Non-biostratigraphical Methods
of Dating and Correlation

Geological Society Special Publications
Series Editor A. J. FLEET

GEOLOGICAL SOCIETY SPECIAL PUBLICATION NO. 89

Non-biostratigraphical Methods of Dating and Correlation

EDITED BY

R. E. DUNAY
Mobil North Sea Ltd,
London

AND

E. A. HAILWOOD
Core Magnetics,
Sedbergh, Cumbria

1995
Published by
The Geological Society
London

THE GEOLOGICAL SOCIETY

The Society was founded in 1807 as The Geological Society of London and is the oldest geological society in the world. It received its Royal Charter in 1825 for the purpose of 'investigating the mineral structure of the Earth'. The Society is Britain's national society for geology with a membership of 7500 (1993). It has countrywide coverage and approximately 1000 members reside overseas. The Society is responsible for all aspects of the geological sciences including professional matters. The Society has its own publishing house which produces the Society's international journals, books and maps, and which acts as the European distributor for publications of the American Association of Petroleum Geologists and the Geological Society of America.

Fellowship is open to those holding a recognized honours degree in geology or cognate subject and who have at least two years' relevant postgraduate experience, or who have not less than six years' relevant experience in geology or a cognate subject. A Fellow who has not less than five years' relevant postgraduate experience in the practice of geology may apply for validation and, subject to approval, may be able to use the designatory letters C. Geol (Chartered Geologist).

Further information about the Society is available from the Membership Manager, The Geological Society, Burlington House, Piccadilly, London W1V 0JU, UK.

Published by The Geological Society from:
The Geological Society Publishing House
Unit 7
Brassmill Enterprise Centre
Brassmill Lane
Bath BA1 3JN
UK
(*Orders:* Tel. 01225 445046
        Fax 01225 442836)

First published 1995

British Library Cataloguing in Publication Data
A catalogue record for this book is available from the British Library
ISBN 1-897799-30-6

Typeset by Bath Typesetting Ltd
Bath, England

Printed in Great Britain by
Alden Press, Oxford

Distributors

USA
  AAPG Bookstore
  PO Box 979
  Tulsa
  Oklahoma 74101-0979
  USA
  (*Orders:* Tel. (918) 584-2555
          Fax (918) 584-0469)

Australia
  Australian Mineral Foundation
  63 Conyngham Street
  Glenside
  South Australia 5065
  Australia
  (*Orders:* Tel. (08) 379-0444
          Fax (08) 379-4634)

India
  Affiliated East-West Press PVT Ltd
  G-1/16 Ansari Road
  New Delhi 110 002
  India
  (*Orders:* Tel. (11) 327-9113
          Fax (11) 326-0538)

Japan
  Kanda Book Trading Co.
  Tanikawa Building
  3-2 Kanda Surugadai
  Chiyoda-Ku
  Tokyo 101
  Japan
  (*Orders:* Tel. (03) 3255-3497
          Fax (03) 3255-3495)

Contents

Non-biostratigraphical methods of dating and correlation: an introduction

R. E. DUNAY[1] & E. A. HAILWOOD[2]

[1]Mobil North Sea Ltd, 3 Clements Inn, London, WC2A 2EB, UK

[2]Core Magnetics, Sedbergh, Cumbria, LA10 5JS, UK and Department of Oceanography, University of Southampton, SO9 5NH, UK

Since the foundation of the science of geology, the description and comparison of preserved organic remains has been the fundamental means to determine the relative ages of sedimentary rock sequences and for correlating these sequences in different areas. Significant problems have arisen in dating and correlating those stratigraphic sequences which are impoverished in, or barren of, fossil remains. As a result, when compared to the plethora of stratigraphic data available for Phanerozoic marine sequences, stratigraphic information available for terrestrial sequences is sparse indeed. It is no surprise, therefore, that major questions in stratigraphy still remain for such periods as the Lower Permian, uppermost Carboniferous and Triassic, which are dominated by non-marine formations.

These problems are exacerbated in correlating and dating barren sequences encountered in offshore exploration and appraisal drilling, as the geoscientist has available normally only drill cuttings and occasional core samples to augment log interpretation. With this sparse dataset, the geoscientist must often make technical judgements on the correlation and continuity of potential hydrocarbon reservoir sections in different wells which may affect exploration and field development decisions. Correlation of biostratigraphically impoverished sequences therefore can be of great significance, as Permian, Triassic and uppermost Carboniferous hydrocarbon reservoirs are economically extremely important, with about 11 000 million barrels of oil equivalent being thus far discovered in these sequences on the UK Continental Shelf alone.

Improvements in analytical techniques during the past few decades have led to the development of geological subdisciplines, some of which, in turn, have potential in providing alternative means for dating and correlating sedimentary sequences. These techniques may prove particularly useful in the dating and correlation of those strata devoid of chronostratigraphically useful fossil remains.

The purpose of this volume is to bring together many of these diverse techniques and disciplines and to explore their potential to solve problems in stratigraphy. It is also the intention to introduce these techniques, each of which is familiar to specialist researchers, to a wider audience of petroleum geologists who may find them useful in the resolution of specific correlation problems.

The techniques represented by the 11 papers in this volume can be broadly grouped into five categories of analysis – mineral, chemical, isotopic, luminescence and cyclicity. Of the papers presenting mineralogical techniques, Morton & Hurst (Correlation of sandstones using heavy minerals: an example from the Statfjord Formation of the Snorre Field, northern North Sea) employ analysis of heavy mineral suites to correlate sandstone sequences of the basal Jurassic–uppermost Triassic Statfjord Formation in the Snorre Field area. Eleven heavy mineral zones have been recognized, which the authors conclude reflect changes in sand provenance, and which can be correlated between wells. Mange-Rajetzky (Subdivision and correlation of monotonous sandstone sequences using high-resolution heavy mineral analysis, a case study: the Triassic of the Central Graben) utilizes heavy mineral techniques in the correlation of normally unfossiliferous Triassic sequences in the central North Sea. Jeans (Clay mineral stratigraphy in Palaeozoic and Mesozoic red-bed facies, onshore and offshore UK) indicates caveats to the use of clay mineralogy in stratigraphic studies, and presents two case studies in this field, the first on the Devonian Old Red Sandstone of Scotland and adjacent offshore areas, and the second on the Triassic of the Southern and Central North Sea and the New Red Sandstone of the south Devon coastal region. Carter et al. (The application of fission track analysis to the dating of barren sequences: examples from red beds in Scotland and Thailand) report on the use of fission track analysis of detrital zircon in determining maximum geological age. In examples from Scotland and Thailand, biostratigraphically barren strata of hitherto equivocal age have been dated as

From Dunay, R. E. & Hailwood, E. A. (eds), 1995, *Non-biostratigraphical Methods of Dating and Correlation* Geological Society Special Publication No. 89, pp. 1–2

Permo-Triassic and Lower Cretaceous, respectively.

Two papers detail chemostratigraphic studies. Racey *et al.* (The use of chemical element analyses in the study of biostratigraphically barren sequences: an example from the Triassic of the central North Sea (UKCS)) present the results of their element analysis study to correlate unfossiliferous Triassic Skagerrak Formation sequences from central North Sea wells using ICP–AES (inductively coupled plasma–atomic emission spectrophotometry). ICP–AES techniques are employed also by Pearce & Jarvis (High-resolution chemostratigraphy of Quaternary distal turbidites: a case study of new methods for analysis and correlation of barren sequences) in their correlation of barren late Quaternary distal turbidites. The authors indicate that this technique has potential for correlating older biostratigraphically barren sequences.

Three contributions deal with the application of isotope analyses to age dating and correlation. Roberts *et al.* (SHRIMP zircon age control of Gondwanan sequences in Late Carboniferous and Early Permian Australia) detail their age assignment of Late Carboniferous volcanic units in Australia to the European Late Carboniferous Stages through $^{40}Ar/^{39}Ar$ ages and the newly developed zircon U–Pb dating technique using the sensitive high resolution microprobe (SHRIMP). Russell (Direct Pb/Pb dating of Silurian macrofossils from Gotland, Sweden) presents his study on direct Pb/Pb dating of stromatoporoid carbonates. The resulting Pb/Pb date is in good agreement with independent chronometric determinations. This study was conducted on fossiliferous strata to provide control, and the results indicate that the technique has potential in dating non-fossiliferous carbonates. Dalland *et al.* (The application of samarium–neodynmium (Sm–Nd) provenance ages to correlation of biostratigraphically barren strata) detail the utility of Sm–Nd isotope analyses in determining provenance ages of Statfjord Formation sediments in the Gullfaks Field, Norwegian North Sea. Provenance age profiles, integrated with other information, facilitate correlation of discrete channel sands with overbank mudstone deposits.

Rendell (Luminescence dating of Quaternary sediments) outlines the use of luminescence dating techniques, which date the last exposure to light of grains of sediment before burial, in dating Quaternary sediments. This technique is currently limited to Late Pleistocene sequences. However, the potential of extending luminescence dating to older deposits certainly exists.

In the final paper of the volume, Yang & Kouwe (Wireline log-cyclicity analysis as a tool for dating and correlating barren strata: an example from the Upper Rotliegendes of the Netherlands) report the results of a cyclicity analysis study in the usually unfossiliferous Upper Rotliegendes Group of offshore Netherlands. Five supersequences and 12 sequences have been recognized, which are useful for regional correlation and field-scale correlation, respectively.

Whilst the papers presented in this special publication provide an excellent cross-section of non-biostratigraphical methods in current use, there have been important developments in other relevant techniques, which are not represented in the volume, including magnetic polarity stratigraphy and high-resolution magnetic susceptibility logging (e.g. Hauger *et al.* 1994; and various papers in Turner & Turner in press). It is to be expected that the continuing demand for improvement in stratigraphic resolution in biostratigraphically-barren strata will lead to significant developments in many of the techniques presented in this volume, and to the introduction of new techniques and approaches in the future. Most importantly, we expect to see increasing emphasis on the integration of a range of different techniques in the study of individual formations, in order to exploit fully the strengths and to identify the weaknesses of the individual techniques.

References

HAUGER, E., LØVLIE, R. & VAN VEEN, P. 1994. Magnetostratigraphy of the Middle Jurassic Brent Group in the Oseberg oil field, northern North Sea. *Marine and Petroleum Geology*, **11**, 375–388.

TURNER, P. & TURNER, A. in press. *Palaeomagnetic Applications in Hydrocarbon Exploration and Production*. Geological Society, London, Special Publication.

Correlation of sandstones using heavy minerals: an example from the Statfjord Formation of the Snorre Field, northern North Sea

ANDREW MORTON[1] & ANDREW HURST[2]

[1]*British Geological Survey, Keyworth, Nottingham, NG12 5GG, UK*
[2]*Department of Geology and Petroleum Geology, University of Aberdeen, Aberdeen, AB9 2UE, UK*

Abstract: Correlation of sandstones using heavy minerals is dependent on the recognition and quantification of provenance-sensitive parameters. Suitable criteria include ratios of minerals that have similar hydraulic and diagenetic behaviour, and properties of populations of individual mineral species. Sandstones of the Statfjord Formation in the Snorre Field display marked variations in their heavy mineral assemblages. Although hydraulic and diagenetic processes have played some part in generating these variations, major changes in the nature of the source material can be detected on the basis of heavy mineral ratio and varietal data. These changes in sand provenance can be used to produce a detailed stratigraphic breakdown of the sequence and to correlate from well to well. Eleven heavy mineral zones have been recognized on the basis of variations in monazite/zircon, garnet/zircon and rutile/zircon ratios, on the incoming of chloritoid, and on the composition of garnet populations. The marked changes in provenance result from the proximity of the depositional location to the hinterland and it is considered unlikely that the detailed correlation made in the Snorre Field area applies to Statfjord sandstones in other parts of the northern North Sea. However, there is evidence for some degree of allogenic control on the heavy mineral variations and therefore some events may have regional significance.

Correlation of many sandstone sequences is commonly difficult to achieve by biostratigraphic methods alone. In marine depositional environments sedimentation rates may be so rapid that biostratigraphic events lack sufficient resolution for detailed sandbody correlation. Greater problems are encountered in non-marine and paralic environments, particularly in red-bed settings, where biostratigraphic control may be virtually absent. Heavy mineral analysis is one of a group of non-biostratigraphic approaches to correlation of sand-rich sequences and can be used either in isolation (as in the case of biostratigraphically-barren strata) or as a complement to traditional biostratigraphic correlation. Heavy minerals are high-density components of siliciclastic sediments, the boundary between the 'light' and heavy' minerals generally being taken at s.g. 2.8. Because they generally form < 1% of sandstones, their analysis involves separation from the less-dense framework component, usually by the use of high-density liquids. Analytical procedures have been reviewed recently by Morton (1985) and Mange & Maurer (1992).

Correlation using heavy minerals

The principle behind the use of heavy mineral analysis and analogous petrographic, mineralogical and geochemical approaches for correlation is that changes in provenance offer a basis for the subdivision of sedimentary sequences. Although heavy minerals are minor constituents of siliciclastic sequences in volumetric terms, a large number of detrital species have been found in sandstones: for example, Mange & Maurer (1992) illustrate *c.* 50 minerals of relatively common occurrence. Because of this diversity, and the restricted paragenesis of many species, heavy-mineral assemblages are sensitive indicators of changing provenance. Changes in sand provenance may be relatively gradual, occurring through unroofing of fresh lithologies in a single source area through continued erosion, or they may be sudden, resulting from changes in basin configuration which introduced sediment from different source terrains. These different styles of changing provenance are manifested in different ways by the heavy mineral assemblages, the former being identified by evolutionary trends and the latter by sudden, abrupt differences.

There are two major considerations that must be taken into account when using heavy minerals for correlation purposes. Firstly, it must be established whether observed variations in heavy mineral suites are caused by changes in provenance, because several other processes can influence the composition of heavy mineral suites. Secondly, changes in sediment provenance are not necessarily chronostratigraphic: they may well be diachronous. Furthermore, changes in provenance are not necessarily unique, because, unlike biostratigraphic events,

From Dunay, R. E. & Hailwood, E. A. (eds), 1995, *Non-biostratigraphical Methods of Dating and Correlation*
Geological Society Special Publication No. 89, pp. 3–22

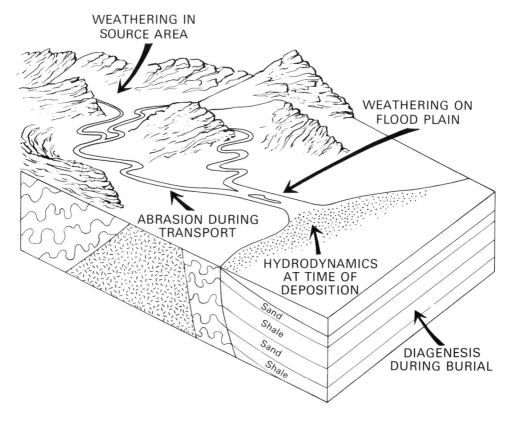

WEATHERING IN
SOURCE AREA

WEATHERING ON
FLOOD PLAIN

ABRASION DURING
TRANSPORT

HYDRODYNAMICS
AT TIME OF
DEPOSITION

Sand
Shale
Sand
Shale

DIAGENESIS
DURING BURIAL

Fig. 1. Schematic representation of the processes that affect heavy mineral assemblages during the sedimentary cycle. The combined effects of these processes may heavily overprint the original source rock mineralogy.

they can be repeated. This must be taken into account when correlating sedimentary sequences on the basis of changes in mineralogy.

Processes governing the composition of heavy mineral assemblages

Although heavy mineral assemblages are sensitive indicators of sediment provenance and valuable for discriminating between sandbodies derived from different sources the composition of the assemblages is not only a function of source rock mineralogy. The relative abundances of heavy minerals within assemblages can be affected at all stages of the sedimentation process (Fig. 1), from weathering in the source area, through transport and deposition, into diagenesis, at both the early, surface-related (eodiagenetic) stage and the late, burial-related (mesodiagenetic) stage (terminology of Burley *et al.* 1985). For correlation purposes, it is im-

portant to distinguish between the processes that affect mineralogy prior to deposition from those that affect mineralogy during and after deposition. Although the composition of heavy mineral assemblages may be altered during the weathering and transport stages, these variations are introduced prior to the ultimate site of deposition, and therefore have potential value for sediment correlation. In contrast, variations caused at the site of deposition, either at the time of deposition itself (by varying hydrodynamic conditions) or subsequently (during diagenesis), may adversely affect the value of heavy mineral suites for correlation purposes. The value of heavy minerals in establishing the nature of the sediment provenance, however, is affected by all of these processes.

Weathering in the source area introduces differences between the mineralogy of the source rocks and the mineralogy of the sediment released into the transport system through the destruction of minerals in soil profiles (see

Bateman & Catt 1985). Despite this, present-day rivers contain rich and diverse heavy mineral suites (e.g. Russell 1937; Shukri 1949; van Andel 1950; Morton & Johnsson 1993) indicating that, on a gross scale, weathering does not significantly affect the diversity of mineral assemblages incorporated into transport systems. It is possible that weathering may cause profound changes in mineralogy in transport-limited weathering regimes (Johnsson *et al.* 1991), but detailed information on the response of heavy minerals to such environments is not available.

During transport itself, minerals are subjected to abrasion processes that could affect the relative abundances of minerals with different mechanical stabilities. Although experimental work has established that detrital minerals display a range of mechanical stabilities (Friese 1931; Thiel 1940, 1945; Dietz 1973), there is little evidence that abrasion processes play a significant role in natural situations (Morton & Smale 1990). However, weathering during periods of alluvial storage during transport in fluvial systems may be more important, particularly in humid tropical environments. Johnsson *et al.* (1991) showed the importance of this process on bulk sand mineralogy, and subsequent work has shown that the process may also be effective in modifying aspects of the heavy mineral suite (Morton & Johnsson, 1993).

The factors which have greater influence on the use of heavy mineral assemblages for correlation are hydraulics and diagenesis; hydraulic processes influence mineralogy at the time of deposition and diagenesis subsequently modifies sediment composition. Hydrodynamic conditions at the time of deposition control the relative abundances of minerals with different hydraulic properties. Rubey (1993) was the first to demonstrate conclusively that changes in hydraulic conditions introduce marked variations to heavy mineral suites, and subsequent work has demonstrated that density and size are the main factors influencing the hydraulic behaviour of heavy mineral grains (Rittenhouse, 1943). Grain shape may also be an important influence (Briggs *et al.* 1962): this is particularly well demonstrated by mica, which, although it has the density of a heavy mineral, actually behaves as a light mineral (Doyle *et al.* 1983). Thus, a sandbody that shows variations in grain size, resulting from differences in hydraulic conditions during deposition, will display concomitant variations in its heavy mineral assemblage. Therefore, correlations made by the uncritical use of heavy mineral data in sequences strongly influenced by varying hydrodynamic conditions will be of doubtful value.

The effects of diagenesis may be even more profound, because post-depositional processes can cause partial or complete loss of many heavy mineral species. During burial, heavy mineral suites become less diverse because of the dissolution of unstable minerals in response to increasing pore-fluid temperature and associated changes in pore-fluid composition (Morton 1984; Milliken 1988; Cavazza & Gandolfi 1992). For example, Late Palaeocene sandstones of the central North Sea contain a diverse assemblage of apatite, amphibole, epidote, garnet, kyanite, rutile, staurolite, titanite, tourmaline and zircon at shallow depths, but as burial increases they progressively lose amphibole, epidote, titanite, kyanite and staurolite, so that at the basin centre they contain an assemblage of apatite, corroded garnet, rutile, tourmaline and zircon. Dissolution of heavy mineral suites also takes place under the influence of low temperature acidic groundwaters (Friis 1976; Morton, 1984, 1986), e.g. in fluvial sandstones or below subaerial unconformities. If the extent to which diagenesis has affected heavy mineral suites is not appreciated correlation using heavy mineral data is likely to lack stratigraphic significance.

Correlation-sensitive components of heavy mineral suites

To ensure that mineralogical correlations have stratigraphic significance it is crucial that the criteria used for correlation purposes are essentially unaffected by variations in hydrodynamic and diagenetic processes. For this to be achieved the features that are quantified must minimize differences in hydraulic and diagenetic behaviour within the suite. The parameters that fulfil this requirement fall into two classes – those acquired by conventional analysis (optical quantification of heavy mineral abundances) and those acquired by varietal studies (quantification of variations displayed by a single mineral species).

Conventional heavy mineral data can be used to identify changes in sediment source by the determination of ratios of stable minerals with similar densities, as these are not affected by changes in hydraulic conditions during sedimentation or by diagenetic processes (Morton & Hallsworth 1994). Ratios such as apatite/tourmaline, rutile/zircon, chrome spinel/zircon and monazite/zircon (Table 1) are useful indicators of changing provenance in deeply-buried sandstones, and the garnet/zircon ratio is also

Table 1. *Determination of provenance-sensitive index values*

Index	Mineral pair	Index determination
ATi	Apatite/tourmaline	$100 \times$ apatite count/(total apatite plus tourmaline)
GZi	Garnet/zircon	$100 \times$ garnet count/(total garnet plus zircon)
RZi	TiO_2 group/zircon	$100 \times TiO_2$ group count/(total TiO_2 group plus zircon)
CZi	Chrome spinel/zircon	$100 \times$ chrome spinel count/(total chrome spinel plus zircon)
MZi	Monazite/zircon	$100 \times$ monazite count/(total monazite plus zircon)

RZi is determined here using the total of the TiO_2 polymorphs, because many analysts do not distinguish the three minerals during the counting procedure, particularly if identification is made using energy-dispersive X-ray analysis rather than by optical methods. If separate counts of rutile, anatase and brookite are made the ratio could be determined using rutile alone. This would have some advantages over the use of the entire TiO_2 group (see Morton & Hallsworth 1994).

useful providing that garnet can be demonstrated to be stable within the sequence under investigation. Variations in the apatite/tourmaline ratio are suspect in sequences affected by acidic groundwater circulation. The grain size characteristics of individual heavy minerals in sandstones is not only controlled by hydraulic conditions but also by the grain size distribution inherited from the parent rock. For this reason it is crucial that the mineral ratios are not determined on the entire heavy mineral population but on a restricted grain size fraction (Morton & Hallsworth 1994).

As varietal studies concentrate on variations seen within one mineral group the range of density and stability across the data set is strongly diminished. The classical approach to varietal studies is to distinguish types on the basis of their optical properties, such as colour, shape and habit. Optical differentiation of zircon and tourmaline types has been used extensively to assess differences in sediment provenance (Krynine 1946; Poldervaart 1955). However, this approach is commonly subjective and boundaries between the classes seen within the population tend to be somewhat arbitrarily drawn. For example, categorization on the basis of grain shape involves quantification of classes such as euhedral, subhedral, angular, subangular, rounded and subrounded: inevitably, problems arise in classifying grains which lie close to class boundaries. Similarly, many colour series are gradational. The problems with such subjective techniques are partly concerned with internal consistency on behalf of the analyst, but become almost insurmountable when comparing data from different researchers. Thus, although optical differentiation of mineral varieties remains an important aspect of heavy mineral analysis, accurate correlation using varietal data should depend on more objectively-acquired numerical data. The single most

important advance in this area has been the determination of the compositional range of one or more heavy mineral species using single-grain geochemical methods, most notably by the electron microprobe. Many detrital heavy minerals lend themselves to this type of analysis, both translucent (Morton 1991) and opaque (Basu & Molinaroli 1991; Grigsby 1990). The most useful minerals are those that show a wide range in chemical composition and are relatively stable, such as garnet (Morton 1991) and tourmaline (Jeans *et al.* 1993).

Stratigraphic significance of correlations made using heavy mineral data

If sandstone sequences show mineralogical variations related to changes in provenance these can be used to establish stratigraphic correlations. Heavy mineral data are used to correlate individual sandbodies or sandbody packages on a variety of scales, from a small-scale, local level to a large-scale regional level. In a hydrocarbon reservoir context it is useful to think of heavy mineral correlations as correlations of volume. It is important to recognize that heavy mineral correlations do not necessarily have chronostratigraphic significance because in many depositional settings changes in provenance may be diachronous.

If heavy minerals record changes in provenance within a single sediment dispersal system these changes will be recorded at essentially the same time throughout the system, at least in geological terms, and can therefore be regarded as having chronostratigraphic value. The areal extent over which such correlations can be applied depends on the geographical limitation of the sediment dispersal system. For example, integration of heavy mineral and biostratigraphic data shows that heavy mineral variations have chronostratigraphic significance

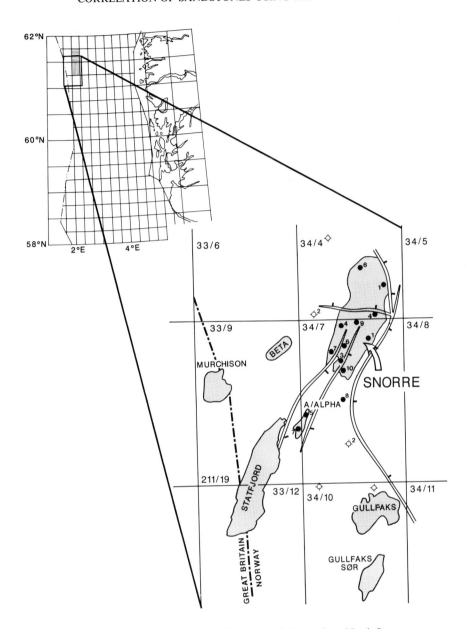

Fig. 2. Location of the Snorre Field in the Norwegian sector of the northern North Sea.

within the Palaeocene submarine fan systems of UK North Sea Quadrant 9 (Morton *et al.* 1993). However, although comparable heavy mineral events occur within the submarine fan system to the south, in Quadrant 16, they occur at a different stratigraphic level, illustrating that the geographic extent of the sediment dispersal system governs the area over which heavy mineral correlations can be made.

Many depositional systems involve sediment input from more than one source direction. In such circumstances changes in mineralogy are less likely to be chronostratigraphic. For example, heavy mineral analysis of the Middle Jurassic Rannoch, Etive and Ness formations (Brent Group) of the northern North Sea (Morton 1992) has shown that sediment supply to this coastal–deltaic system had two major components. There was a fluvial component, which was broadly derived from the south and had a transport direction essentially perpendicular to the shoreline, and a longshore component derived from the east with a transport direction parallel to the shoreline. Although the boundary between the underlying shoreface sequence and the overlying delta-top sequence is clearly defined on a mineralogical basis, the correlations made on this basis are markedly diachronous on a regional basis because of the gradual northward progradation of the coastal–deltaic complex. Despite this, mineralogical correlations identified in such circumstances have important lithostratigraphic value and this may have equal importance, especially for hydrocarbon reservoir characterization purposes.

These examples show that the significance and regional extent of correlations made using heavy mineral, or analogous petrographical, mineralogical or geochemical data must be assessed by integrating the data with sedimentological information, together with biostratigraphy and seismic stratigraphy (if available). Heavy mineral data, of course, have an important part to play in determining the regional extent of sediment dispersal systems, but without involvement of independent sedimentological and other data there is a danger of being drawn into a circular argument.

Case example: the Statfjord Formation of the Snorre Field

Regional setting and sedimentological context

The Snorre Field lies in Blocks 34/4 and 34/7 of the Norwegian sector of the northern North Sea (Fig. 2). The hydrocarbon reservoir units belong to the Statfjord Formation and the Hegre Group (Nystuen et al. 1989). The Hegre Group (Scythian–Rhaetian) comprises a continental red-bed sequence, including channelized and non-channelized fluvial sandstones, lacustrine and flood-basin mudrocks, and possibly aeolian or reworked aeolian sandstones (Nystuen et al. 1989). The Hegre Group is subdivided into three formations: the Teist Formation, at the base, is overlain by the Lomvi Formation, with the Lunde Formation at the top. The overlying Statfjord Formation comprises sandstone units up to 22 m thick, interbedded with mudstone units between 0.5 and 12 m thick (Fig. 3). The sandstone units comprise stacked individual channel-fill sandstone bodies with erosional

lower contacts. The dominant lithology is trough cross-bedded medium- to very coarse-grained or gravelly, moderately- to poorly-sorted sandstone: planar cross-bedded, horizontally-bedded and ripple cross-laminated facies are less common (Høimyr et al. 1993). The intercalated mudstones are grey to reddish brown and contain stacked palaeosols of flood-plain origin (Høimyr et al. 1993). The Statfjord Formation is considered to be of Rhaetian to Sinemurian age (Deegan & Scull 1977) and is overlain by Pliensbachian-age marine mudrocks of the Dunlin Group. Biostratigraphic resolution within the sequence is poor because of the continental facies. Four wells have been investigated (34/7-3, 34/7-4, 34/7-6 and 34/7-10): of these, 34/7-3, 34/7-6 and 34/7-10 have been extensively cored, allowing good stratigraphic coverage, but in Well 34/7-4 data are available only from the upper part of the sequence.

Heavy mineral assemblages

The heavy mineral assemblages in sandstones from the uppermost Lunde and Statfjord Formation of the Snorre Field are rich and diverse. Thirteen detrital non-opaque minerals have been identified: these are apatite, chloritoid, chrome spinel, epidote group minerals, garnet, gahnite (the blue–green zinc spinel), kyanite, monazite, TiO_2 minerals (virtually exclusively rutile, with only minor amounts of anatase and brookite), staurolite, titanite, tourmaline and zircon (Table 2). Many of these species show wide fluctuations in relative abundance: apatite ranges from 0–21.5%, epidote from 0–10%, garnet from 0.5–94.5%, kyanite from 0–69.5%, monazite from 0–16.5%, rutile from 0.5–37%, titanite from 0–14.5%, staurolite from <0.5–16.5%, tourmaline from <0.5–14%, and zircon from 0.5–56%. The remaining minerals (chloritoid, chrome spinel and gahnite) are scarce throughout and fail to become volumetrically significant components of the assemblages.

The stratigraphic variations in the heavy mineral assemblages appear to lack any consistent pattern, as shown by Well 34/7-3 (Fig. 4). This is because the observed variations in the assemblages are not simply related to changes in provenance: hydraulic and diagenetic processes are also heavily involved. Of the two, diagenetic controls are the more significant. The sandstones under investigation are buried to depths in excess of 2400 m: Palaeocene sandstones at comparable depths in the central North Sea have lost epidote, titanite and kyanite through dissolution processes. Epidote, titanite and kyanite dissolution in the Statfjord sandstones

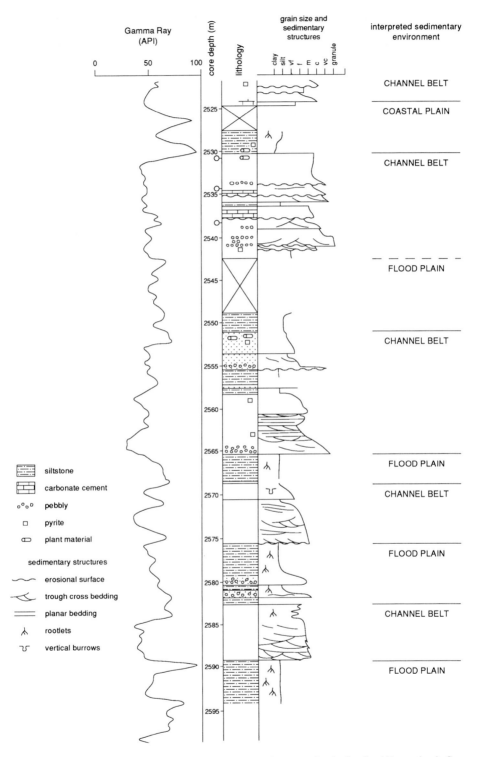

Fig. 3. Summary core description and interpreted depositional environment for the Statfjord Formation in Snorre Field Well 34/7-6.

Table 2. *Heavy minerals in Statfjord Formation sandstones from the Snorre Field*

Well	Depth (m)	Zone	AP	CT	CR	EP	GT	GH	KY	MO	RU	SP	ST	TO	ZR	GZi	RZi	MZi
34/7-3	2416.3	11	1.0	0.5		10.0	48.0		11.0	R	6.0	14.5	5.0	1.5	2.5	95.1	56.7	4.8
	2418.6	10					0.5		69.5	R	9.0		0.5	R	20.5	2.4	32.4	1.0
	2424.9	9		R		R	36.5	R	38.0	R	5.0		4.0	2.0	12.0	75.0	38.9	4.4
	2441.8	7					71.5		6.0	R	5.0		0.5	R	17.0	80.8	22.5	2.0
	2447.5	7					46.0		32.5	R	7.5		4.0	0.5	9.5	82.9	30.1	4.8
	2458.5	6					1.0		9.0	3.0	17.5		9.5	3.5	56.0	1.8	24.8	4.8
	2463.4	6			R		3.5		3.0	7.5	37.0		12.5	4.5	32.0	9.9	63.0	22.5
	2467.0	6					4.0		15.0	5.5	24.0		16.5	2.5	32.5	11.0	33.3	16.7
	2477.3	6					2.0		0.5	6.5	28.5		10.0	3.5	49.0	3.0	34.2	9.1
	2511.2	3					92.0			1.5	3.5		1.0	R	2.0	97.9	48.2	44.0
34/7-4	2534.8	8				0.5			6.5	3.0	17.5	R	7.5	5.5	58.0	0.0	22.6	6.6
	2545.5	7					60.5	0.5	5.5	0.5	7.0		1.5	R	24.5	71.1	32.9	9.1
	2547.3	7			R		63.0	R	0.5	0.5	7.5		4.5	0.5	23.5	72.8	37.5	6.5
	2548.4	7					85.5		0.5	1.0	4.0		0.5	0.5	8.0	91.4	30.1	12.3
	2551.6	7	2.0				69.5		6.0	1.0	3.5		6.5		11.5	85.8	40.0	14.3
	2554.7	6	2.0				2.0	R		8.0	13.5		9.0	1.0	61.5	3.2	24.2	17.4
34/7-6	2521.0	9					61.0		0.5	R	10.0		3.0	2.0	23.5	72.2	33.1	1.2
	2523.6	9					66.5		7.5	R	9.0		2.5	0.5	14.0	82.6	40.8	5.7
	2534.1	8					3.5	R	14.0	1.0	18.0		13.5	2.5	47.5	6.9	29.6	2.0
	2534.8	8					2.5	0.5	29.5	1.0	5.0		14.0	R	35.5	1.4	31.0	3.9
	2538.0	7					60.0	R	21.0	R	15.0		5.5	2.0	8.5	87.6	35.9	3.3
	2553.7	6			0.5		1.0	R	0.5	7.0	29.5		8.0	3.5	50.0	2.0	21.3	7.4
	2563.9	6					1.5		0.5	3.0	26.0		12.5	4.0	52.5	2.8	28.6	4.8
	2571.4	5	R				37.0			6.5	28.5		2.5	2.5	23.0	61.7	46.5	16.7
	2572.4	5	1.0				19.5			9.0	24.0		5.0	4.0	37.5	34.3	40.0	25.4
	2574.0	5	7.0				15.0			9.0	21.5		3.0	2.5	42.0	25.6	37.5	25.9
	2586.0	4	21.5		0.5		3.0			9.0	25.5		6.0	3.0	31.5	8.7	27.0	25.9
	2587.2	4	16.0				0.5			13.5	27.0		5.0	2.0	36.0	1.4	35.5	23.7
	2589.0	4	1.0				2.0			16.5	23.5		6.0	0.5	50.0	3.9	32.9	32.0
	2600.0	3	21.0		R		55.0			6.5	10.5		0.5	1.0	5.5	90.9	48.9	38.2
	2604.7	3	3.0				80.5			1.0	6.5		0.5	0.5	8.0	91.0	59.2	35.1
	2615.5	2	5.5				87.0			R	4.5		R	0.5	2.5	97.2	82.1	*20.9*
34/7-10	2532.0	11	6.5	0.5	0.5		66.5		5.0	0.5	5.5	4.0	0.5	2.0	8.5	88.7	54.5	*3.2*
	2533.0	11	3.0	1.0	0.5		76.5		3.0	0.5	5.5	1.5	1.0	1.5	6.0	92.7	32.0	1.0
	2534.0	11	2.5	0.5			64.5	0.5	6.0	0.5	9.5	2.0	1.0	1.5	11.5	97.7	38.3	1.0
	2549.9	8	R			0.5	16.5	R	2.0	7.0	19.0		4.0	1.5	49.5	25.0	39.8	17.4
	2552.9	7					80.0			1.0	5.5		1.5	7.0	5.0	94.1	53.5	18.5
	2562.0	7					64.0		R	R	11.0		5.0	1.5	18.5	77.6	35.5	5.7
	2565.2	7	0.5				40.0		11.0	2.5	7.0		1.5	0.5	37.0	52.0	18.7	9.9
	2593.0	4					3.5			6.0	36.0		9.5	14.0	31.0	10.1	69.5	20.0
	2604.2	4	R				8.0		2.0	14.5	21.0		4.5	1.0	47.0	14.0	36.7	27.5
	2612.0	3	1.5				64.0			5.5	7.5		2.5	R	19.0	72.1	32.9	29.6
	2615.0	3	0.5				91.0			1.0	2.5		0.5	R	4.5	95.3	48.5	21.3
	2629.0	2	3.0				90.5			0.5	3.0		0.5	R	2.5	97.3	79.1	29.9
	2636.0	2	2.0				94.5			R	0.5		1.0	0.5	0.5	99.5	84.1	*35.3*
	2645.0	1	0.5				93.5			0.5	1.5		0.5	R	3.5	96.4	47.1	21.9

Data for heavy mineral species, in frequency (%), from a count of 200 non-opaque detrital grains in the 63–125 μm (3ϕ–4ϕ) fraction. Index values (defined in Table 1) were calculated on a minimum count of 100 grains, except data in italics which denote a low grain count (< 50). AP, apatite; CT, chloritoid; CR, chrome spinel; EP, epidote; GT, garnet; GH, gahnite; KY, kyanite; MO, monazite; RU, TiO$_2$ minerals; SP, titanite; ST, staurolite; TO, tourmaline; ZR, zircon. R denotes rare (< 0.5%).

of the Snorre Field can be inferred from their sporadic distribution, their tendency to be less common in the more deeply-buried sections, and the development of severe corrosion features on grain surfaces (Fig. 5). Staurolite also shows moderate to extreme etching, suggesting that some grains may have been lost through dissolution. In contrast, garnet surfaces show

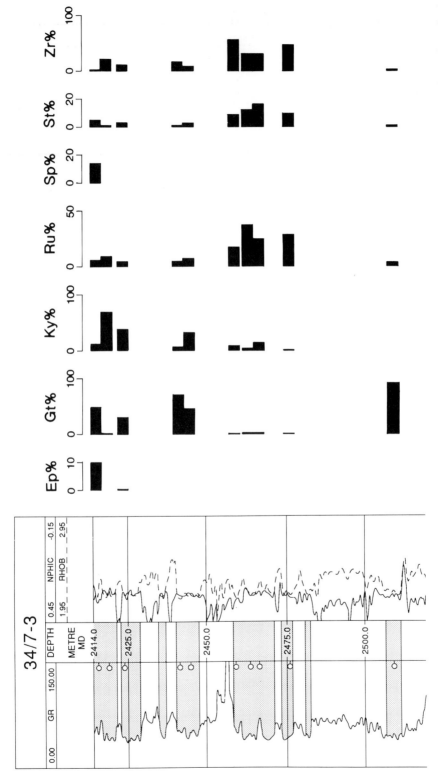

Fig. 4. Stratigraphic variation in abundance of the most common heavy minerals in the Statfjord Formation of Well 34/7-3. Data refer to the frequency (in %) determined on a count of 200 non-opaque detrital grains in the 63–125 μm fraction. GR, natural gamma-ray emission (in API units); NPHIC, neutron porosity log (in porosity units); RHOB, formation density (in g cm^{-3}).

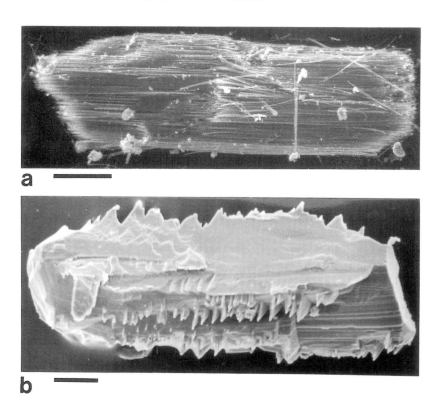

Fig. 5. Scanning electron micrographs showing surface textures of heavy minerals from the Statfjord Formation in the Snorre Field. Scale bar is 20 μm in both cases. (**a**) Epidote grain from Well 34/7-3, 2416.0 m, showing fibrous development as a result of corrosion. (**b**) Kyanite grain from Well 34/7-4, 2551.5 m, showing typical platy habit modified by corrosion generating hacksaw terminations.

only limited surface corrosion features: therefore, proportions of this mineral are unlikely to have been affected by burial-related dissolution. The variations in abundance of apatite are also largely related to diagenesis: this mineral shows evidence of both dissolution and overgrowth. Apatite is consistently present in the uppermost Lunde Formation, but is absent in many Statfjord Formation samples. Apatite dissolution takes place by the action of acidic groundwaters (Morton 1986), and the difference between the distribution of apatite in the Lunde and Statfjord Formations may therefore be related to the change in climatic conditions, from the semi-arid environment prevalent during Lunde Formation times to the humid environment prevalent during Statfjord Formation deposition (Nystuen *et al.* 1989). Further apatite dissolution and reprecipitation may have taken place during the Late Jurassic tectonic phase, as the Snorre structure is one of the few in the northern North Sea that was subaerially exposed through footwall uplift at this time

(Yielding *et al.* 1992).

Hydraulic controls are less significant, but may have had some influence in a number of samples. The anomalously high kyanite content (69.5%) at 2418.6 m in Well 34/7-3 (Fig. 4) may be a hydraulic effect: similar, although less marked, kyanite concentrations occur at 2447.5 m in Well 34/7-3 (32.5%) and at 2534.8 m (29.5%) in Well 34/7-6. Kyanite is readily concentrated because of its platy habit, which has a marked effect on its hydraulic behaviour (Briggs *et al.* 1962). Flores & Shideler (1978) also noted that minerals with bladed or platy habits tend to be hydraulically equivalent to smaller quartz grains compared with their equant or prismatic counterparts. The generally low abundance of the relatively light heavy mineral tourmaline reflects the coarse-grained facies that characterizes the Statfjord Formation. By analogy, apatite, which has a similar density to tourmaline, was probably also relatively scarce prior to its extensive dissolution through acidic groundwater circulation.

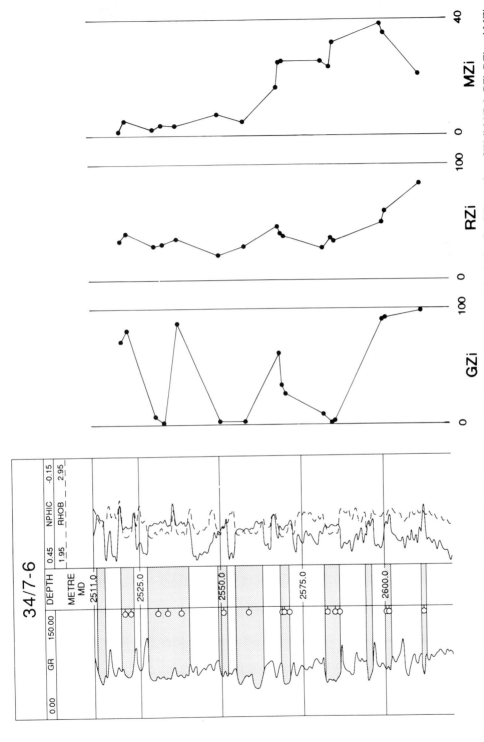

Fig. 6. Variations in ratio of garnet/zircon (GZi), TiO$_2$ minerals/zircon (RZi) and monazite/zircon (MZi) in the Statfjord Formation of Well 34/7-6. GZi, RZi and MZi are determined as shown in Table 1. As far as heavy mineral recovery allows, each ratio is based on a count of 200 grains of the mineral pair. GR, natural gamma-ray emission (in API units); NPHIC, neutron porosity log (in porosity units); RHOB, formation density (in g cm^{-3}).

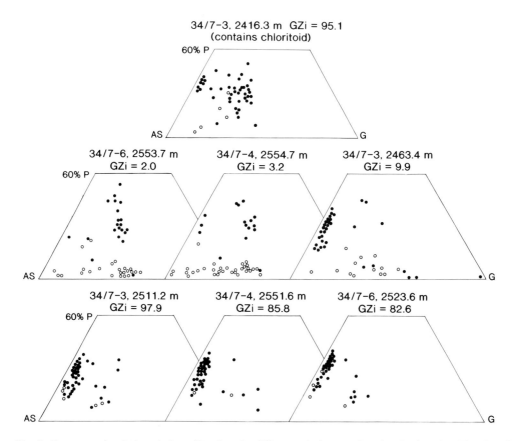

Fig. 7. Garnet geochemical analysis verifies that the differences in heavy mineral ratio data (see Fig. 5) and distribution of chloritoid are related to differences in sediment provenance. Samples with a low GZi, or high GZi and those containing chloritoid all have markedly different garnet populations. Fifty garnet grains were analysed per sample using a Link Systems energy-dispersive X-ray analyser attached to a Microscan V electron microprobe. Compositions are expressed in terms of the abundances of the almandine plus spessartine (AS), pyrope (P) and grossular (G) end-members. ○, Grains with > 5% spessartine; ●, grains with < 5% spessartine.

Provenance-sensitive parameters

Of the provenance-sensitive heavy mineral indices (Table 1) proposed by Morton & Hallsworth (1994), the apatite–tourmaline index (ATi) cannot be used because of the evidence of extensive apatite dissolution and reprecipitation. The chrome spinel–zircon index (CZi) remains very low throughout and has no value as a stratigraphic indicator in this example. The indices that are considered useful in the stratigraphic subdivision of the Statfjord Formation in the Snorre Field are the garnet–zircon index (GZi), the monazite-zircon index (MZi) and the TiO$_2$ group–zircon index (RZi). Of these, GZi and MZi show major variations, with RZi providing useful support.

The variations shown by these parameters are illustrated by reference to Well 34/7-6 (Fig. 6). This shows that the sequence can be divided into two major parts, a lower unit with high MZi values (> 20) and an upper unit with low MZi values (< 20). Within each unit there are major variations in the GZi value, which vary from very low (c. 1) to very high (c. 99). RZi values show less striking variations, but tend to be highest in the lowermost part of the sequence. In one zone near to the base of the analysed succession RZi values are extremely high (c. 80).

In addition to the variations shown by GZi, MZi and RZi, the distribution of the relatively scarce mineral chloritoid is considered to be another useful stratigraphic indicator. The topmost sandbody in two wells, 34/7-3 and 34/7-10, is distinctive in containing small, but significant, amounts of this mineral, which is entirely absent in the underlying sequence and throughout wells 34/7-4 and 34/7-6. Although the stability of chloritoid has been questioned by Mange & Maurer (1992), it displays no evidence of dissolution in the Statfjord Formation of the Snorre Field. Therefore, the appearance of chloritoid is considered to mark a change in provenance, indicating that the chloritoid-bearing sandstone is a correlatable unit.

Garnet geochemistry

The major differences in GZi values observed in the Statfjord Formation have two possible origins. One is that they are related to differences in provenance, but an alternative is that they have resulted from garnet dissolution by acidic groundwaters, either at the depositional site or previously, during periods of alluvial storage on the floodplain. Garnets are known to be unstable in acidic groundwaters (Grimm 1973; Friis 1976; Friis et al. 1980), although appear to be more stable than apatite under such circumstances (Morton 1984). Although there is little evidence for corrosion on garnet grain surfaces in the low-garnet zones, garnet compositions of both garnet-rich and garnet-poor sandbodies were studied in order to evaluate the cause of the variations. Garnets were also studied from the chloritoid-bearing sandstone unit to test the hypothesis that it had a different provenance.

Sandstones with high GZi values are characterized by abundant low-grossular almandine–pyrope garnets, with grossular < 10% and pyrope between 20 and 40% (Fig. 7). A small number of more grossular-rich garnets are also found, but there are very few garnets with < 10% pyrope, and virtually none with > 5% spessartine. In contrast, sandstones with very low GZi values (< 5) have very small amounts of the low-grossular almandine–pyrope component. They are dominated by low-pyrope almandine–spessartines, with variable grossular contents: pyrope is < 10%, spessartine > 5% and grossular ranges up to 50%. A subsidiary grouping of grossular- and pyrope-rich almandines lacking significant spessartine contents is also conspicuous. Sandstones with slightly higher GZi values (> 5) have similar types of garnet, but also contain a large number of the low-grossular almandine–pyrope group that characterizes the high GZi sandstones. The chloritoid-bearing sandstone is dominated by a group of grossular- and pyrope-rich almandines (grossular 10–30%, pyrope 10–40%), with a subsidiary group of low-grossular almandine–pyropes.

The sandstones with low GZi contain a component that is entirely absent from the high GZi sandstones, and therefore cannot represent weathered high GZi material. Furthermore, the garnet suites in the low GZi sandstones are more diverse than those associated with the high GZi material, the reverse of what would be expected if weathering processes were involved. The chloritoid-bearing sandstones have similar GZi values to the high GZi type, but have entirely different garnet assemblages. Therefore, the garnet geochemical data confirm that the variations in GZi value and distribution of chloritoid are primary, and are related to differences in provenance. They can therefore be used with confidence to establish a correlation framework.

Stratigraphic subdivision

The variations in provenance-sensitive parameters described above permit the subdivision of the Statfjord Formation and immediately underlying Lunde Formation into 11 zones, which can be correlated between the four wells with varying degrees of confidence (Fig. 8).

Zone 1
This has been identified only in the lowermost sample from 34/7-10, the well with the greatest stratigraphic coverage. It is characterized by high MZi (> 20), very high GZi (96), and high RZi (47).

Zone 2
This has been identified in two wells, 34/7-3 and 34/7-10. Its most obvious characteristic is the extremely high RZi (79–84): the other indices are comparable to those of Zone 1. The top of the zone is close to the boundary between the Lunde and Statfjord Formations, and may be a useful indicator of this boundary in cases where it is difficult to identify on the basis of other characteristics, such as geophysical log signatures.

Zone 3
This has similar characteristics to Zone 1, the boundary between Zones 2 and 3 being identified by a decrease in RZi from the very high levels of Zone 2 to more normal values of 48–59. The zone has been identified in 34/7-3, 34/7-6 and 34/7-10. In 34/7-10 the GZi values decline upwards, marking the initiation of the trend seen in the subsequent zone.

Zone 4
This differs from Zone 3 in having low GZi (1–14). RZi also tends to be lower than in Zone 3, generally between 27 and 37, although one sample has an anomalously high value of 69.

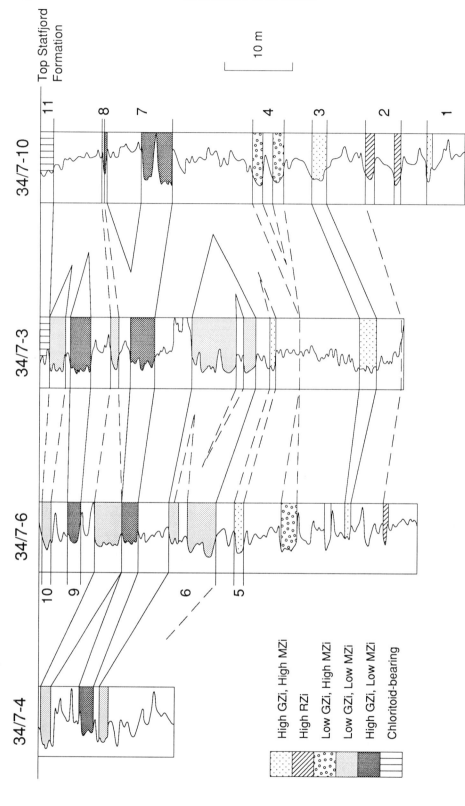

Fig. 8. Subdivision of the Statfjord Formation in the Snorre Field into 11 sandstone zones on the basis of variations in heavy mineral ratios GZi, MZi and RZi (see text), and proposed correlation between Wells 34/7-4, 34/7-6, 34/7-3 and 34/7-10. Dashed line indicates uncertainty because of (1) lack of data for Zones 5 and 8 in 34/7-3; (2) lack of information on the base of Zone 6 in 34/7-4; and (3) the absence of sand with Zone 4 characteristics in 34/7-3. All log curves are natural gamma emissions (GR), as shown in Figs 4 & 6.

MZi remains at the same high levels (> 20), as in Zones 1–3. Zone 4 has been identified in 34/7-6 and 34/7-10, but sandstones with similar characteristics were not found in 34/7-3, where they appear to have been replaced by a thick mudstone unit.

Zone 5

This has been identified only in 34/7-6, and therefore appears to be laterally impersistent. However, there is a possible equivalent in 34/7-3 from which no data are available. It is characterized by a trend of upward-increasing GZi values from 26 to 62, paralleled by the RZi trend from 37 to 47. MZi values are close to 20.

Zone 6

This continues the upward decreasing trend in MZi values initiated in the underlying Zone 5: with one exception, the samples have MZi values between 5 and 17. The zone has low GZi values, which range from 2 to 11: RZi is relatively low, between 21 and 34, although one anomalously high value of 63 was recorded. Zone 6 is well developed in 34/7-3, 34/7-4 and 34/7-6, but the equivalent interval in 34/7-10 is represented by floodplain mudstones.

Zone 7

This marks the return of garnet-rich heavy mineral suites, with GZi between 52 and 91. MZi and RZi values are comparable to those of Zone 6, with MZi between 2 and 18 and RZi between 19 and 53. Zone 7 is well developed in all four studied wells: in 34/7-10 it appears to include an intercalated mudstone unit.

Zone 8

This has a low GZi (0–25). MZi and RZi are similar to those of Zones 6 and 7, with MZi between 2 and 17 and RZi between 23 and 40. It has been identified in 34/7-4, 34/7-6 and 34/7-10, and a possible equivalent occurs in an unanalysed zone in 34/7-3.

Zone 9

This has a high GZi (72–83), RZi between 33 and 41, and MZi between 1 and 6. It has been identified in 34/7-3 and 34/7-6, but is absent in 34/7-4 because of erosion of the uppermost part of the Statfjord Formation. It appears to have passed into a mudstone interval in 34/7-10.

Zone 10

This is marked by a return to low GZi values, with RZi and MZi comparable to the underlying zone. It has only been identified in 34/7-3, but may be present in the unanalysed

zone at the very top of the Statfjord Formation in 34/7-6. It appears to have passed into mudstone in 34/7-10 and is absent in 34/7-4, probably through erosion.

Zone 11

This is characterized by the consistent presence of chloritoid. Chrome spinel is also present in very minor amounts. The GZi is high (81–95), MZi low (1–5), and RZi is relatively high (32–57). This sand is also distinctive by the consistent presence of apatite, the grains of which lack evidence for significant dissolution or reprecipitation. This suggests that this topmost sand was less strongly influenced by acidic groundwaters than the underlying sands: either it was not subject to subaerial exposure following deposition on the floodplain, or it was actually deposited in brackish or marine conditions. In either case, it heralds the subsequent marine transgression that led to the deposition of the marine Dunlin Group mudrocks. Zone 11 is found in 34/7-3 and 34/7-10, but is absent, probably through erosion, in 34/7-4 and 34/7-6.

Zonation within sandstone bodies

In Well 34/7-6 the sandstone body between c. 2350 and 2542 m contains two distinct heavy mineral assemblages, belonging to Zones 7 and 8 (Fig. 9). In Well 34/7-4, to the north, these zones are separated by a fine-grained mudstone unit, whereas in 34/7-10, to the south, Zone 7 is composite, comprising a thick lower sandstone, overlain by a thick mudstone and capped by a very thin sandstone: this is directly overlain by a sandstone assigned to Zone 8. From the sedimentological description of Well 34/7-6, the sandstone appears to be a multistorey channel belt sandstone body with several distinct erosion surfaces. The samples were taken from three different fining-upwards units, two above and one below a prominent carbonate-cemented sandstone with a siltstone cap (Fig. 9). The siltstone may be interpreted as a locally-developed drape.

From a reservoir geological viewpoint, it is important to ascertain whether this sandstone should be described as a single reservoir unit or as a composite unit. Given that no conclusive sedimentological evidence exists to make this distinction, the heavy mineral data provide important criteria for understanding the structure of the sandstone body. By comparison with 34/7-4 and 34/7-10, it appears likely that up to 12 m of section may have been eroded in 34/7-6. In a fluvial depositional environment, erosion of underlying strata by autogenic processes such as

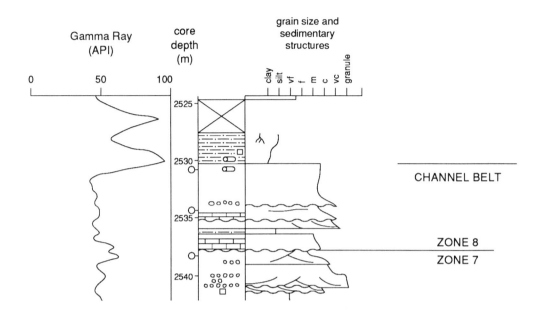

Fig. 9. Detailed sedimentological log of the sandstone body in Well 34/7-6 containing heavy mineral Zones 7 and 8. Notations and symbols are as in Fig. 4.

the migration of channels and channel belts, is expected. However, if 12 m of erosion occurred prior to deposition of Zone 8 in 34/7-6, it is unlikely to have been autogenetically driven, and it is therefore considered to be a response to more regionally significant allogenic processes.

Although the present sample coverage does not allow precise positioning of the major erosion surface within the sandstone body, two erosion surfaces were identified during sedimentological description (Fig. 9). Either of these surfaces could be the sequence boundary (*sensu* van Wagoner *et al.* 1991), representing a regionally significant surface within the Statfjord Formation. In Wells 34/7-4 and 34/7-3 the same surface is preserved as a sandstone-on-mudstone contact (Fig. 8) Both erosion surfaces in Well 34/7-6 are overlain by carbonate-cemented sandstones. The carbonate cement is believed to be derived from the redistribution of carbonate present as reworked calcrete eroded from interfluve areas. It is considered likely that the lower, thicker carbonate-cemented sandstone reflects a more important erosive event, and is therefore preferred as the sequence boundary, but this cannot be verified on the basis of existing data.

Comparison with the Sm–Nd zonation scheme

Mearns *et al.* (1989) produced a zonation of the Statfjord and upper Lunde sequence using the Sm–Nd isotopic method. This enabled subdivision of the sequence into four cycles. Comparison is hampered because different sample sets were used, but the two subdivisions have some features in common (Fig. 10). For example, the top of cycle 1 correlates well with the top of heavy mineral Zone 5, which marks the change from relatively monazite-rich to relatively monazite-poor assemblages. Sandstones below this marker horizon tend to have older Sm–Nd model ages than those above. Superimposed on this, sandstones with a low GZi have older model ages than those with a high GZi. This suggests that variations in abundance of monazite, garnet and zircon may play important roles in controlling the Sm–Nd model age. This needs further investigation, firstly to acquire heavy mineral and whole-rock Sm–Nd data on the same samples, and secondly to determine individual mineral model ages. Such constraints

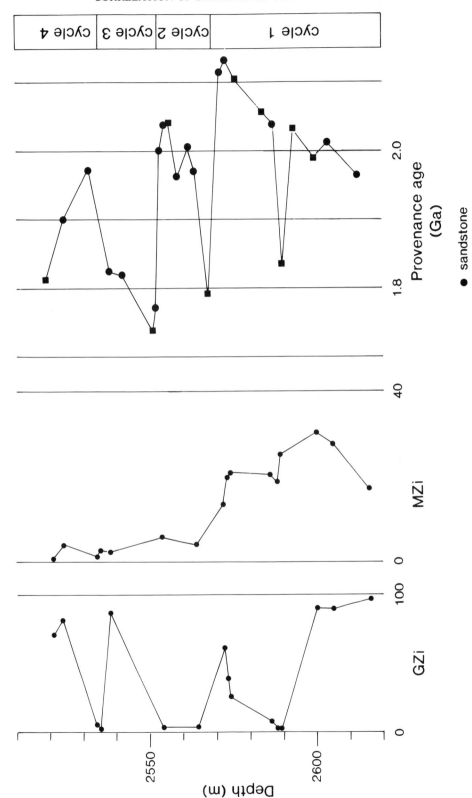

Fig. 10. Relationship between variations in GZi and MZi and the Sm–Nd provenance age (Mearns *et al.* 1989) of the Statfjord Formation in Well 34/7-6.

will also prove useful in determining the ultimate provenance of the Statfjord Formation sandstones, which remains enigmatic, although the Sm–Nd data indicate the strong involvement of Archaean crust.

Conclusions

Changes in sediment provenance provide a basis for stratigraphic correlation of sandstone sequences. Such changes can be detected by means of heavy mineral analysis. In order for heavy mineral correlations to have stratigraphic significance it is essential that the parameters used reflect changes in sediment source and not changes in depositional conditions or diagenetic environment. Two possible approaches are available, one that makes use of aspects of the entire heavy mineral suite, as obtained by conventional optical analysis, and one that concentrates on varietal data, acquired through investigations of the varieties shown by an individual mineral species. Suitable parameters obtained from conventional optical analysis include ratios of minerals with similar hydraulic and diagenetic behaviour, as described by Morton & Hallsworth (1994). Quantification of varietal characteristics is best achieved by the use of objective techniques, most notably single-grain geochemical analysis.

Correlations made on the basis of detecting changes in provenance should only be applied across a single depositional system. Similar events may occur in other depositional systems but need not necessarily occur in the same stratigraphic level. It is therefore important that provenance-based correlations are considered within the overall sedimentological framework. Furthermore, unlike biostratigraphic datums, heavy mineral events are not necessarily unique: repeated shifts in sediment source may produce a series of similar mineralogical events during deposition of a sandstone sequence. This aspect should be carefully considered when using mineralogical data to establish stratigraphic subdivision.

By combining provenance-sensitive aspects of conventional heavy mineral data with geochemical varietal data, a detailed correlation of the fluvial sandstone sequence forming the hydrocarbon reservoir in the Snorre Field has been achieved. There are a number of notable features. One important point is that heavy mineral data demonstrate the lateral imperistence of a number of units, such as Zones 4–6. These lie within the lower and middle reservoir zones as defined by Høimyr et al. (1993), who comment that the sandstones of the upper reservoir unit have wider lateral extent. However, the data indicate that Well 34/7-10 has a poor development of the upper reservoir unit, with heavy mineral Zones 9 and 10 apparently passing into a thick mudstone unit. In general, the zones with a high GZi have greater lateral extent than those with a low GZi.

Another important feature is the identification of zone boundaries at sandstone-sandstone contacts within multistorey sandstone units, such as the Zone 7–8 boundary in 34/7-6 and the Zone 10–11 boundary in 34/7-3. In other wells these zones are separated by mudstone units. This has important implications for reservoir volumetrics and sandbody connectivity. It must be emphasized, however, that the existence of correlatable sandstone zones does necessarily imply that the individual sandbodies correlate directly from well to well.

There is evidence that allogenic processes have played a part in the introduction of different heavy mineral suites, and consequently some of the events may be regionally correlatable. Preliminary data from Statfjord Formation sequences elsewhere in the northern Viking Graben show that similar variations, particularly in the GZi, are present, and it is possible that these will prove to be regionally correlatable features.

The authors are grateful to Statoil and Saga Petroleum a.s. for permission to publish this paper. The paper is published with the approval of the Director of the British Geological Survey (NERC).

References

BASU, A. & MOLINAROLI, E. 1991. Reliability and application of detrital opaque Fe–Ti oxide minerals in provenance determination. *In:* MORTON, A. C., TODD, S. P. & HAUGHTON, P. D. W. (eds) *Developments in Sedimentary Provenance Studies.* Geological Society, London, Special Publication, **57**, 55–65.

BATEMAN, R. M. & CATT, J. A. 1985. Modification of heavy mineral assemblages in English coversands by acid pedochemical weathering. *Catena,* **12**, 1–21.

BRIGGS, L. I., McCULLOCH, D. S. & MOSER, F. 1962. The hydraulic shape of sand particles. *Journal of Sedimentary Petrology,* **32**, 645–656.

BURLEY, S. D., KANTOROWICZ, J. D. & WAUGH, B. 1985. Clastic diagenesis. *In:* BRENCHLEY, P. J. & WILLIAMS, B. P. J. (eds) *Sedimentology: Recent Developments and Applied Aspects.* Geological Society, London, Special Publication, **18**, 189–226.

CAVAZZA, W. & GANDOLFI, G. 1992. Diagenetic processes along a basin-wide marker bed as a function of burial depth. *Journal of Sedimentary*

Petrology, **62**, 261–272.

DEEGAN, C. E. & SCULL, B. J. 1977. *A standard lithostratigraphic nomenclature for the central and northern North Sea*. Report of the Institute of Geological Sciences, **77/25**.

DIETZ, V. 1993. Experiments on the influence of transport on shape and roundness of heavy minerals. *Contributions to Sedimentology*, **1**, 103–125.

DOYLE, L. J., CARDER, K. L. & STEWARD, R. G. 1983. The hydraulic equivalence of mica. *Journal of Sedimentary Petrology*, **53**, 643–648.

FLORES, R. M. & SHIDELER, G. L. 1978. Factors controlling heavy mineral variations on the south Texas outer continental shelf. *Journal of Sedimentary Petrology*, **48**, 269–280.

FRIESE, F. W. 1931. Untersuchung von Mineralen auf Abnutzbarkeit bei Verfractung im Wasser. *Tschermaks Mineralogisches und Petrographisches Mitteilungen*, **41**, 1–7.

FRIIS, H. 1976. Weathering of a Neogene fluviatile fining upwards sequence at Voervadsbro, Denmark. *Bulletin of the Geological Society of Denmark*, **25**, 99–105.

——, NIELSEN, O. B., FRIIS, E. M. & BALME, B. E. 1980. Sedimentological and palaeobotanical investigations of a Miocene sequence at Lavsbjerg, central Jutland, Denmark. *Danmarks Geologiske Undersøgelse*, **Årbog 1979**, 51–67.

GRIGSBY, J. D. 1990. Detrital magnetite as a provenance indicator. *Journal of Sedimentary Petrology*, **60**, 940–951.

GRIMM, W. D. 1973. Stepwise heavy mineral weathering in the Residual Quartz Gravel, Bavarian Molasse (Germany). *Contributions to Sedimentology*, **1**, 103–125.

HØIMYR, Ø., KLEPPE, A. & NYSTUEN, J. P. 1993. Effects of heterogeneities in a braided stream channel sandbody on the simulation of oil recovery: a case study from the Lower Jurassic Statfjord Formation, Snorre Field, North Sea. *In:* ASHTON, M. (ed.) *Advances in Reservoir Geology*. Geological Society, London, Special Publication, **69**, 109–134.

JEANS, C. V., REED, S. J. B. & XING, M. 1993. Heavy mineral stratigraphy in the UK Trias of the North Sea Basin. *In:* PARKER, J. R. (ed.) *Petroleum Geology of Northwest Europe: Proceedings of the 4th Conference*. Geological Society, London, 609–624.

JOHNSSON, M. J., STALLARD, R. F. & LUNDBERG, N. 1991. Controls on the composition of fluvial sands from a tropical weathering environment: Sands of the Orinoco drainage basin, Venezuela and Colombia. *Bulletin of the Geological Society of America*, **103**, 1622–1647.

KRYNINE, P. D. 1946. The tourmaline group in sediments. *Journal of Geology*, **54**, 65–87.

MANGE, M. A. & MAURER, H. F. W. 1992. *Heavy Minerals in Colour*. Chapman & Hall, London.

MEARNS, E. W., KNARUD, R., RAESTAD, N., STANLEY, K. O. & STOCKBRIDGE, C. P. 1989 Samarium-neodymium isotope stratigraphy of the Lunde and Statfjord Formations of Snorre Oil Field, northern North Sea. *Journal of the Geological Society of London*, **146**, 217–228.

MILLIKEN, K. L. 1988. Loss of provenance information through subsurface diagenesis in Plio-Pleistocene sediments, northern Gulf of Mexico. *Journal of Sedimentary Petrology*, **58**, 992–1002.

MORTON, A. C. 1984. Stability of detrital heavy minerals in Tertiary sandstones of the North Sea Basin. *Clay Minerals*, **19**, 287–308.

—— 1985. Heavy minerals in provenance studies. *In:* ZUFFA, G. G. (ed.) *Provenance of Arenites*. Reidel, Dordrecht, 249–277.

—— 1986. Dissolution of apatite in North Sea Jurassic sandstones: implications for the generation of secondary porosity. *Clay Minerals*, **21**, 711-733.

—— 1991. Geochemical studies of detrital heavy minerals and their application to provenance studies. *In:* MORTON, A. C., TODD, S. P. & HAUGHTON, P. D. W. (eds) *Developments in Sedimentary Provenance Studies*. Geological Society, London, Special Publication, **57**, 31–45.

—— 1992. Provenance of Brent Group sandstones: heavy mineral constraints. *In:* MORTON, A. C., HASZELDINE, R. S., GILES, M. R. & BROWN, S. (eds) *Geology of the Brent Group*. Geological Society, London, Special Publication, **61**, 227–244.

—— & HALLSWORTH, C. R. 1994. Identifying provenance-specific features of detrital heavy mineral assemblages in sandstones. *Sedimentary Geology*, **90**, 241–256.

——, —— & WILKINSON, G. C. 1993. Evolution of sand provenance during Paleocene deposition in the northern North Sea. *In:* PARKER, J. R. (ed.) *Petroleum Geology of Northwest Europe: Proceedings of the 4th Conference*. Geological Society, London, 73–84.

—— & JOHNSSON, M. J. 1993. Factors influencing the composition of detrital heavy mineral suites in Holocene sands of the Apure River drainage basin, Venezuela. *In:* BASU, A. & JOHNSSON, M. J. (eds) *Processes Controlling the Composition of Clastic Sediments*. Geological Society of America, Special Paper, **284**, 171–185.

—— & SMALE, D. 1990. The effects of transport and weathering on heavy minerals from the Cascade River, New Zealand. *Sedimentary Geology*, **68**, 117–123.

NYSTUEN, J. P., KNARUD, R., JORDE, K. & STANLEY, K. O. 1989. Correlation of Triassic to Lower Jurassic sequences, Snorre field and adjacent areas, northern North Sea. *In:* COLLINSON, J. (ed.) *Correlation in Hydrocarbon Exploration*. Graham & Trotman, London, 273–289.

POLDERVAART, A., 1955. Zircon in rocks, 1: sedimentary rocks. *American Journal of Science*, **253**, 433–461.

RITTENHOUSE, G. 1943. Transportation and deposition of heavy minerals. *Bulletin of the Geological Society of America*, **54**, 1725–1780.

RUBEY, W. W. 1933. The size distribution of heavy minerals within a water-lain sandstone. *Journal of Sedimentary Petrology*, **3**, 3–29.

RUSSELL, R. D. 1937. Mineral composition of Mississippi river sands. *Bulletin of the Geological Society of America*, **48**, 1308–1348.

SHUKRI, N. M. 1949. The mineralogy of Nile sediments. *Quarterly Journal of the Geological Society of London*, **105**, 511–529.

THIEL, G. A. 1940. The relative resistance to abrasion of mineral grains of sand size. *Journal of Sedimentary Petrology*, **10**, 103–124.

—— 1945. Mechanical effects of stream transportation in mineral grains of sand size. *Bulletin of the Geological Society of America*, **56**, 1207.

VAN ANDEL, T. H. 1950. *Provenance, Transport and Deposition of Rhine Sediments*. Veenman en Zonen, Wageningen.

VAN WAGONER, J. C., MITCHUM, R. M., CAMPION, K. M. & RAHMANIAN, V. D. 1991. *Siliciclastic Sequence Stratigraphy in Well Logs, Cores and Outcrops*. American Association of Petroleum Geologists, Methods in Exploration Series, **7**.

YIELDING, G., BADLEY, M. E. & ROBERTS, A. M. 1992. The structural evolution of the Brent province. *In:* MORTON, A. C., HASZELDINE, R. S., GILES, M. R. & BROWN, S. (eds) *Geology of the Brent Group*. Geological Society, London, Special Publication, **61**, 27–43.

Subdivision and correlation of monotonous sandstone sequences using high-resolution heavy mineral analysis, a case study: the Triassic of the Central Graben

MARIA A. MANGE-RAJETZKY

Department of Earth Sciences, Oxford University, Parks Road, Oxford, OX1 3PR, UK

Abstract: The subdivision and correlation of monotonous siliciclastic sequences with poor biostratigraphic control and few lithostratigraphic markers has long been a problem. The need to find an efficient approach is even greater for subsurface sequences where information from cores and wireline logs has proved inadequate. In this short paper a novel approach, high-resolution heavy minerals analysis (HRHMA), is introduced, which is particularly suitable for the subdivision of problematical clastic sediments where other techniques are inconclusive. A full paper, which will present the technique in more detail, will be published shortly.

High-resolution heavy mineral analysis (HRHMA) is based on the recognition that the majority of rock-forming and accessory minerals show a wide diversity of size, habit, colour, internal structure, chemistry and optical properties, controlled principally by specific conditions during crystallization. Since a wide range of lithologies provide detritus to siliciclastic sediments, their heavy mineral assemblages are complex and an individual heavy mineral species may comprise several kinds of varieties, each preserving a different genetic and/or sedimentological history.

The technique

Principles and practice

HRHMA records diagnostic varieties of individual heavy minerals, thus the number of variables is significantly higher compared to those of a simple, species-level, analysis. Distinguishing the varietal types of the chemically highly resistant minerals, such as zircon, tourmaline and, when suitable, apatite prove most rewarding since they are ubiquitous and remain stable during diagenesis. Therefore, HRHMA is especially useful in the study of sediments from which the diagnostic but chemically unstable species were eliminated by post-depositional dissolution, leaving a residue of ultrastable zircon + tourmaline + rutile ± apatite (Mange & Maurer 1992). An important advantage of this technique is that, because it deals with each particular species, the influence of modifying factors [especially hydraulic and diagenetic (Morton 1985)] is considerably reduced. In addition to core samples, HRHMA can be successfully applied to cuttings.

Variables are based on grain morphology, colour, internal structure, etc. The measurement of grain shape (degree of roundness and sphericity) is a standard sediment petrological procedure and is an important part in the description of sedimentary textures. It has been in general use since the 1930s (Wadell 1935; Cailleux 1947, 1952; Powers 1953, 1982; Pryor 1971; Pettijohn *et al.* 1973). Colour varieties, especially those of zircon, can be provenance-diagnostic; for example, Mackie (1923) and Tomita (1954) successfully traced sediments to respective parent lithologies by using purple zircon varieties. Structural types of zircon (showing either euhedral zoning or overgrowth, etc.) are significant since they are linked to specific parageneses (Speer 1980).

In the present case, zircon, tourmaline and apatite varieties were distinguished. The most informative morphological categories range from sharp euhedral (such as zircon) or prismatic crystals with sharp terminations (e.g. tourmaline, apatite) through rounded prisms, subrounded morphologies to well-rounded and spherical grains. Zircons exhibiting zoning or overgrowth were recorded separately. The colourless and pink to purple zircon types were also allocated into separate categories. During analysis the total of 100 varietal grains for each species were point counted and then recalculated into relative percentages.

Application

The Triassic continental red beds of the Central Graben of the North Sea – Quadrants 22 and 29, UK, and Quadrants 6 and 7, Norway (Fig. 1) – are lithologically monotonous and comprise alternations of dominantly red mudstones, siltstones and sandstones, representative of alluvial fan, fluvial, sabkha and lacustrine

From Dunay, R. E. & Hailwood, E. A. (eds), 1995, *Non-biostratigraphical Methods of Dating and Correlation*
Geological Society Special Publication No. 89, pp. 23–30

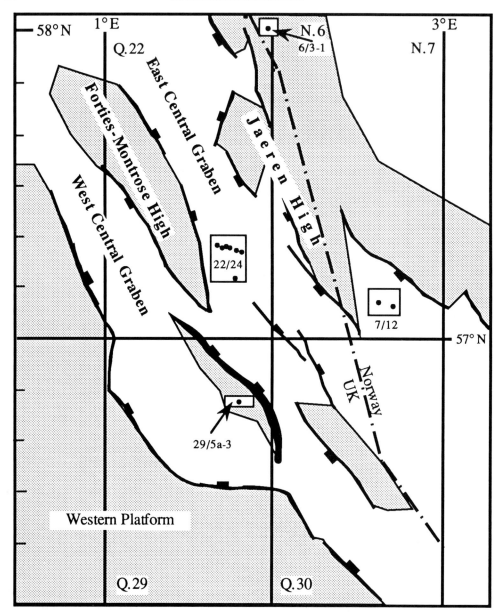

Fig. 1. Map of study area, showing major structural elements of the Central Graben and generalized location of the analysed wells.

environments, with occasional marine incursions near to the base. The succession is rarely complete and thickness is extremely variable. This is due both to variation in the initial depositional thickness as a result of accommodation space available at a particular location, controlled by salt-related tectonics (Hodgson *et al.* 1992), and to erosion due to extensive middle Jurassic uplift. Their subdivision and correlation

have long been a problem, compounded by their poor biostratigraphic record.

Triassic stratigraphy of the central and northern North Sea is based on the lithostratigraphy established for geographical areas (Deagan & Skull 1977) and subsequently modified by well data (Vollset & Doré 1984; Lervik *et al.* 1989; Fisher & Mudge 1990; Cameron 1993). Dating the sequences still involves a great deal of

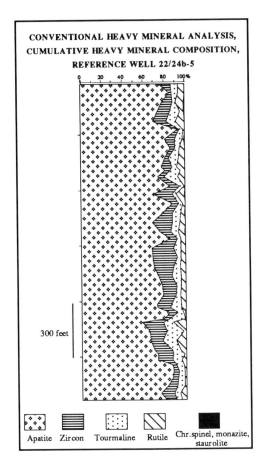

CONVENTIONAL HEAVY MINERAL ANALYSIS,
CUMULATIVE HEAVY MINERAL COMPOSITION,
REFERENCE WELL 22/24b-5

Apatite   Zircon   Tourmaline   Rutile   Chr.spinel, monazite, staurolite

Fig. 2. Heavy mineral spectrum of reference Well 22/24b-5 using conventional heavy mineral analysis.

assemblages (using the 0.063–0.210 mm size fraction) is depicted in Fig. 2: the apatite-dominated heavy mineral spectrum of reference Well 22/24b-5 shows no visible mineralogical markers and negligible overall variations in the relative proportions of individual species. HRHMA, on the same sequence of samples, has provided a strikingly different picture when the cumulative percentages of zircon, tourmaline and apatite varieties and ratios of heavy mineral pairs were plotted against depth (Fig. 3). This figure reveals that the heavy mineral varieties of these seemingly monotonous red beds change systematically, thus permitting the establishment of subdivisions, termed heavy mineral zones.

Subdivisions

The Triassic sequences in Quadrant 22 comprise three major, well-constrained heavy mineral zones (SB, M and A). These can be subdivided into subzones. Figure 4, showing the occurrence of tourmaline varieties, illustrates these zones and their lateral continuity: (1) The lower (Zone SB) contains monotonous, polycyclic suites typified by the extreme roundness of apatite. It is laterally extensive and corresponds to the Smith Bank Formation. The overlying sequences represent the Skagerrak Formation; (2) The middle zone (Zone M) is generally transitional and commences with a laterally widespread lower part, characterized by uniform mineralogy (subzone q), but the grains are less rounded than below. The simple, uniform suites of the lower portion of this zone evolve into higher diversity assemblages, substantiating further subdivisions (e.g. subzones d, c, etc.). The varietal composition of the uppermost part (subzone b), distinguished by the appearance of sharp prismatic tourmaline and further reduced apatite roundness, is particularly important. These assemblages are precursors of a significant change, detected in the top interval. (3) The upper zone (Zone A) is highlighted by a contrasting varietal spectrum. At its lower boundary all varietal patterns change dramatically with the abrupt increase in abundance of the first cycle grains. The horizon at the base of Zone A can be traced from well to well, thus providing a basin-wide Marker Horizon (MH).

The well analysed in Quadrant 29 and wells in Quadrant 7 in the Norwegian sector comprise only two major divisions. The upper interval (Zone A), bounded by the MH, corresponds to that of Quadrant 22, but in the sequences below Zone A only individual heavy mineral zones can be delineated. Their mineralogy differs from that in Quadrant 22 since they frequently contain

uncertainty, though recently in the south Central Graben palynomorphs have been successfully integrated into a refined lithostratigraphy (Goldsmith *et al.* 1995).

Earlier heavy mineral analyses are found mostly in confidential oil company reports. These were employed to provide the clues on provenance and to establish some kind of heavy mineral stratigraphy. Jeans *et al.* (1993) attempted to construct a heavy mineral stratigraphy in the UK Triassic (Western Approaches, onshore England and the central North Sea). They included several wells from Quadrants 22 and 29 in order to facilitate correlation and to detect the regional extent of particular heavy mineral assemblages. Since all these analyses were limited to species level, the potential of HRHMA was not tested.

The low resolution of conventional, species-level analysis, as applied to highly mature

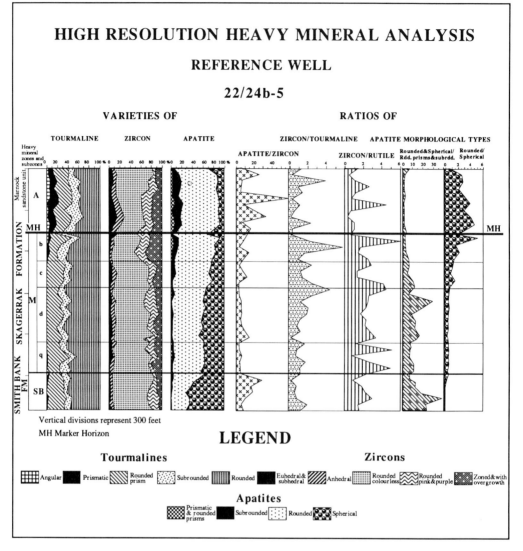

Fig. 3. Results of HRHMA, reference Well 22/24b-5, indicating heavy mineral zones and subzones and the position of the Marker Horizon.

first-cycle suites.

In the Marnock Field of the north Central Graben high-quality reservoir sandstones, deposited at the top of the Skagerrak Formation during the late Triassic (Norian to Rhaetian), are informally termed the Marnock Sandstone unit (Hodgson *et al.* 1992; Cameron 1993). Varietal signatures of the analysed sandstones from this field unequivocally identify this unit as the mineralogically distinct and laterally widespread Zone A, with highly significant implications, since its base is defined by mineralogical criteria as the MH. Although the base of this

unit can be determined by detailed logging of core, it is not easily distinguishable on wireline logs (Cameron 1993).

The recognition of Zone A, with the Marker Horizon at its base, in and beyond the Marnock Field, amplifies the value of HRHMA. This method, by revealing mineralogically distinct intervals, has thus provided both a framework for lithostratigraphic subdivision and a potential aid for the correlation of relevant zones, either on a local or on a basin-wide scale. Provenance-controlled fundamental changes in mineralogy also pinpoint correlatable mineralogical markers.

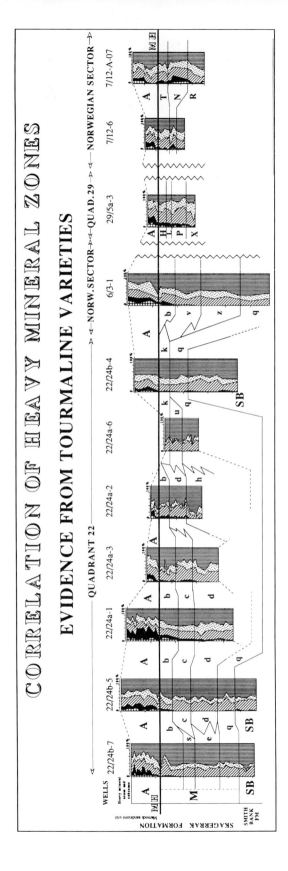
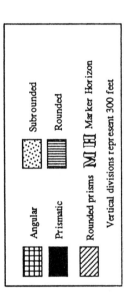

Fig. 4. Regional correlation of the heavy mineral zones based on tourmaline varieties.

Whereas in the south Central Graben the Skagerrak Formation is subdivided into widely correlatable sandstone and mudstone members (Goldsmith *et al.* 1995), this has not been possible in the north Central Graben. As recounted above, however, results of HRHMA now provide a framework within which subdivisions can be made and this is particularly important for the thick Zone M. Until HRHMA has been carried out in the south Central Graben the provenance and lithostratigraphic relationships of the two groups of Skagerrak Formation sediments are unlikely to be resolved.

Correlation

Well-to-well correlation of particular zones was achieved by comparing the varietal patterns and changes in the distributional trends of particular heavy mineral ratios with the wireline log motifs. An example for confident correlation using heavy mineral varieties is shown by the tourmaline patterns in Fig. 4.

Two distinct correlations are evident; the first relates to the upper zone (Zone A) whereas the second connects the underlying intervals: (1) Zone A has been recognized in the majority of wells and forms a mineralogically fully-constrained interval. It can be correlated basin-wide. In Quadrant 22 and in Well 6/3-1 subzone q of Zone M (evolving from Zone SB) and subzone b (underlying the MH) can be correlated across several wells. Other subzones (e.g. c and d) have a more limited extent. (2) The lower zones in Quadrant 22 and in Well 6/3-1 differ from those in Quadrant 29 and Quadrant 7 (Norwegian sector), and consequently cannot be correlated with them.

Evolution and provenance

The threefold character of the most complete sequences in Quadrant 22 indicates three phases of sedimentation:

(1) The Smith Bank Formation was deposited in a low energy environment. The wide spatial distribution and homogenization of the fine to very fine grained detritus suggests occasional distal sheet floods. Its polycyclic aspect indicates reworking of mature pre-existing sediments. High sphericity of the grains, many with a 'frosted' appearance, signal that aeolian (probably marginal Permian) deposits could be important contributors.

(2) In the lower subzones of Zone M of the Skagerrak Formation sedimentation commenced with limited internal drainage which progressively evolved into a higher energy fluvial regime. This is deduced from the upwards increase in mineralogical diversity and enhanced spatial variations. In the lower part of the sequence the main source of the detritus was intrabasinal, probably derived from fault scarps, intrabasinal highs and, occasionally, from marginal areas such as the East Shetland Platform and adjacent contemporary exposures (containing pre-Triassic sediments, metasediments and ophiolites of the Caledonides). The variations observed in the different wells can be interpreted in terms of separate systems with varying energies operating at different periods, frequent intrabasinal reworking and local tectonic movements. Mineral signatures of subzone b (Fig. 4) and its broader spatial extent point to an expanding drainage network, reaching distant primary lithologies. This stage is seen as being the precursor of an event that caused widespread modifications in palaeogeography.

(3) The dramatic change at the MH and the broad basin-wide continuity of Zone A represent an event which occurred on a regional scale. By the late Triassic the northward drift of continents caused a change from a semi-arid to a semi-humid climate. Tectonic pulses initiated uplift in Scandinavia and Greenland (Ziegler 1978; Jacobsen & van Veen 1984). Prior to these changes transport of material from Scandinavia was inhibited by low energy conditions, the existence of intrabasinal highs and marginal sub-basins adjacent to the Fenno–Skandian Shield. By Zone A time, increased run-off and basin configuration permitted the expansion of transport paths as far as the Fenno–Scandian complexes, thereby diluting the previously dominant intrabasinal polycyclic suites with abundant first-cycle minerals. The broad spatial dispersal of the first-cycle detritus, shown by the lateral continuity of Zone A, was promoted by a well-established fluvial system.

Sand provenance in the lower zones in Quadrant 29 and Quadrant 7 (Norwegian sector) share common characteristics, indicated by the frequent presence of first-cycle material. However, there is a dissimilarity between Quadrants 29 + Norwegian sector 7 and Quadrants 22 + Well 6/3-1 in the interval below the MH, which suggests that the two groups of sediments were the products of separate depositional systems. These may have existed at different times and were confined to different structural settings.

Concluding remarks on the applicability of HRHMA

The use of this technique has provided remarkably good results for a variety of sedimentary formations, the majority of which were problematical. In massive structureless turbidite sequences (Witch Ground Graben, North Sea), a heavy mineral stratigraphy and well-to-well correlation has been successfully established. In addition, varietal signatures have revealed the existence of different fan systems, fed from contrasting sources (Mange *et al.*, unpublished work). A similarly good resolution has been achieved for Rotliegend sediments (Mange, confidential data). Here boundaries of heavy mineral zones, erected independently, were found not only to coincide accurately with lithostratigraphic units, identified by the interpretation of electric logs, but also formed discrete zones within these major divisions. Facies changes were also highlighted by differences in the varietal spectra.

The wealth of informative data obtained by using HRHMA ensures that the extent of subjectivity involved in this technique (mainly the definition of morphological and colour categories) is minimal, and is negligible when compared to the complexity of information it yields. HRHMA is internally consistent, adds versatility and sophistication to conventional heavy mineral analysis, permits subdivision and correlation of monotonous, biostratigraphically barren siliciclastic sequences and complements facies analysis. It can also indicate changes in sedimentary processes, in suitable cases it provides information on dating, assists in differentiation and/or identification of submarine fans and always helps reconstructing provenance.

This work benefited from stimulating discussions with Harold G. Reading, Philip A. Allen and David T. Wright. Steve Wyatt's continued help and assistance in the laboratory is highly appreciated.

Permission to use heavy mineral data was granted by Shell UK Exploration and Production, Esso Exploration and Production UK Limited, BP Exploration, BP Norway, Agip UK, Mobil North Sea Limited, Monument Oil and Gas and ARCO British Limited, all of whom the author gratefully acknowledges. The opinions expressed here are those of the author and do not necessarily reflect those of companies involved in supplying material.

References

CAILLEUX, A. 1947 L'indice d'emoussé des grains de sable et gres. *Revue Geomorphologique Dynamique*, **3**, 78–87.

—— 1952. Morphoskopische Analyse der Geschiebe und Sandkörner und ihre Bedeutung für die Paläoklimatologie. *Geologische Rundschau*, **40**, 11–19.

CAMERON, T. D. J. 1993. Triassic, Permian and pre-Permian of the central and northern North Sea. *In:* KNOX, R. W. O'B. & CORDEY, W. G. (eds) *Lithostratigraphic Nomenclature of the UK North Sea.* British Geological Survey Publication.

DEAGAN, C. E. & SCULL, B. J. 1977. *A standard lithostratigraphic nomenclature for the central and northern North Sea.* Institute of Geological Sciences Report *77/25*.

FISHER, M. J. & MUDGE, D. C. 1990. Triassic. *In:* GLENNIE, K. W. (ed.) *Introduction to the Petroleum Geology of the North Sea.* Blackwell, Oxford, 191–218.

GOLDSMITH, P., RICH, B. & STANDRING, J. 1995. Triassic correlation and stratigraphy in the south Central Graben, United Kingdom North Sea. *In:* BOLDY, S. A. R. (ed.) *Permian and Triassic Rifting in NW Europe.* Geological Society, London, Special Publication, **91**.

HODGSON, N. A., FARNSWORTH, J. & FRASER, A. J. 1992. Salt-related tectonics, sedimentation and hydrocarbon plays in the Central Graben, North Sea, UKCS. *In:* HARDMAN, R. F. P. (ed.) *Exploration Britain: Geological insights for the next decade.* Geological Society, London, Special Publication, **67**, 31–63.

JACOBSEN, V. W. & VAN VEEN, P. 1984. The Triassic offshore Norway north of 62°N. *In:* BROOKS, J. & GLENNIE, K. W. (eds) *Petroleum Geology of the North European Margin.* Graham & Trotman, London, 317–327.

JEANS, C. V., REED, S. J. B. & XING, M. 1993. *In:* PARKER, J. R. (ed.) *Petroleum Geology of Northwest Europe: Proceedings of the 4th Conference.* The Geological Society, London, 609–624.

LERVIK, K. S., SPENCER, A. M. & WARRINGTON, G. 1989. Outline of Triassic stratigraphy and structure in the central and northern North Sea. *In:* COLLINSON, J. D. (ed.) *Correlation in Hydrocarbon Exploration.* Graham & Trotman, London, 173–189.

MACKIE, W. 1923. The source of purple zircons in the sedimentary rocks of Scotland. *Transactions of the Edinburgh Geological Society*, **11**, 200–213.

MANGE, M. A. & MAURER, H. F. W. 1992. *Heavy Minerals in Colour.* Chapman & Hall, London.

MORTON, A. C. 1985. Heavy minerals in provenance studies. *In:* ZUFFA, G. G. (ed.) *Provenance of Arenites.* Reidel, Dordrecht, 249–277.

PETTIJOHN, F. J., POTTER, P. E. & SIEVER, R. 1973. Sand and Sandstone. Springer-Verlag, New York.

POWERS, M. C. 1953. A new roundness scale for sedimentary particles. *Journal of Sedimentary Petrology*, **23**, 117–119.

—— 1982. Comparison chart for estimating roundness and sphericity. *AGI Data Sheet 18*. American Geological Institute.

PRYOR, W. A. 1971. Grain Shape. *In:* CARVER, R. E. (ed.) *Procedures in Sedimentary Petrology*. Wiley-Interscience, New York, 131–150.

TOMITA, T. 1954. Geologic significance of the colour of granite zircon and the discovery of the Precambrian in Japan. *Kyushu University Memoir*, Faculty of Science, Series D, Geology, **4**, 135–161.

SPEER, J. A. 1980. Zircon. *In:* RIBBE, P. H. (ed.) *Reviews in Mineralogy. Vol. 5. Orthosilicates*. Mineralogical Society of America, Washington DC, 67–112.

WADELL, H. A. 1935. Volume, shape, and roundness of quartz particles. *Journal of Geology*, **43**, 250–280.

VOLLSET, J. & DORÉ, A. G. (eds) 1984. *A revised Triassic and Jurassic lithostratigraphic nomenclature for the Norwegian North Sea*. Norwegian Petroleum Directorate, Bulletin 3.

ZIEGLER, P. A. 1978. North-western Europe: Tectonics and basin development. *Geologie en Mijnbouw*, **57**, 509–626.

Clay mineral stratigraphy in Palaeozoic and Mesozoic red bed facies onshore and offshore UK

C. V. JEANS

Department of Earth Sciences, Cambridge University, Downing Street, Cambridge CB2 3EQ, UK

Abstract: Different methods of clay mineral correlation within sedimentary sequences provide varying degrees of stratigraphical resolution. These range from: (1) precise time correlation by the recognition of clay-forming events; (2) levels of general stratigraphical significance based upon changes in clay mineral assemblages reflecting widespread climatic, topographic or sea-level changes; (3) correlations of no stratigraphical significance where the distribution of clay minerals reflect patterns of detrital dispersal, pore-fluid flow in buried sediments, or metamorphism. The main use of clay mineral stratigraphy is to provide an alternative means of correlating effectively unfossiliferous sediments by tying them into biostratigraphically dated standard sequences in which the clay mineral variations are well established. Since 1980 it has been used in offshore UK to help correlation in the Old Red Sandstone (Devonian–Lower Carboniferous) of the North Sea and west of Shetland, and the New Red Sandstone (late Carboniferous–Triassic) of the Southwest Approaches, southern North Sea Basin and the central North Sea. Results of these studies are described and critically assessed.

Clay stratigraphy involves the identification and correlation of sediments using the distribution of clay mineral assemblages. It can be applied to any sediments; however, its main use is with sedimentary rocks which are either intrinsically or effectively barren of biostratigraphically useful fossils. This paper describes the use of clay mineral stratigraphy in the study of the Old Red Sandstone (Devonian–Lower Carboniferous) and New Red Sandstone (late Carboniferous–Triassic) in the offshore areas of the UK. Both these facies are of continental origin and are characterized by restricted floras and faunas, as well as highly oxidizing conditions of preservation. Neither the Old Red Sandstone nor New Red Sandstone is barren of fossils, although there are considerable sections that are completely unfossiliferous. In the offshore, both formations are effectively barren in respect to routine micropalaeontology. In 1980, when clay stratigraphical methods were first applied to the offshore Old Red Sandstone and New Red Sandstone, there were sufficient onshore data to indicate that their clay mineralogies in relation to lithofacies were fundamentally different and consequently could be differentiated in the subsurface. Since then a considerable range of clay mineral correlations have been achieved. The first part of this paper deals with the principles underlying different styles of clay mineral correlation and their stratigraphical significance, whereas the second and main part discusses examples from the Southwest Approaches,

southern North Sea Basin, central North Sea and areas West of the Shetland (Fig. 1).

Principles and types of clay mineral stratigraphy

The mineral components of a clay mineral assemblage ($< 2\,\mu$m) separated from a sediment may have the following genetic origins: (1) *Detrital.* Either recycled from earlier sediments or representing clay neoformed in coeval sediments and soils. The type of mineral component will be controlled by climate, topography and source rocks. (2) *Intrinsic neoformation (Jeans 1984 for discussion).* Clay components formed within the sediment by reaction of its autochthonous pore fluids with its unstable components. The nature of the mineral components will be controlled by the pore-fluid chemistry and the availability of the sediment's reactive components: these factors are ultimately influenced by climate and topography. (3) *Non-intrinsic neoformation (Jeans 1984 for discussion).* These components represent the reaction of allochthonous pore fluids with the sediment. The immediate controlling factors are the chemistry of the pore fluids and the reactivity of the host sediment. The distribution pattern will be controlled by the pore-fluid pathways during burial diagenesis. (4) *Burial metamorphism (Jeans 1984 for discussion).* This part of the clay mineral assemblage develops as the result of increased temperature or pressure during burial.

From Dunay, R. E. & Hailwood, E. A. (eds), 1995, *Non-biostratigraphical Methods of Dating and Correlation* Geological Society Special Publication No. 89, pp. 31–55

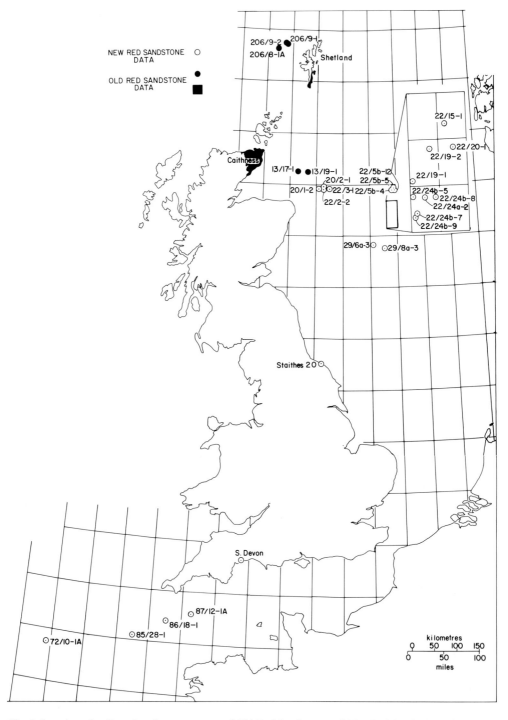

Fig. 1. Location of wells and onshore exposures of Old Red Sandstone and New Red Sandstone.

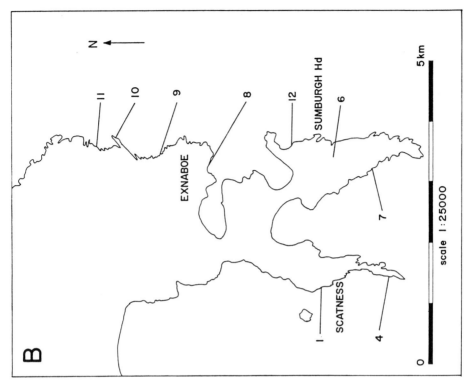

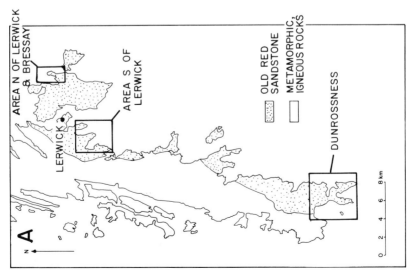

Fig. 2. (**A**) Old Red Sandstone outcrop, SE Shetland. (**B**) Sampled sedimentary sequences, Dunrossness, Shetland. (**C**) Sampled sedimentary sequences, areas south and north of Lerwick and Bressay, Shetland.

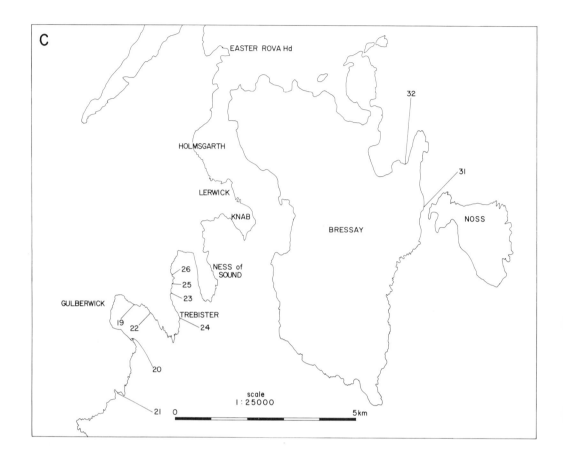

The overall control is by rates of heat flow and the burial history.

By using the different genetic components of clay assemblages, five types of clay mineral correlation can be achieved, each with different stratigraphical significance. (1) Local correlation of detrital clay variation within a single or closely related deposition system. This has poor stratigraphical significance and correlation may be limited by differential settling of various clay minerals controlled by particle size and/or flocculation characteristics. (2) Use of combined detrital and intrinsically neoformed clay assemblages. Good general stratigraphy based upon regional changes in climate and topography. Standard biostratigraphically-dated sequences are needed for comparison. (3) Clay mineral forming events. These can be very widespread with excellent stratigraphical significance. There may be difficulties of recognition as the clay mineral assemblages are likely to vary with the chemistry of the depositional environment.

Perhaps the best known examples are argillized ash bands representing the widespread fallout of ash from subaerial volcanic eruptions: correlation difficulties may include the lack of specific identification characters, varying preservation in different depositional settings (best preserved in sites of rapid, fine-grained sedimentation), and environmentally controlled variations in the mineraology of the argillization products (typically kaolin-rich in freshwater/brackish environments and smectite-rich in the marine). (4) Correlation of non-intrinsic neoformed clay components reflecting patterns of allochthonous pore-fluid reaction. This has no stratigraphical significance. (5) Correlation of degree of meta- mor- phism. This is generally of no stratigraphical significance. In certain circumstances, the degree of metamorphism can be used to distinguish between strata of different ages. Glasmann & Wilkinson (1993) have used this to differentiate between Devonian and Permo-Triassic sediments in Quadrant 9, North Sea.

Sampling and mineral analysis

It is important when deciding on the type of correlation required to select a suitable sampling style. Three points are particularly important. Variations in clay mineralogy are, with few exceptions, cryptic to either the naked eye or light microscopy; the clay mineralogy of a sedimentary sequence may vary both with and independently of lithofacies. The clay assemblages of argillaceous sediments are dominantly detrital with lesser contribution from intrinsic neoformation and burial metamorphism. In sandstones, the clay assemblages are dominantly neoformed either intrinsically or non-intrinsically, depending upon whether the sandstone has been protected from allochthonous pore-fluids.

It is important that consistent methods of mineralogical analysis are maintained, otherwise comparison between individual samples or different wells is difficult. The results described in this paper have been obtained over a 20-year period, the majority under commercial conditions with temporal and financial limitations at a premium. Consequently, fashionable niceties of academic studies (details of mixed layering, crystallinity) have been intentionally avoided because in an initial examination of the X-ray traces there was no evidence that these characters had an important role in clay stratigraphy. General details of the analytical methods used are as follows. Semi-quantitative clay mineral analysis has been based upon X-ray diffractograms ($CuK\alpha$) obtained from untreated (Ca^{2+} saturated), glycerolated, 400°C (or 440°C) heated and 550°C heated oriented aggregates of the $<2\,\mu m$ e.s.d. fraction prepared by the procedure of Jeans (1978, p. 625). Semi-quantification was based upon a modified version of Griffin's (1971) method, semi-quantification of non-clay minerals was based upon X-ray analysis of unoriented bulk sediment samples using either peak heights of characteristic X-ray lines (e.g. calcite 104, dolomite 104, halite 200) or the external standard binary mixture method.

The following points concerning identification should be noted. Smectite and smectite/mica refer to glycerol sensitive minerals giving, respectively, 0 0 1 peaks at 17 Å and in the 10–17 Å region in the glycerolated X-ray trace and collapsing to c. 10 Å on heating. Corrensite refers to glycerol-sensitive minerals giving a c. 32 Å 0 0 1 peak in the glycerolated X-ray trace and exhibiting partial collapse on heating. Chlorite/smectite refers to glycerol sensitive minerals with a 0 0 1 peak in the 17–14 Å region

in the glycerolated X-ray trace, which collapses to 10–14 Å on heating. Regular mica/chlorite refers to a glycerol insensitive mineral, with peaks at 12.5 Å (0 0 2) and 7.9 Å (0 0 3), which on heating show relative changes of intensity but no appreciable alteration in position. Mica/chlorite refers to a glycerol insensitive mineral with an 0 0 1 peak in the 10–14 Å region which may exhibit intensity changes on heating but no appreciable change in position.

In Fig. 8, the proportion of Mg-rich chlorite has been estimated from variations of the 0 0 2 : 0 0 1 intensity values of total chlorite in the X-ray diffractograms of the 550°C heated clay sample. It has been assumed that the higher the intensity value, the greater the contents of Mg-rich chlorite.

Old Red Sandstone

In 1979–1980 correlation based upon clay stratigraphy was achieved between onshore biostratigraphically-dated sections in NE Scotland and SE Shetland (Fig. 2), and undated offshore sequences penetrated in the North Sea and west of Shetland. The onshore-type sequence of most use was from NE Scotland (Fig. 3) and was based upon the work of Ferrero (in Burolett *et al.* 1969) with additional data from Wilson (1972) and Donovan (1971). Type sections were also established in SE Shetland for the Middle Old Red Sandstone. The overall variation in stratigraphical clay mineralogy exhibited by the type sections was assumed to reflect changes in climate and topographic control related to the history of the Orcadian Basin.

Onshore sections: NE Scotland and SE Shetland

Type sections were established for the Middle Old Red Sandstone sediments of Dunrossness (Fig. 4) and the areas north and south of Lerwick (Fig. 5). These were based upon 168 samples from 21 measured sequences (Fig. 2) collected by P. A. Allen (Oxford University) and J. A. E. Marshall (Southampton University). The clay mineral assemblages of the Middle Old Red Sandstone sediments of these two areas are dominated by mica and chlorite, with other clay phases (smectite, mixed-layer chlorite–mica) being occasional minor components. The mica and chlorite are coarsely crystalline and exhibit petrographic evidence of their authigenic development in the sediment, probably by the recrystallization of detrital and earlier authigenic

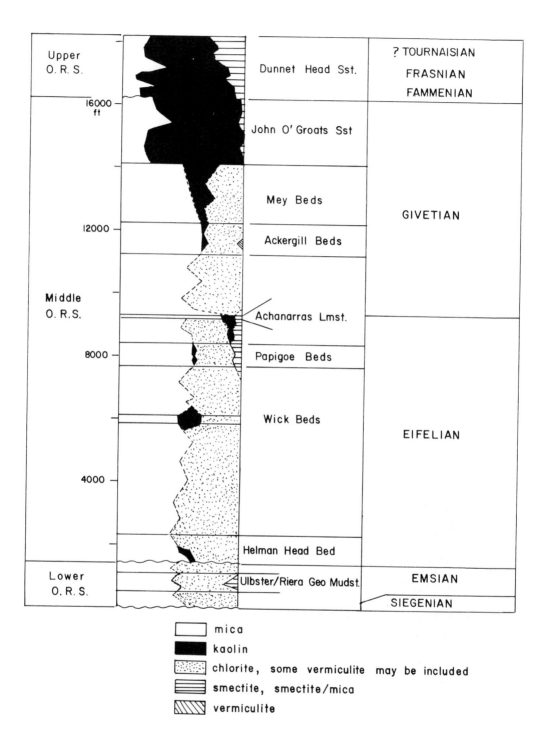

Fig. 3. Clay stratigraphy of the Caithness Old Red Sandstone. It is not clear how the clay mineral data of Hillier & Clayton (1989) and Hillier (1993) fits into or modifies this scheme.

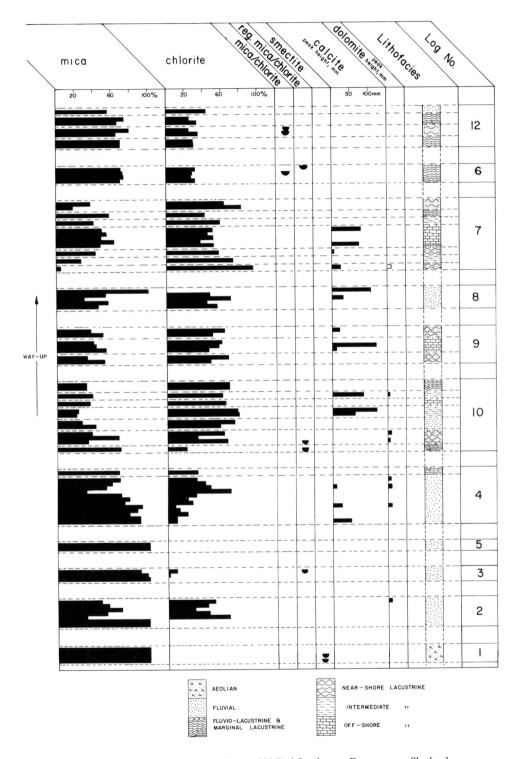

Fig. 4. Clay stratigraphy and depositional lithofacies, Old Red Sandstone, Dunrossness, Shetland.

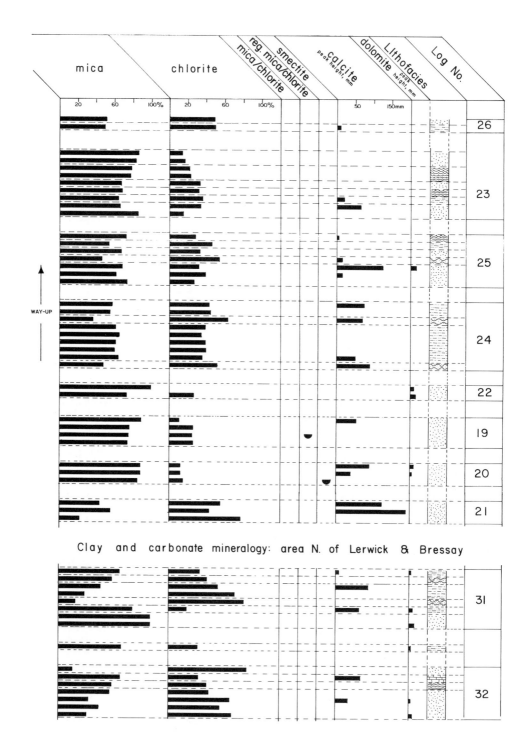

Fig. 5. Clay stratigraphy and depostional lithofacies, Old Red Sandstone, areas south and north of Lerwick and Bressay, Shetland.

Table 1. *Average and range of clay mineral/lithofacies relations in the Old Red Sandstone of SE Shetland*

Lithofacies (sample no. in parentheses)	Mica (%)	Chlorite (%)	Chlorite glycerolated 002:001	Chlorite 550°C heated 002:001
Aeolian (4)	100	—	—	—
Fluvial (61)	69 (15–100)	31 (0–85)	1.97 (1.46–3.88)	0.13 (0–1.84)
Lacustrinal subfacies 0 (18)	66 (32–79)	34 (21–68)	2.10 (1.41–2.97)	0.06 (0–0.36)
1 (18)	42 (5–75)	58 (25–95)	2.20 (1.63–3.21)	0.08 (0–0.32)
2 (19)	51 (21–80)	49 (20–79)	2.78 (1.97–4.26)	0.09 (0–0.21)
3 (9)	49 (36–63)	51 (37–64)	3.09 (2.30–6.00)	0.29 (0.06–0.88)

Lacustrine lithofacies: subfacies 0, fluvial–lacustrine, marginal lacustrine; subfacies 1, near-shore lacustrine; subfacies 2, intermediate lacustrine; subfacies 3, offshore lacustrine.

clays. Variations in the height ratio of the 0 0 1 and 0 0 2 peaks of chlorite (glycerolated and 550°C heated samples) from different samples are thought to reflect variation in the Fe and Mg contents. Figures 4 & 5 demonstrate a lack of consistent stratigraphical variation in clay mineralogy; however, fairly consistent relationships exist between the mineral assemblages and the interpreted depositional lithofacies (Table 1). The clay assemblages of the aeolian lithofacies consist only of mica. The fluvial lithofacies has a high average mica content (69%) and this decreases markedly as one passes through the marginal lacustrine subfacies (average 64%) to the near-shore subfacies (average 42%); there is a consistent increase in mica content again from the near-shore lacustrinal subfacies (average 42%) to the offshore lacustrinal subfacies (average 50%). A similar, but reverse, pattern occurs with the chlorite content of the clay assemblages from the different lithofacies. The 0 0 2 : 0 0 1 values of the 550°C X-ray diagrams of chlorite show a consistent pattern with high values occurring in the fluvial lithofacies and offshore lacustrine subfacies and low values in the marginal, near-shore and intermediate lacustrine subfacies.

Calcite and dolomite are present in the Old Red Sandstone sediment of SE Shetland, and show no consistent pattern of distribution in relation to stratigraphy or lithofacies. Calcite is dominant; dolomite occurs only in minor amounts, tending to be restricted to the fluvial lithofacies.

Offshore wells, North Sea

Well 13/17-1 (Fig. 6)

Fine-grained sediments (probably overbank deposits) from a red-bed sequence were investigated between 6688 and 8257 ft. They are characterized by mica (69–95%) and an Fe-rich chlorite (5–31%) with a marked decrease in chlorite content below 8150 ft. The clay assemblages are compared to those parts of the Caithness sequence (parts of the Middle Old Red Sandstone below the John O'Groats Sandstone or the Lower Old Red Sandstone; Fig. 3) where kaolin is absent. There is no microfloral evidence as to the age of the interval.

Well 13/19-1 (Fig. 6)

Fine-grained sediments (probably overbank deposits) of a red-bed sequence were investigated between 3995 and 7053.5 ft. Five clay mineral zones are recognized: (1) Zone I (3995–5532 ft) contains major mica (32–79%) and kaolin (21–68%): minor chlorite occurs in the uppermost three samples (3995–4558 ft): dolomite and calcite (dominant) occur in most samples; (2) Zone II (5628–5759 ft) contains major mica (83–94%) and minor kaolin (6–17%): both dolomite (dominant) and calcite occur; (3) Zone III (5772–6108 ft) contains major mica (66–83%) and moderate kaolin (17–34%): dolomite and calcite (dominant) are present; (4) Zone IV (6230–6500 ft) contains major mica (88–97%) and minor kaolin (3–12%): both dolomite (dominant) and calcite

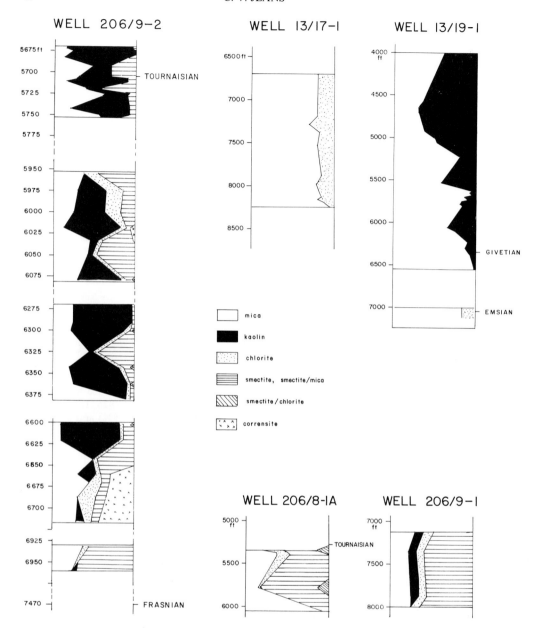

Fig. 6. Old Red Sandstone clay stratigraphy in Quadrant 13 (North Sea) and in Quadrant 206 (west of Shetland). Biostratigraphically dated horizons are shown.

occur; (5) Zone V (7033–7053 ft) is characterized by major mica (81–84%) and minor chlorite (16–19%): dolomite is present.

There is correlation between the relative abundance of kaolin and the dominant carbonate; dolomite characterizes the low kaolin zones, whereas calcite dominates the high-kaolin zones. Comparison with the Caithness sequence

(Fig. 3) suggests Zone I is perhaps equivalent to the John O'Groats Sandstone, whereas the alternating kaolin contents of Zones II–IV could be equivalent to the Mey or Ackergill Beds. The absence of chlorite from zones II–IV (cf. presence in the Mey and Ackergill Beds) could suggest local facies variation or that these zones are to be correlated with the John O'Groats

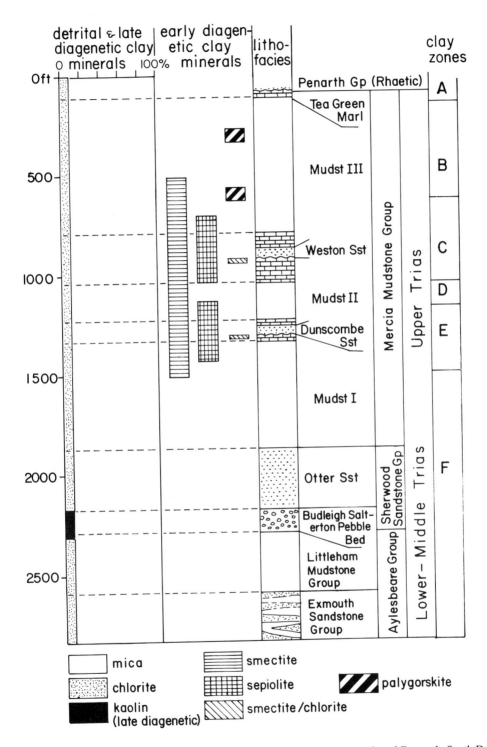

Fig. 7. Generalized clay stratigraphy of the New Red Sandstone between Axmouth and Exmouth, South Devon coast. Based on Jeans (1978) and unpublished data.

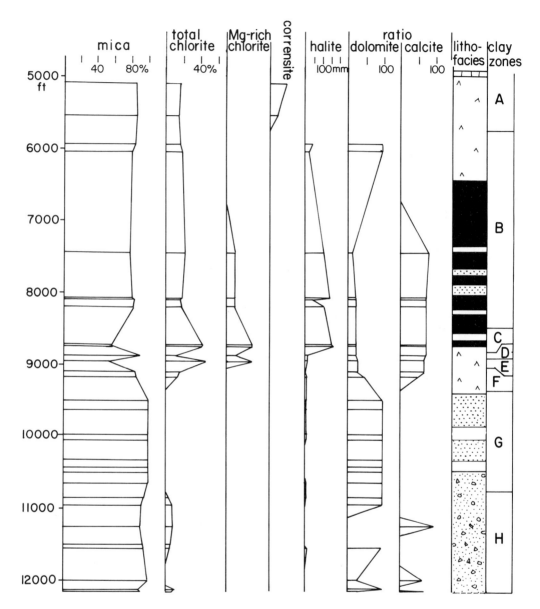

Fig. 8. Clay stratigraphy of the New Red Sandstone, Well 72/10-1A, Western Approaches, based on sidewall cores. Relative abundance of Mg-rich chlorite and corrensite shown schematically. Halite content related to 2.82 Å peak height (mm) in the bulk sample X-ray diffractogram.

Sandstone. This general correlation is substantiated by a Givetian microfloral assemblage at *c.* 6250 ft. Zone V is matched to the Lower Middle or Lower Old Red Sandstone of the Caithness sequence, and this is confirmed by an Emsian palynomorph assemblage.

Offshore wells west of Shetland

Well 206/8-1A (Fig. 6)
Fine-grained sediments (probably overbank deposits) from a red-bed sequence were investigated in three short intervals (5346–5389 ft,

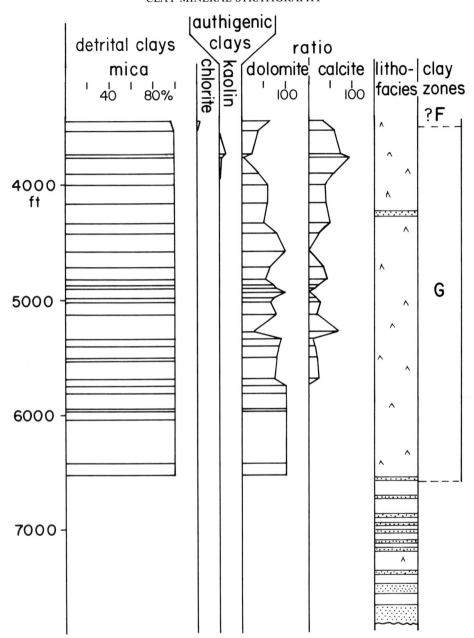

Fig. 9. Clay stratigraphy of the New Red Sandstone, Well 86/18-1, Western Approaches, based on sidewall samples. It is assumed that mica and chlorite are detrital in origin, whereas kaolin is of diagenetic origin. Kaolin is shown as a percentage of total clay assemblage.

5786.7–5787.2 ft and 6038–6056 ft). The upper and middle intervals are dominated by smectitic and mixed-layer minerals with lesser amounts of mica and chlorite. The lower interval is dominated by poorly crystalline mica (with smectite or vermiculite interlayers) and minor chlorite.

Kaolin is absent in all three intervals. An Upper Old Red Sandstone age was tentatively assigned, based on comparison to the Caithness section. A Tournaisian palynofloral assemblage occurs just above the upper interval.

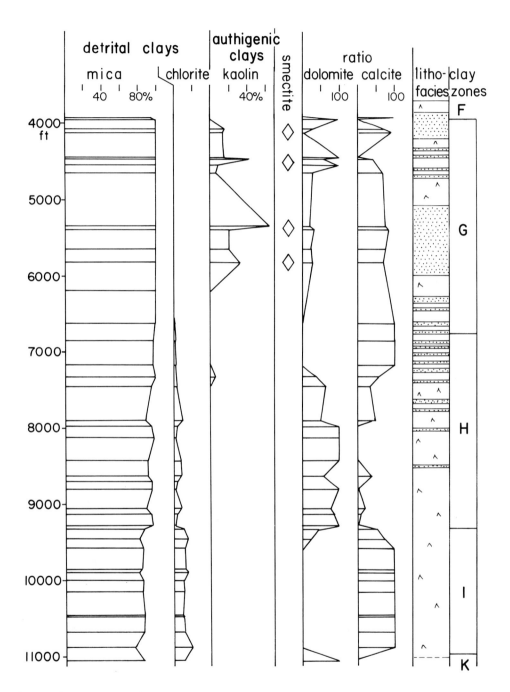

Fig. 10. Clay stratigraphy of the New Red Sandstone, Well 85/28-1, Western Approaches, based on sidewall samples. Smectite is shown schematically. It is assumed that mica and chlorite are detrital in origin, whereas kaolin is of diagenetic origin. Kaolin is shown as a percentage of total clay assemblage. Chlorite and kaolin exhibit antipathetic distribution patterns.

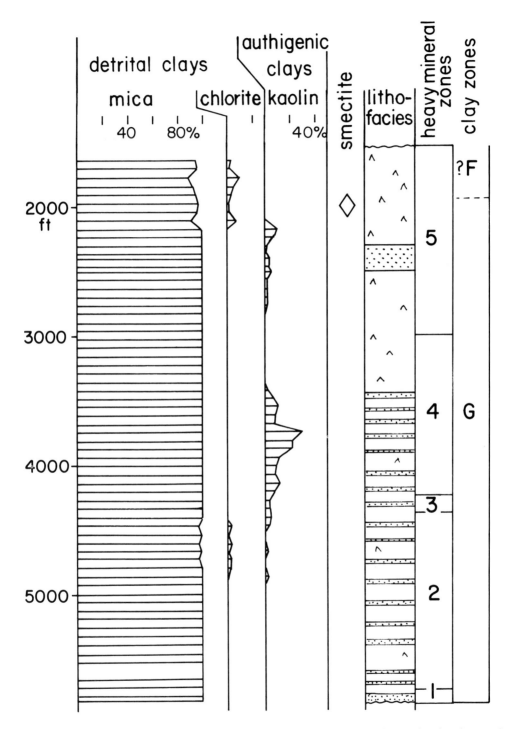

Fig. 11. Clay stratigraphy of the New Red Sandstone, Well 87/12-1A, Western Approaches, based on cutting samples. The lithological sequence is from Evans (1990). The heavy mineral zones of Jeans *et al.* (1993) are shown. It is assumed that mica and chlorite are of detrital origin, whereas kaolin is of diagenetic origin. Kaolin is shown as a percentage of total clay assemblage. Chlorite and kaolin exhibit antipathetic distribution patterns.

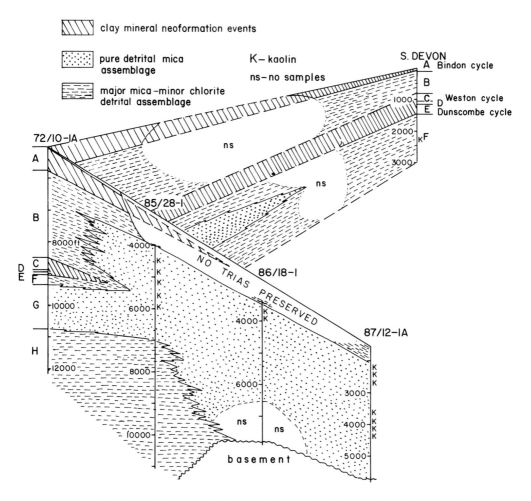

Fig. 12. Clay mineral correlation within the New Red Sandstone of the Western Approaches and South Devon. Kaolin is assumed to be of diagenetic origin. There are two types of correlation, one based upon widely felt clay-forming events, the other on the distribution of detrital clay minerals.

Well 206/9-1 (Fig. 6)
Fine-grained cuttings from a red-bed sequence contained mica, smectite, irregular mixed-layer smectite/mica and occasionally kaolin as major components of the clay assemblage. Fe-rich chlorite is a minor component. An Upper Old Red Sandstone age was indicated based on general similarity to the upper part of the Caithness sequence. No palynomorph assemblages were recovered.

Well 206/9-2 (Fig. 6)
Four cored sections in this predominantly sandstone red-bed sequence were investigated. There is good correlation between lithology and clay assemblages. The sandstones are dominated by kaolin–mica assemblages associated with minor smectite or irregular mixed-layer minerals, whereas the shales are characterized by mica, smectite and irregular mixed-layer smectite/mica with only minor kaolin. An Upper Old Red Sandstone age is suggested on comparison to the Caithness sequence (Fig. 3). Tournaisian and Frasnian microfloral assemblages confirmed this assignment.

New Red Sandstone

The long-established facies term New Red Sandstone is used here to include the continental

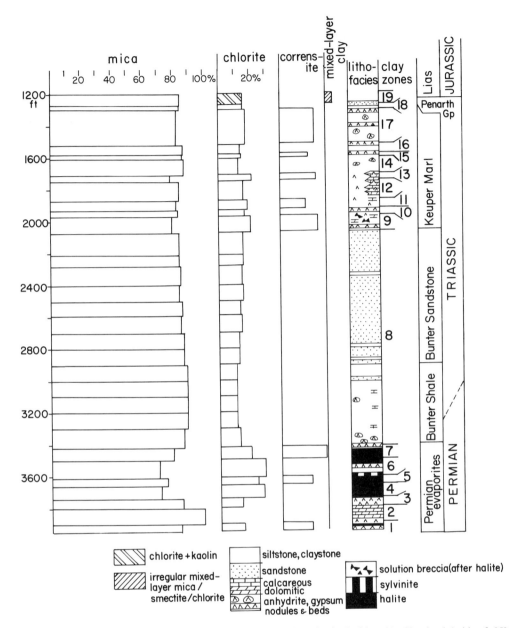

Fig. 13. Summary of Triassic and Upper Permian clay stratigraphy in Staithes 20, Cleveland (grid ref: N2 47603451800) based on core samples at *c.* 1.5 m intervals. Lithological and stratigraphical sequence based on Wood (1973). Relative abundances of corrensite and mixed-layer clay shown schematically.

red-bed facies, often containing major evaporite deposits, which may range in age from late Carboniferous to latest Trias. These sediments are particularly barren of fossils and throughout most of the sections biostratigraphical dating and correlation is minimal.

Southwest Approaches

The first attempt to use clay mineral stratigraphy in the New Red Sandstone of the Western Approaches provided excellent correlation of the Upper Triassic clay mineral cycles on the South

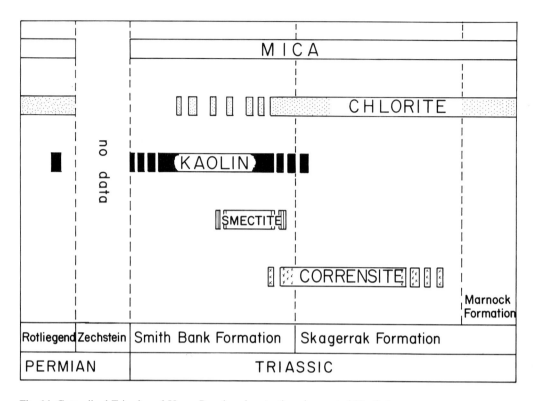

Fig. 14. Generalized Triassic and Upper Permian clay stratigraphy, central North Sea.

Devon coast (Fig. 7) with matching cycles in Well 72/10-lA (Fig. 8), some 500 km distant (Fisher & Jeans 1982). It was also suggested that the clay mineral zones (G, H) occurring below these cycles in 72/10-lA might be matched with clay mineral variation described by Henson (1973) from the Lower New Red Sandstone in the cliffs between Sidmouth and Exmouth. The author's subsequent investigations indicate that this is unlikely. Investigations of Wells 86/18-1 (Fig. 9), 85/28-1 (Fig. 10) and 87/12-lA (Fig. 11) in the Western Approaches demonstrate that the lower part of the zonal clay mineral sequence in 72/10-lA can be recognized over a wide area. However, the stratigraphical significance of these zones is slight. For example, detailed work (Jeans et al. 1994) on the pure mica assemblage of Zone G indicates its origin in coeval desert soils and that much of the clay mica in the detrital background assemblage of mica and minor chlorite is of similar origin: a correlation line differentiating between pure mica clay assemblages and a mica–chlorite assemblage will only be mapping out the distribution of detrital chlorite relative to mica. Figure 12 summarizes present clay mineral stratigraphy of the New

Red Sandstone in the Western Approaches and its correlation with the South Devon coastal sections.

Southern North Sea Basin

Little is known about the regional clay mineralogy of the Permo-Triassic sediments in the southern part of the North Sea Basin. This reflects the successful application of electric log, lithofacies and limited biostratigraphical data to correlation within this area and with onshore sections in Western Europe and the UK. Published data (Rossel 1982) has been concerned with the distribution of clay minerals in the Rotliegend Sandstone reservoirs, where the variation has been related to burial depth, the thickness of the gas-generating underlying Carboniferous sediments and the facies of the overlying Zechstein. The only stratigraphical clay mineralogy available for this area is from the Staithes No 20 borehole on the Cleveland coast which penetrated and cored the complete Triassic sequence, as well as the upper part of the Permian. Figure 13 summarizes the clay mineral sequence based upon the clay mineral

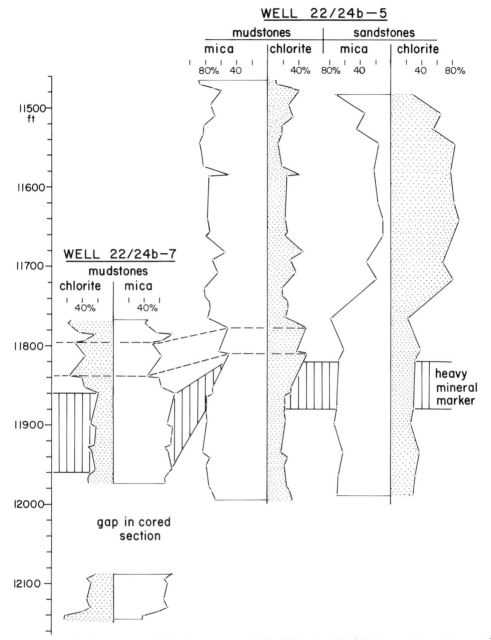

Fig. 15. Correlation between cored Triassic sequences, Wells 22/24b-5 and 22/24b-7, central North Sea, using heavy minerals and variations in detrital clay assemblages. Marked differences in the clay mineral assemblages of mudstones and sandstones reflect the more extensive development of authigenic clays in the latter.

analysis of over 500 samples: Zone 2, characterized by a pure clay mica assemblage, has been studied in detail by Jeans *et al.* (1994). It is not clear whether this sequence is representative of the more central parts of the Southern North Sea Basin.

Central North Sea

The Triassic sediments of the central North Sea differ from those of the southern North Sea Basin and onshore UK by containing no major evaporite deposits. However, their clay mineral

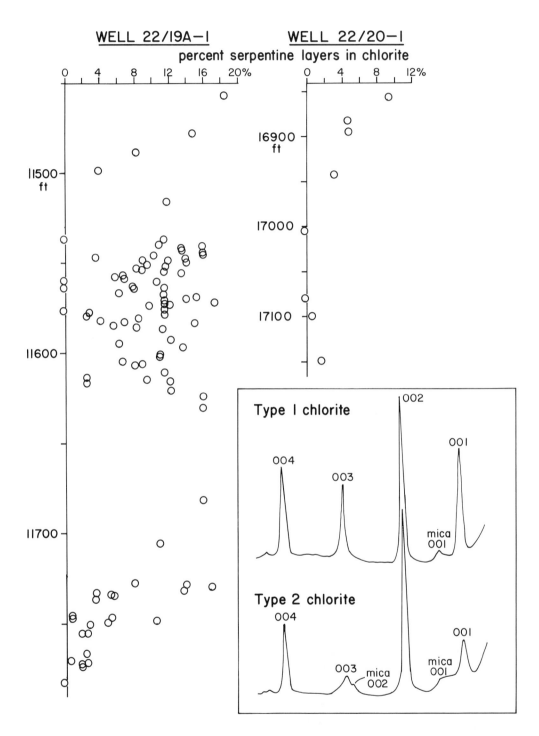

Fig. 16. Percentage of serpentine interlayers within the average chlorite sandstone cement, topmost Triassic zone of Wells 22/19-1 and 22/20-1. Calculations based upon the methods of Reynolds *et al.* (1992). Detailed petrography suggests that the pattern of values results from a mechanical mixture of a diagenetically early chlorite (Type 1 with little or no serpentine interlayers) and a later chlorite (Type 2 with 16–18% serpentine interlayers).

assemblages are similar, consisting essentially of two suites. A detrital one is dominated by mica with some chlorite, while an authigenic suite of Mg-rich clay minerals is linked to the evaporation status of the sediments. In onshore UK and at least parts of the status of the southern North Sea Basin there is relatively little difference between the clay assemblages of the argillaceous and aren-aceous sediments. When deep burial with enhanced temperature, pore-fluid reactivity and, perhaps, pore-fluid movement, has occurred, such as in the central North Sea, later diagenetic suites of clay minerals tend to be restricted to the sandstones: these are the products of the chemical interaction between the less stable components of the sandstones (certain feldspar types, biotite, etc.) and their own or allochthonous pore fluids.

Generalized sequence
Figure 14 summarizes the generalized clay mineral stratigraphy of the latest Permian and Triassic sediments of the central North Sea. It is based upon the study of largely core material from the following 19 wells: 20/1-2, 20/2-1, 20/2-2, 20/3-1, 22/5b-4, 22/5b-5, 22/5b-12, 22/12a-2, 22/15-1, 22/19-1, 22/19-2, 22/l0-1, 22/24a-2, 22/24b-5, 22/24b-7, 22/24b-8, 22/24b-9, 29/6a-3 and 29/8a-3. The limitations of this clay stratigraphy must not be overlooked. The stratigraphical framework upon which correlation between wells is based is poorly substantiated. The cored sections examined are often only a small part of the total Triassic section. There is clear evidence of facies control on the type of clay assemblage and, consequently, changes in clay mineralogy shown to occur vertically in Fig. 14 may occur laterally as well. Correlation between the clay assemblages and the evaporitic status of the sediment is not well developed (cf. on-shore UK, southern North Sea Basin). Whether this reflects the complex, changing Triassic palaeogeography of the central North Sea area or results from the redeposition of Permian Mg-rich clays, generally similar to the Triassic ones, is not clear. Such Permian clays could have been released by the erosion and dissolution of evaporites as they were brought to the suface by halokinetic tectonics, of which there is evidence (Smith *et al.* 1993).

Detailed detrital clay correlation
The Permo-Triassic red-bed sequence penetrated in Wells 22/24b-5 and 22/24b-7, some 7 km apart, consist of alternations of mudstones and sandstones. The clay assemblage of the mudstone consists predominantly of mica with lesser amounts of chlorite, assumed to be essentially detrital in origin. In contrast, the assemblages of the sandstones are dominated by chlorite with lesser amounts of mica, reflecting the extensive precipitation of chlorite cements in their pore space. In spite of detailed sampling of the mudstones throughout the sequence, it was not possible to match uniquely the patterns of detrital clay mineral variation between the two wells. However, this was achieved (Fig. 15) when the two sequences were linked stratigraphically by a marked change in the detrital heavy mineral pattern. However, it is clear that no matter how good the sampling opportunities, major difficulties may be encountered because of the very limited range of detrital clay components, the presence of considerable gaps in the sample coverage representing the sandstone zones, and the modification of the detrital assemblage by the local penetration of authigenic clay cements.

Clay mineral zones, topmost Triassic
Two types of authigenic clay cement may be associated with the upper part of the Triassic section. Figure 16 shows the distribution of a zone of randomly interstratified serpentine/chlorite in the sandstones at the uppermost part of the Triassic in two wells from Quadrant 22. The abundance of this cement relative to the normal clay cements is greatest at the top and decreases downsection: its distribution often coincides with the unit of greenish-grey or grey sediments occurring at the top of the Triassic sequence known informally as the Marnock Formation. Jeans *et al.* (1993) have suggested that the Marnock Formation reflects climatic amelioration at the end of the Triassic [equivalent onshore to the Tea Green Marl (Blue Anchor Formation)] and the effects of the inclusion of increased organic matter in the sediment. It is clear from the petrography of the serpentine–chlorite that it post-dates earlier clay cements including a relatively Mg-rich chlorite. However, its close spatial association with this facies suggests that serpentine–chlorite could be an intrinsic component of the diagenesis of the Marnock Formation in this area.

In Well 29/6a-3 a thick zone of authigenic kaolin cement occurs in the top part of the Triassic section immediately beneath the overlying marine Jurassic (Fig. 17). The kaolin-bearing Triassic sediments are grey or greenish-grey and these pass down into normal reddish-brown Triassic mudstones and sandstones beneath containing typical Triassic assemblages and no kaolin. The petrography of the kaolin-bearing sandstones indicates that the early diagenetic clay cements were mica, corrensite and chlorite occurring as grain-coating skins,

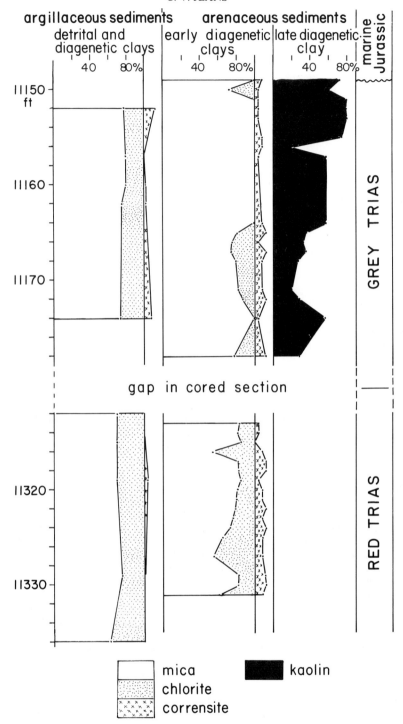

Fig. 17. Well 29/6a-3, central North Sea. Clay mineral distribution in the grey and reddish-brown sediment at the top of the Triassic sequence showing the relative abundance of: the detrital and diagenetic clay minerals in the argillaceous sediments; the early diagenetic clay minerals of the arenaceous sediments (corrensite is shown schematically); and kaolin (late diagenetic cement related to the erosion surface at the top of the Triassic) as a percentage of the total clay assemblage of the arenaceous sediments.

whereas the kaolin has developed at a later stage after or during the extensive dissolution of detrital feldspar. Dolomite cement is also present and this has been corroded during feldspar dissolution and kaolin precipitation. The author's interpretation is that this kaolin-rich zone represents part of a deep subaerial chemical weathering profile under good draining conditions developed on the Triassic sediments prior to the deposition of the overlying marine Middle Jurassic sediments. An alternative explanation is that the kaolin cement is intimately associated with the ingress of meteoric water through the subaerially exposed sub-Jurassic erosion surface, and the precipitation of the kaolin took place after the burial of Jurassic sediments. Macaulay *et al.* (1993) have put forward this hypothesis for the kaolinite of the Magnus Oilfield.

Concluding discussion

The clay mineral patterns within the Old Red Sandstone (Devonian–Lower Carboniferous) of northern Scotland and the adjacent offshore areas and within the Permo-Triassic sediments of the North Sea and Western Approaches are useful in supplementing biostratigraphical and lithological subdivision on a local basinal scale and can, in rare instances, provide exact intrabasinal correlation by matching widespread clay forming events. The successful application of clay stratigraphy depends particularly on the understanding of the origin of the clay assemblages in relation to their host sediment and within a stratigraphical framework. It is important to realize that clay mineral assemblages may undergo considerable changes during diagenesis which are controlled by chemistry of the pore fluids reflecting the original depositional fluids and the breakdown of unstable minerals within the sediment. Generally, throughout the subsurface sediments of the North Sea there is good correlation between clay assemblages and lithofacies, indicating intrinsic diagenesis (defined by Jeans 1984), and consequently conditions are suitable for the application of clay stratigraphy. Exceptions occur in the porous sediments acting as the main plumbing of the North Sea Basin, through which a large volume of allochthonous fluids have moved. Some of the changes occurring in the clay mineral assemblages during diagenesis are correlated with the depth of burial or the degree of organic metamorphism. For example, Hillier & Clayton (1989) and Hillier (1993) have related the clay mineralogy of the Devonian sediments of the Orcadian Basin (NE Scotland) to the degree of organic metamorphism and conclude that the pattern of mineral

assemblages is temperature controlled and suggest that this may also apply to the Permo-Triassic clay assemblages. If true, this means that the patterns of clay assemblages used for stratigraphy in both the Old Red Sandstone and New Red Sandstone have little significance other than correlating similar degrees of metamorphism. Such suggestions do not bear critical evaluation because the heat stability of the main clay mineral groups is well known (MacKenzie 1957). With very few exceptions (e.g. halloysite, allophane, berthierine), these do not show heat instability under inert conditions at the temperatures (up to c. 300°C) suggested by the organic metamorphism in the Old Red Sandstone sediments of the Orcadian Basin or at temperatures normally experienced in hydrocarbon wells (<200°C). The initial structural breakdown of clay minerals is by dehydroxylation. This takes place over the following temperatures for the main clay mineral groups; kaolins, 500–700°C; chlorites, 450–650°C; micas, 350–700°C; smectites, 500–750°C; sepiolite, 350–750°C; palygorskite, 350–650°C. At lower temperatures, non-structural water (sorbed, interlayer, or zeolitic in sepiolite and palygorskite) is lost but this reaction is reversible under inert burial conditions.

However, if clay minerals are in chemical environments where they are unstable, they are likely to alter and the rates of these reactions may be wholly or partially temperature/time controlled. The effects of these reactions range from recrystallization with little or no compositional changes to the complete replacement of earlier clay assemblages by new ones. When considered in isolation, and outside a properly constrained geological and stratigraphical framework, such clay mineral changes are often related to burial metamorphism. For example, Hillier (1993) claims that the occurrence of authigenic chlorite, corrensite and chlorite/smectite mixed-layer phases in the Orcadian Devonian mudstones is related to temperature variations reflected in the vitrinite reflectance and spore coloration, and suggests that the same temperature control is responsible for the occurrences of these minerals in the Permo-Triassic sediments of the UK. The spatial relationships of clay assemblages in Staithes No. 20 borehole (Fig. 12) and the regional variation in the central North Sea (Fig. 13), onshore UK (Jeans 1978) and in the Western Approaches (Fisher & Jeans 1982) clearly contradicts this and indicates that the pattern of assemblages is intrinsically linked to the facies and has been 'set' at the time of deposition: this does not mean that the clay assemblages have

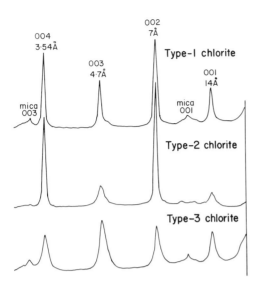

Fig. 18. X-ray diffractograms (CuKα, glycerolated) of common varieties of chlorite occurring as cement in Triassic sandstones, central North Sea. Analysis by ATEM (W. J. McHardy, MLURI, Aberdeen) gives the following compositions. Type 1, $(Fe_{2.8} Mg_{5.2} Al_4) (Si_{6.4} Al_{1.6}) O_{20}(OH)_{16}$; Type 2, $(Fe_{3.8} Mg_{4.4} Al_{3.8}) (Si_{6.2} Al_{1.8}) O_{20}(OH)_{16}$; Type 2 chlorite is similar in X-ray characteristics to the randomly interstratified serpentine/chlorite described by Reynolds *et al.* (1992).

remained unaltered. Until Hillier *et al*'s (1989) and Hillier's (1993) results from the Devonian Orcadian sediments are put into and tested within a detailed regional, stratigraphical and lithofacies framework, his hypothesis is open to question.

There is room for considerable improvement in our understanding of the clay-mineral lithofacies relationships in the Old Red Sandstone of Scotland. With present available data, including the considerable additions of Hillier & Clayton (1989) and Hillier (1993), it is not possible to establish a predictive clay mineral–lithofacies model. Results (see above) from the SE Shetland suggest that good lithofacies/clay mineral correlations are still preserved even at high vitrinite reflectance values (average 5.7, range 3.2–6.6, suggesting temperatures well in excess of 300°C according to Hillier & Clayton (1989)) and it would be surprising if good lithofacies/clay mineral correlations are not well developed in NE Scotland.

A tested clay mineral–lithofacies model is available for the Triassic (Lucas 1962; Jeans 1978; Fisher & Jeans 1982), although there are still two major problems concerning components of this model.

Kaolin–detrital chlorite relationships

Kaolin is generally a rare component of the intrinsic clay assemblages of the Permo-Triassic sediments of western Europe. It is typically associated with low salinity, continental sediments characterized by calcite as the evaporitic phase. The kaolin usually exhibits an antipathetic distribution pattern relative to the detrital chlorite of the background clay component (Figs 7, 10 & 11) or in rare instances may occur together with chlorite (Figs 14 & 17). Until further detailed investigations are carried out three interpretations have to be considered.

1. Kaolin and chlorite are of detrital origin but from different sources. Supporting evidence is poor as these minerals usually have an antipathetic relationship, rarely occurring together.
2. Kaolin has replaced detrital chlorite under particular pore-fluid conditions. There is possible evidence of this in the kaolin weathering zone of Well 29/6a-3 (Fig. 17) and in the Budleigh Salterton Pebble Bed, S. Devon (Fig. 7).
3. Both kaolin and the assumed detrital chlorite have been neoformed within the sediment, but under different pore-fluid conditions. Observations suggest that kaolin is associated with a lower salinity environment than chlorite.

Triassic chlorites, central North Sea

At least three varieties of chlorite (Fig. 18) occur in the neoformed clay cements of the Triassic sandstones from the central North Sea area. Their pattern of stratigraphical and regional distribution is generally unclear as a result of poor stratigraphical control, but it is likely that their development reflects variation in the pore-fluid chemistry and not of temperature or pressure.

I wish to thank the following companies for permission to publish this paper: British Gas, BP Exploration Ltd, Cleveland Potash Ltd, Elf Enterprise Caledonia Ltd, Esso Exploration & Production UK Ltd, Mobil North Sea Ltd, and Shell UK Exploration and Production. The interpretations and conclusions presented are mine alone and are not necessarily those of the companies mentioned above. Many people have been most helpful in various aspects of this project; in particular I thank P. A. Allen, A. F. Atherton, S. J. C. Cannon, R. E. Dunay, M. J. Fisher, J. A. E. Marshall, P. F. Rawson and M. Scherer for their support and interest.

References

BUROLETT, P. F., BYRAMJEE, J. & COUPPEY, C. 1969. Contribution a l'étude sedimentologique des terrains Devonien du Nord-est l'Écosse. *Notes et memoires Companie Française des Pétroles, Paris*, **9**, 1–85

DONOVAN, R. N. 1971. *The geology of the coastal tract near Wick, Caithness.* PhD thesis, University of Newcastle upon Tyne, UK.

EVANS, C. D. R. 1990. *The Geology of the Western English Channel and its Western Approaches.* HMSO for the British Geological Survey, London.

FISHER, M. J. & JEANS, C. V. 1982. Clay mineral stratigraphy in the Permo-Triassic red bed sequence of BNOC 72/10-1A, Western Approaches, and the South Devon coast. *Clay Minerals*, **17**, 79–89.

GLASMANN, J. R. & WILKINSON, G. C. 1993. Clay mineral stratigraphy of Mesozoic and Paleozoic red beds, Northern North Sea. *In:* PARKER, J. R. (ed.) *Petroleum Geology of Northwest Europe: Proceedings of the 4th Conference.* The Geological Society, London, Vol. 1, 625–636.

GRIFFIN, G. M. 1971. Interpretation of X-ray diffraction data. *In:* CARVER, R. E. (ed.) *Procedures in Sedimentary Petrology.* Wiley Interscience, New York, 541–569.

HENSON, M. R. 1973. Clay minerals from the lower New Red Sandstone of South Devon. *Proceedings of the Geologists Association*, **84**, 429–445.

HILLIER, S. 1993. Origin, diagenesis and mineralogy of chlorite minerals in Devonian lacustrine mudrocks, Orcadian Basin, Scotland. *Clays & Clay Minerals*, **41**, 240–259.

—— & CLAYTON, T. 1989. Illite/smectite diagenesis in Devonian lacustrine mudrocks from northern Scotland and its relationship to organic maturity indicators. *Clay Minerals*, **24**, 181–196.

JEANS, C. V. 1978. The origin of the Triassic clay asemblages of Europe with special reference to the Keuper Marl and Rhaetic of parts of England. *Philosophical Transactions of the Royal Society of London*, **289**, 549–639.

—— 1984. Patterns of mineral diagenesis: an introduction. *Clay Minerals*, **19**, 263–270.

——, MITCHELL, J. G., SCHERER, M. & FISHER, M. J. 1994. Origin of Permo-Triassic clay mica assemblages. *Clay Minerals*.

——, REED, S. J. B. & XING, M. 1993. Heavy mineral stratigraphy in the UK Trias: Western Approaches, onshore England and the Central North Sea. *In:* PARKER, J. R. (ed.) *Petroleum Geology of Northwest Europe: Proceedings of the 4th Conference.* The Geological Society, London, 609–624.

LUCAS, J. 1962. *La transformation des minéraux argileux dans la sédimentation: Études sur les argiles du Trias.* Mémoires du Service de la Carte Géologique d'Alsace et de Lorraine, **23**.

MACAULAY, C. I., FALLICK, A. E. & HASZELDINE, R. S. 1993. Textural and isotopic variations in diagenetic kaolinite from the Magnus Oilfield sandstones. *Clay Minerals*, **28**, 625–639.

MACKENZIE, R. C. (ed.) 1957 *The Differential Thermal Analysis of Clays.* Mineralogical Society (Clay Minerals Group), London.

REYNOLDS, R. C., DISTEFANO, M. P. & LAHANN, R. W. 1992. Randomly interstratified serpentine/chlorite: its detection and quantification by powder X-ray diffraction methods. *Clays & Clay Minerals*, **40**, 262–272.

ROSSEL, N. C. 1982. Clay mineral diagenesis in Rotliegend aeolian sandstones of the southern North Sea. *Clay Minerals*, **17**, 69–77.

SMITH, R. I., HODGSON, N. & FULTON, M. 1993. Salt control on Triassic reservoir distribution, UKCS Central North Sea. *In:* PARKER, J. R. (ed.) *Petroleum Geology of Northwest Europe: Proceedings of the 4th Conference.* The Geological Society, London, Vol. 1, 547–557.

WILSON, M. J. 1972. Clay mineralogy of the Old Red Sandstone (Devonian) of Scotland. *Journal of Sedimentary Petrology*, **41**, 995–1005.

WOOD, P. J. E. 1973. Potash exploraion in Yorkshire: Boulby mine pilot borehole. *Transactions (Section B) Institution of Mining and Metallurgy*, **82**, 99–106.

The application of fission track analysis to the dating of barren sequences: examples from red beds in Scotland and Thailand

ANDREW CARTER, CHARLES S. BRISTOW & ANTHONY J. HURFORD

Research School of Geological and Geophysical Sciences, Birkbeck College and University College London, Gower Street, London WC1E 6BT, UK

Abstract: In continental red-beds facies, which often contain little or no organic material for biostratigraphic dating, fission track (FT) analysis of detrital zircon crystals represents one method by which a maximum stratigraphic age may be determined. In addition, FT analysis of detrital apatite provides a means of palaeotemperature estimation, especially in areas where there is insufficient, or unsuitable, organic matter for vitrinite reflectance. Two examples of the application of FT analysis to the solution of stratigraphic problems in red-bed sequences are described. In Scotland, the age of the Tongue Outlier is disputed, previous studies having suggested either a Devonian or Permo-Triassic age. Comparison of FT data from both detrital apatite and zircon indicates a Permo-Triassic age. In Thailand, the 3 km thick continental clastic Khorat Group has a well-defined lithostratigraphy but is poorly constrained biostratigraphically. The lower Nan Phong Formation has a Rhaetic age whilst the Khok Kruat Formation at the top of the group has been assigned an Aptian age. In the absence of any biostratigraphic control, the intermediate Phu Kradung and Phra Wihan Formations have been assumed to be of Middle and Upper Jurassic age. FT data presented here for detrital zircons indicate a Lower Cretaceous age for Phra Wihan Formation of the Khorat Group.

Heavy mineral assemblages have been used as provenance indicators and for stratigraphic correlation within sedimentary rocks for many years (see Morton & Hurst, this volume; Mange-Rajetzky, this volume). By dating certain of these detrital minerals, using isotopic (including fission track) techniques, chronological information may be derived for otherwise barren sediments. If the dated crystals come from syn-depositional volcanic eruptions then an actual age of sedimentation may be derived. However, measured isotopic ages may reflect a variety of factors including the ages of the hinterland rocks, weathering, transport and diagenetic processes which can modify the detrital mineral suite, and post-depositional heating which can reset the isotopic signature.

Fission track (FT) analysis depends upon the accumulation within the crystal lattice of linear radiation damage effects, some $10-16\,\mu m$ long, produced by the natural fission decay of ^{238}U. The density of these spontaneous fission tracks is a function of the time during which they have been accumulating, of the rate at which fission occurs and also of the uranium content of the host crystal. Sample uranium content is determined using a neutron activation technique. As with other isotopic dating methods, the FT system can be reset by increases in ambient temperature, when spontaneous FTs are repaired or annealed. The annealing process can

be quantified in part by track length measurement, newly-formed tracks with uniform length being systematically shortened by increases in temperature. At sufficiently high temperatures, all tracks are removed and the FT age is totally reset. A more detailed summary of the FT method can be found in Hurford & Carter (1991).

Two commonly-occurring heavy minerals, apatite and zircon, are ideally suited to FT analysis. Apatite, with its lower temperature for the annealing of FTs, 60–120°C over geological time, is well suited to low temperature thermal history analysis. Detailed understanding of the track annealing process in apatite means that track length measurement can be used to decipher the integrated thermal history of a sample, and that numerical modelling can be used to predict probable time–temperature histories for any measured fission track age and length parameters (see Green *et al.* 1989 and refs therein). In contrast, zircon has the inherently higher track annealing temperature of 200–250°C over geologic time (Hurford 1986), although a precise measurement and description of the annealing parameters in zircon is still in preparation (see e.g. Tagami *et al.* 1990). Despite this lack of a detailed annealing model, zircon, with its relatively high closure temperature, is less prone than apatite to post-depositional annealing at those temperatures most frequently

From Dunay, R. E. & Hailwood, E. A. (eds), 1995, *Non-biostratigraphical Methods of Dating and Correlation*
Geological Society Special Publication No. 89, pp. 57–68

Table 1. Fission track apatite and zircon analytical data determined for Scottish and Thai samples

Field number	Lithology	Mineral	No. of crystals	Dosimeter ρd	Nd	Spontaneous ρs	Ns	Induced ρi	Ni	Age dispersion χ^2	Age dispersion %	Age (Ma) $\pm 1\sigma$	Mean track length (μm)	s.d.	No. of tracks
Tongue outlier northern Scotland															
SCOT-1181	Conglomerate	Apatite	20	1.208	8368	1.014	1047	0.762	787	<1	36.6	267±26	13.04±0.15	1.51	100
SCOT-1181	Conglomerate	Zircon	40	0.425	2942	12.59	5568	1.113	492	<1	18.9	293±17		1.44	
HUR-306*	Conglomerate	Apatite	20	1.341	3718	1.863	1678	1.798	1620		8	231±10	12.86±0.14		101
Phra Wihan Formation, Khorat Group, Thailand															
T90/12	Sandstone	Apatite	35	1.467	9884	0.624	1124	2.830	5096	<1	38	58±5	13.85±0.20	1.55	100
T90/12	Sandstone	Zircon	19	0.476	3301	1.079	2063	0.134	256	<1	25	247±22			
T90/13	Sandstone	Zircon	18	0.429	3301	1.503	3480	0.249	578	5	11	158±9			
T90/15	Sandstone	Zircon	20	0.430	3301	1.675	3724	0.193	378	<1	28	236±21			
T90/16	Sandstone	Zircon	12	0.432	3301	1.454	907	0.229	139	25	3	172±16			
T90/17	Sandstone	Apatite	20	1.264	8759	0.475	340	1.938	1386	<1	36	50±5	12.84±0.28	2.11	57
T90/17	Sandstone	Zircon	28	0.432	2400	0.878	3151	0.145	522	<1	30	163±13			
T90/18	Sandstone	Zircon	30	0.433	2400	1.114	6081	0.176	961	<1	30	162±11			
T90/25	Sandstone	Zircon	16	0.435	2400	1.243	1620	0.206	268	<1	32	156±17			
T90/31	Sandstone	Zircon	11	0.437	2400	1.802	1673	0.191	177	<1	28	257±31			
T90/33	Sandstone	Apatite	20	1.264	8759	0.359	195	1.580	855	<1	32	53±6	12.08±1.82	3.14	4
T90/33	Sandstone	Zircon	9	0.438	2400	1.256	1079	0.201	173	<1	32	179±25			

Notes: (1) Track densities are ($\times 10^6 \, tr \, cm^{-2}$), numbers of tracks counted (N); (2) analyses by external detector method using 0.5 for the $4\pi/2\pi$ geometry correction factor; (3) ages calculated using dosimeter glasses CN-5 (apatite) and CN-2 (zircon) – analyst Carter $\zeta_{CN5} = 339 \pm 5$, $\zeta_{CN2} = 124 \pm 5$ and *Hurford $\zeta_{CN5} = 356 \pm 5$; calibrated by multiple analyses of IUGS apatite and zircon age standards (see Hurford 1990); (4) $P\chi^2$ is probability for obtaining χ^2 value for ν degrees of freedom, where ν = no. crystals -1; (5) central age is a modal age, weighted for different precisions of individual crystals (see Galbraith 1992); (6) uncertainties are quoted throughout the text at 1σ level; (7) Tongue conglomerate samples collected at Coldbackie Bay, NC 612 604 by Phil Dolding and AJH.

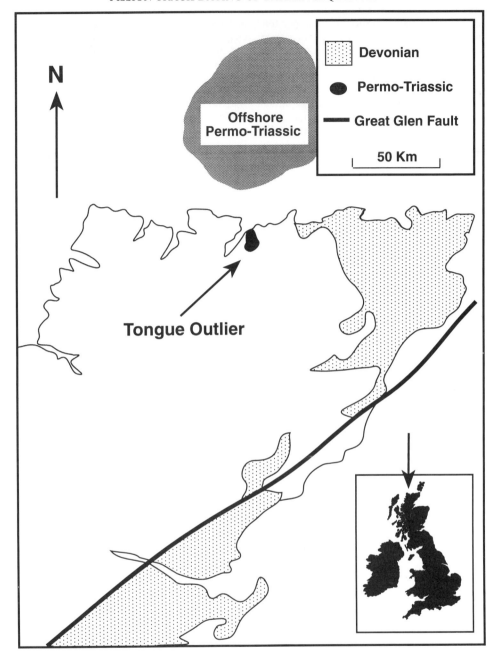

Fig. 1. Sample location map showing the position of the Tongue Outlier with respect to the Devonian rocks of northeast Scotland, and suspected offshore Permo-Triassic rocks.

encountered during upper crustal burial. Zircon is thus well-suited to providing single-crystal FT ages from clastic sedimentary rocks, and frequently offers an indicator of sediment provenance (Hurford & Carter 1991). The measured zircon FT age may reflect the thermal history of the source area (or areas) rather than the time of source area 'formation', be that intrusive or extrusive magmatism, or the time of basement exhumation. Accordingly, in interpreting a FT age consideration must be given to: (1) the relationship of measured ages to the known or suspected age of sedimentation; (2) comparison of measured ages with ages of probable source areas; (3) dispersion of single grain ages within a sample aliquot: a single homogeneous population indicates a single sediment source, or the time of total post-depositional resetting; a spread in single grain ages may reflect either the range in source ages or indicate partial post-depositional resetting (dispersion of ages can be assessed from radial plots of the single grain ages (see Figs 2 & 4), or from the relative error of a sample central age value (see Table 1)); (4) comparison of FT ages from co-existing detrital apatite and zircon, with differing thermal sensitivities, to provide constraints on post-depositional temperatures.

To illustrate this application, two examples of zircon FT dating are given, showing how they can both solve a long-standing stratigraphic problem, and constrain the stratigraphy in a frontier area where both geology and biostratigraphy are poorly understood. Since a sediment cannot be older than its source rocks, the youngest age measured for a detrital mineral provides a maximum age for that sediment, providing there is clear evidence that no significant post-depositional annealing has occurred. The following examples from northern Scotland and Thailand illustrate the application of these interpretative principles to zircon and apatite FT data to derive constraints on the stratigraphic and provenance ages of analysed sediments.

The Tongue outlier, northern Scotland

East of the village of Tongue, on the northern coast of Scotland, is a series of small outliers of red conglomerates (Fig. 1), first described by Sedgwick & Murchison in 1827, and interpreted by them as western extensions of the Old Red Sandstone (Devonian). For over a hundred years the outcrops were repeatedly mapped as Devonian (e.g. Nicol 1861; Geikie 1878; Crampton & Carruthers 1914) until, in 1956, McIntyre *et al.* questioned the age, proposing that the rocks might equally well be Torridonian or Permo-Triassic in age. Discussion continued during the 1980s, with several authors favouring Permo-Triassic deposition (an age strengthened by the discovery of Permo-Triassic rocks several kilometres offshore containing rare biostratigraphic evidence, e.g. Woodland 1979; Evans *et al.* 1982; Johnstone & Mykura 1989), whilst others continued to prefer the Devonian option (Blackbourn 1981; O'Reilly 1983). A Torridonian age for the Tongue conglomerates was

discounted because of the presence of Cambrian clasts.

Uncertainty as to the depositional age of the Tongue rocks has resulted primarily from the complete lack of biostratigraphic evidence, compounded by the limited size and extent of the outcrop such that comparison of general tectonic style and lithologies with more distant exposures is tenuous. An attempt to resolve these uncertainties initiated a FT study which, if unable to define the precise time of deposition, would provide a maximum stratigraphic age. A 3 kg sample collected from the Coldbackie Bay outcrop, 4 km northeast of the village of Tongue, yielded sufficient quantities of both apatite and zircon for analysis. The sample FT data are summarized in Table 1 with individual zircon and apatite grain ages displayed on radial plots (Fig. 2a & b). On such plots (Galbraith 1990) the error is standardized so that each point has the same standard error in the *y* direction. The *x*-scale indicates the precision of an individual grain age, the closer points are to the radial age scale, the more precise the age determination.

Mean central ages of 293 ± 17 Ma ($\pm 1\sigma$) for zircon and 267 ± 26 Ma for apatite are younger than the Devonian but older than the Triassic. Relative errors attached to the central age values (see Table 1) indicate a spread of grain ages for each mineral greater than expected from a homogeneous single age population. This dispersion is also seen in the radial plots (Fig. 2a) where zircon grain ages range from *c.* 220 Ma to *c.* 450 Ma, the younger zircons forming a distinct cluster. More robust analysis of the component ages using methods described by Galbraith & Green (1990) and Sambridge & Compston (1994) reveal a dominant Permian source at 270 ± 20 Ma and a younger late Triassic source at 220 ± 18 Ma, which comprises 45% of the dated grains. Older metamict and semi-metamict zircons with high, uncountable track densities are absent from such analyses, introducing an element of bias towards the younger ages, although, for stratigraphic purposes this is unimportant, since it is the younger end of the age range that is of relevance.

The zircon age distribution (Fig. 2a) can be interpreted in two ways. Firstly, if there has been no post-depositional annealing of tracks in zircons, the youngest single grain ages would represent the ages of the youngest source material and thus provide a maximum depositional age for the sediment; in this scenario the sediment would have a late Triassic stratigraphic age. Alternatively, the detrital zircon population may reflect partial resetting since deposition.

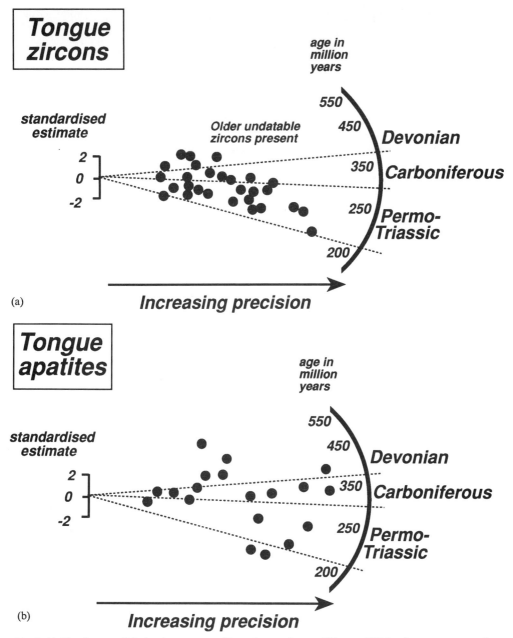

Fig. 2. (a) The zircon radial plot shows a spread in grain ages from *c.* 220 to *c.* 450 Ma, the youngest ages cluster within the range 220 ± 18 Ma. These suggest a Late Triassic age for Tongue rocks. (b) The apatite radial plot shows a similar spread in grain ages (*c.* 150 to *c.* 500 Ma), and clearly shows that the apatites have not been totally reset since deposition, whether Devonian or Triassic, consequently the zircon ages are interpreted as source ages and indicate the Tongue Outlier to be of Triassic age.

Consideration of the apatite FT results (Fig. 2b) provides evidence for the post-depositional thermal history and for testing the alternative interpretations of the zircon data. Apatite single grain ages show a wide dispersion, ranging from *c.* 150 to 500 Ma, with a mean apatite track length of $13.04 \pm 0.15 \, \mu$m and standard deviation of $1.5 \, \mu$m. If the Tongue conglomerate was deposited in the Permo-Triassic, most apatite single grain ages would be equal to, or older than, deposition, indicating only very modest subsequent annealing. If deposition was Devonian then, although some apatite grains have been reset to give Triassic and younger ages, some crystals have retained Devonian and older ages, indicating that annealing was far from total. In each scenario the results are consistent with shallow burial and modest annealing since deposition at maximum temperatures < 100°C, with subsequent gradual exhumation through the Mesozoic and Cenozoic.

The post-depositional temperature limitation of 100°C imposed by the apatite data, irrespective of the timing of the deposition, clearly precludes the possibility of major zircon annealing at temperatures in excess of 200°C, as required if deposition was Devonian. The conclusion is reached, therefore, that the Tongue conglomerate was deposited in the Permo-Triassic, which implies that the Permo-Triassic currently outcropping on the sea bed a few kilometres to the north of the Kyle of Tongue (Evans *et al.* 1982) originally extended further south. The spectrum of zircon ages found for the Tongue conglomerate contains grains with a Late Carboniferous–Early Permian signature, possibly derived from igneous activity that occurred throughout Scotland at that time. The youngest zircon grain ages clustering *c.* 220 ± 18 Ma may record a mid- to late-Triassic volcanic source. Alternatively the Triassic ages may represent cooling ages associated with uplift and erosion on the flanks of major extension and rifting in the North Atlantic and North Sea at this time.

The Khorat Group, Thailand

The Khorat Group, situated in northeast Thailand (Fig. 3), is a sequence of continental rocks up to 3 km thick of Triassic to Cretaceous age with a well-defined lithostratigraphy; biostratigraphically, however, the sediments are poorly constrained. The group was divided into eight formations (see Fig. 5) on the basis of lithostratigraphy (Ward & Bunnag 1964), of which the Huai Hin Lat Formation at the base of the Khorat Group has been dated as Norian (late Triassic) in age on the basis of vertebrate remains (Buffetaut & Ingavit 1986), palynology and floral evidence (Konno & Asama 1973). The overlying Nam Phong Formation is undated (Buffetaut *et al.* 1993) and, while it is usually placed with the latest Triassic to early Jurassic, there is no good evidence to support this age. Sattayarak *et al.* (1991) have suggested that the Nam Phong unconformably overlies the Huia Hin Lat Formation and that the Huai Hin Lat

should be dropped from the Khorat Group. However, recent authors (Buffetaut *et al.* 1993; Mouret *et al.* 1993) have continued to include the Huai Hin Lat in the Khorat Group, and the Nam Phong Formation is usually attributed a latest Triassic age (Buffetaut *et al.* 1993). In the interpretation of Sattayarak *et al.* (1991) there may be a considerable time gap at the unconformity between the Huai Hin Lat and Nam Phong Formations.

The Nam Phong Formation is overlain by the Phu Kradung Formation, which is poorly dated and has been ascribed a middle Jurassic age on the basis of its stratigraphic position and a tenuous correlation of crocodilian remains *Sunosuchus thailandicus* with poorly dated late Jurassic to early Cretaceous rocks in China (Buffetaut *et al.* 1993). The overlying Phra Wihan Formation is undated (the only tentative evidence are dinosaur trackways which are not age diagnostic) and is ascribed a middle Jurassic age on the basis of its stratigraphic position. This agrees with the correlation proposed by Mouret *et al.* (1993), between the lithological and facies changes and a relative sea-level fall *c.* 177 Ma. The Sao Khua Formation has yielded a rich and varied fauna, including freshwater hypodont shark remains, parts of theropod and sauropod dinosaurs, *Lepidotes*-like actinopterigians, turtle plates and a nearly complete *Goniopholis* skeleton. The dinosaurs have been compared with Late Jurassic forms in China and USA, and the crocodilian *Goniopholis* is known in Europe from Late Jurassic and Early Cretaceous rocks (Buffetaut *et al.* 1993). Some authors have suggested a local unconformity between the Sao Khua and Phu Phan Formations (Maranate & Vella 1986), although this has not been recorded on seismic data (Mouret *et al.* 1993). To date, no diagnostic fossils have been recorded from the Phu Phan Formation, which is assumed to be Early Cretaceous because of its stratigraphic position. Mouret *et al.* (1993) suggest that the Phu Phan Formation may correlate with a relative sea-level fall *c.* 128.5 Ma. The overlying Khok Kruat Formation is dated as Aptian to Albian on the basis of an ornithischian dinosaur *Psittacosaurus* and a peculiar species of hypodont shark *Thaiodus ruchae* (Buffetaut *et al.* 1993). The Mahasarakham Formation and the Borabu Formation appear to lie unconformably on the Khok Kruat Formation (Sattayarak *et al.* 1991) and have yielded Albian–Cenomanian pollen. The overlying Phu Tok Formation is undated and is attributed a latest Cretaceous to Tertiary age by Mouret *et al.* (1993). (Figure 5 summarizes the biostratigraphic information.)

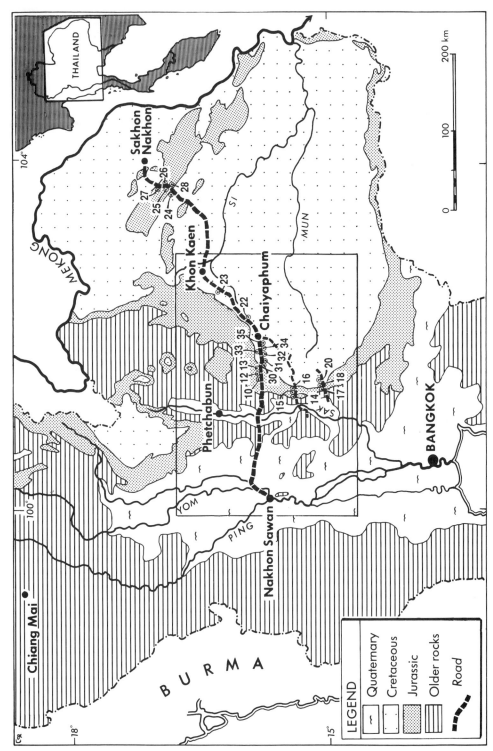

Fig. 3. Geological sketch map of Thailand to show location of samples as listed in Table 1. The boxed area refers to the rocks that contained apatite.

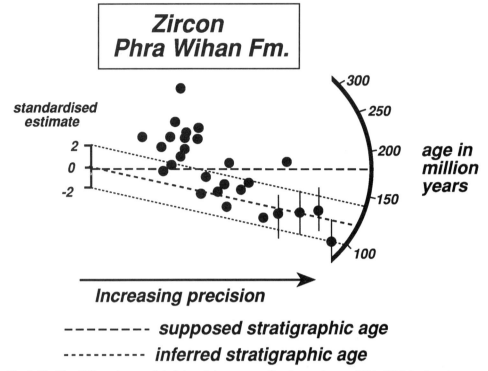

Fig. 4. The Phra Wihan zircon radial plot contains a range of grain ages from *c.* 120 to 350 Ma. A major grouping at 125 Ma is much younger than the supposed stratigraphic age of *c.* 175 Ma. The presence of both a spread in grain ages and old zircons indicates that the sample has not been totally reset since deposition and, therefore, the younger group of zircons provides an indicator of maximum stratigraphic age.

It is clear that while the lower and upper formations within the Khorat Group are well constrained biostratigraphically as Triassic and Cretaceous, respectively, other formations are not well dated. In the past the available stratigraphy has been used to fill the gaps with the Khorat Group spanning the Jurassic, although this presents some palaeogeographic problems since middle Jurassic shallow marine limestones are reported in Western Thailand, Cambodia and Laos (Fontaine *et al.* 1988), and there is no intervening delta. The lack of biostratigraphic control on the age of the Nam Phong, Phu Kradung, Phra Wihan and Phu Phan Formations, and the lack of information on the age of the unconformity between the Huai Hin Lat and Nam Phong Formations, has left a considerable gap in our understanding of the regional stratigraphy and palaeogeographic development. Attempts to correlate the non-marine Khorat Group strata with relative sea-level curves on the basis of basinwards shifts in facies (Mouret *et al.* 1993) have to be treated with caution since there is a wide range of tectonic or climatic factors which can lead to a

change in fluvial style, and there is no direct evidence for incision due to base level changes. In view of the uncertainty over biostratigraphic ages and lack of direct evidence of base level controls on Khorat Group sedimentation, it is reasonable to discount the tentative ages assigned by Mouret *et al.* (1993).

In the absence of reliable dates for much of the Khorat Group the fission track method has been used to date detrital zircon grains. FT analysis was carried out on 20 samples from the Khok Kruat, Sao Khua and Phra Wihan Formations, the latter sampled from nine different locations spread over some 200 km. Apatites were mostly absent, probably due to the effects of diagenesis and weathering. Samples which did contain apatite were confined to the western part of the Khorat plateau. For the purposes of this paper discussion will be confined to the Phra Wihan Formation.

The FT data from the Phra Wihan Formation are listed in Table 1. The four samples with apatite FT data indicate a Cretaceous–early Tertiary resetting event, and clearly are of little direct stratigraphic use. In contrast, the nine zircon samples contain a wide variation in central and grain ages (most samples have a high percentage of relative errors) indicating a potential for yielding useful provenance and stratigraphic information. The spread in central ages ranges from 257 ± 31 to 156 ± 17 Ma, and indicates variable influences of the component source ages throughout the formation. Most interestingly, six of nine central ages are at, or within, error of the assigned stratigraphic age of c. 175 Ma, based on a correlation with the Haque sea-level curve (Mouret et al. 1993). Clearly, the population of zircon grain ages within these samples warrants a closer examination; the younger grains may provide more robust corroborative stratigraphic information.

Figure 4 shows a radial plot for sample T90/17 that is typical of the younger central age group. It is immediately clear that, despite a central age of 163 ± 13 Ma, which is within error of the supposed stratigraphic age, a significant number of grain ages are not only much younger but form a single age population at c. 120 Ma. More robust modelling of the component ages for this sample using methods described by Galbraith & Green (1990) and Sambridge & Compston (1994) places this younger mode at 125 ± 20 Ma. Providing the zircons have not experienced any post-depositional annealing, these younger ages are significant in that they constrain the time that the Phra Wihan Formation must have been deposited. Unfortunately the apatites, which in the Tongue Outlier provided a constraint on the likelihood of zircon annealing, have been totally reset in the Khorat Group samples in latest Cretaceous–early Tertiary times, and therefore cannot be used to exclude the possibility that the zircons have been annealed. Other evidence is required.

The Khorat Group, which has a total thickness of some 3 km, has at its top a major Tertiary unconformity, recorded by the apatite FT data, and from which an unknown quantity of Tertiary overburden has been removed. It is conceivable that this additional cover was sufficient, possibly combined with higher than normal geothermal gradients, to place the Phra Wihan Formation within the zircon partial annealing zone ($> 200°C$). It is clear from the spread in zircon grain ages, shown in the radial plot of Fig. 4, that the zircon age population is heterogeneous, with ages ranging from c. 120 to c. 400 Ma. Total post-depositional annealing of the zircon population can, therefore, be ruled out, since many grains have ages older than Jurassic. Detecting partial annealing, however, is slightly more problematical. In such situations single grain ages more susceptible to annealing

(perhaps resulting from an influence of zircon composition) will be reduced in age to a level close to the time of maximum temperature. If partial resetting of zircon associated with Tertiary burial had taken place then there should be a significant number of grains with late Cretaceous–early Tertiary ages; this is demonstrably not the case. Furthermore, metamict and semi-metamict zircons (very old and/or high uranium crystals) would have been annealed; most analysed samples contain c. 10% metamict crystals.

The likelihood that the zircons of the Phra Wihan Formation have been partially reset at temperatures $> 200°C$ is fairly small; several lines of evidence, whilst not totally ruling out this possibility, do strongly suggest that the grain ages are source ages. By accepting this it is possible, by examining the radial plots for the nine Phra Wihan samples and undertaking more robust analysis of the component ages, to determine the major source age groups, whilst acknowledging that, throughout the period of deposition, one particular sediment source may have been more dominant than another. Including the youngest group already discussed at 125 Ma, other major source ages are from early Jurassic (c. 200 Ma) and Permian (c. 250–280 Ma), with subordinate, smaller clusters of crystals between 300 and 400 Ma. Other older, metamict and undatable crystals were present, but comprised no more than c. 10% of the total zircon population. The adjusted formation ages that are required by the youngest zircon age modes are summarized in Fig. 5, including for comparison the youngest age modes determined for two other formations. Whilst the youngest age mode in the Sao Khua Formation is older (160 ± 6 Ma) than that of the Phra Wihan Formation, indicating a different and older source, the Khok Kruat Formation has a very similar age mode (133 ± 13 Ma). Clearly, the contribution and influence of this younger source varied with time. In terms of provenance, these younger age modes must be from a near to contemporaneous source, the zircons being clear and euhedral imply the minimum of transport. In contrast, the older grains have a greater tendency to be subhedral to well rounded. A potential source for the young zircons is from the igneous activity in Vietnam whilst the remaining major source ages probably came from inliers such as the Khontum Massif. Palynological data from the Phra Wihan Formation suggests a Barremian age (Racey et al. 1994) which provides additional confirmation of our interpretation of the fission track data.

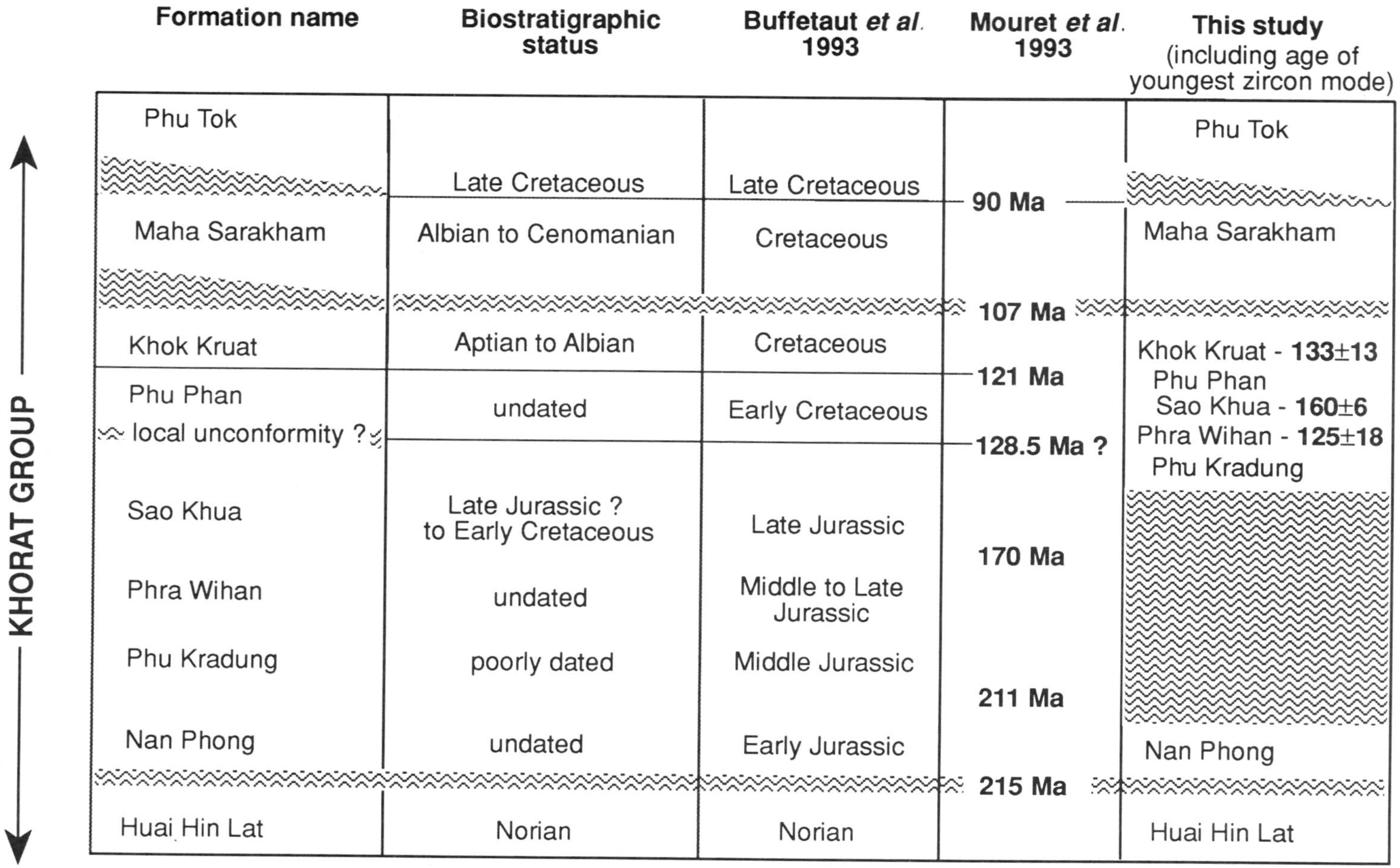

Formation name	Biostratigraphic status	Buffetaut et al. 1993	Mouret et al. 1993	This study (including age of youngest zircon mode)
Phu Tok		Late Cretaceous		Phu Tok
Maha Sarakham	Late Cretaceous	Cretaceous	— 90 Ma	Maha Sarakham
Khok Kruat	Albian to Cenomanian	Cretaceous	— 107 Ma	Khok Kruat - **133±13**
Phu Phan ≈ local unconformity ? ≈	Aptian to Albian	Early Cretaceous	— 121 Ma	Phu Phan
	undated		— 128.5 Ma ?	Sao Khua - **160±6**
Sao Khua	Late Jurassic ? to Early Cretaceous	Late Jurassic		Phra Wihan - **125±18**
Phra Wihan	undated	Middle to Late Jurassic	— 170 Ma	Phu Kradung
Phu Kradung	poorly dated	Middle Jurassic		
Nan Phong	undated	Early Jurassic	— 211 Ma	Nan Phong
Huai Hin Lat	Norian	Norian	— 215 Ma	Huai Hin Lat

KHORAT GROUP

Fig. 5. The diagram summarizes the current biostratigraphic information and the postulated ages for the Khorat Group. The last column includes details of the stratigraphic realignment required by the zircon FT data.

Conclusions

Two case studies illustrate how zircon fission track analysis can provide both stratigraphic constraints and provenance information from otherwise barren rocks. The first example, from the Tongue Outlier, has shown how apatite FT analysis can be used to constrain thermal history and to interpret zircon FT data in resolving a long-standing problem as to the depositional age of a barren sediment. Zircon single grain ages place the deposition of the Tongue conglomerate within the Permo-Triassic. The second example, from Thailand, has illustrated that, in a frontier area where there is the minimum of stratigraphic information, the assigning of stratigraphic ages to barren formations between two correlated/dated formations may be significantly wrong. Zircon single grain ages from the Khorat group place the Phra Wihan Formation within the lower Cretaceous, rather than middle Jurassic as previously suggested. Subsequent unpublished pollen data support this interpretation of the zircon FT data.

Fission track analysis in London is supported by NERC Grants GR3/7086 and 8291. Support from the University of London South East Asia Research Consortium and Total Khorat Ltd is gratefully acknowledged. Thanks to Phil Dolding for a sample from the Tongue Outlier.

References

BLACKBOURN, G. A. 1981. Probable Old Red Sandstone conglomerates around Tongue and adjacent areas, north Sutherland. *Scottish Journal of Geology*, **17**, 103–218.

BUFFETAUT, E. & INGAVIT, R. 1986. The succession of vertebrate faunas in the continental Mesozoic of Thailand. *Journal of the Geological Society of Malaysia*, **19**, 167–172.

———, SUTEETHORN, V., MARTIN, V., CHAIMANEE, Y. & TONG-BUFFETAUT, H. 1993. Biostratigraphy of the Mesozoic Khorat Group of northeastern Thailand: the contribution of vertebrate palaeontology. *International Symposium of Biostratigraphy of Mainland Southeast Asia: Facies and Paleontology*, Chiang Mai, Thailand, 51–62.

CHONGLAKMANI, C. & SATTAYARAK, N. 1978. Stratigraphy of the Huai Hin Lat Formation (Upper Triassic) in northwestern Thailand. *In:* NUTALAYA, P. (ed.) *Proceedings GEOSEA III*. Dept Mineral Resources, Bangkok, Thailand, 739–762.

CRAMPTON, C. B. & CARRUTHERS, R. G. 1914. *The Geology of Caithness*. Memoir of the Geological Survey of Great Britain.

EVANS, D., CHESHER, J. A., DEEGAN, C. E. & FANNIN, N. G. T. 1982. The offshore geology of Scotland in relation to the IGS shallow drilling programme, 1970–1978. *Report of the Institute of Geological Sciences*, No. 79/15.

FONTAINE, H., ALMERAS, Y., BEAUVAIS, L. ET AL. 1988. Jurassic of West Thailand. *Second International Symposium on Jurassic Stratigraphy, Lisboa 1988*. Instituto Nacional de Investigação Cinetífica, Lisboa, 1081–1094.

GALBRAITH, R. F. 1990. The radial plot: graphical assessment of spread in ages. *Nuclear Tracks and Radiation Measurements*, **17**, 207–214.

——— 1992. Statistical models for mixed ages. *Abstracts with Programs, 7th International Workshop on Fission Track Thermochronology*. University of Pennsylvania, July 1992, 7.

——— & GREEN, P. F. 1990. Estimating the component ages in a finite mixture. *Nuclear Tracks and Radiation Measurements*, **17**, 197–206.

GEIKIE, A. 1878. On the Old Red Sandstone of Western Europe. *Transactions of the Royal Society of Edinburgh*, **28**, 345–452.

GREEN, P. F., DUDDY, I. R., LASLETT, K. A., GLEADOW, A. J. W. & LOVERING, J. F. 1989. Thermal annealing of fission tracks in apatite–4. Quantitative modelling techniques and extension to geological timescales. *Chemical Geology (Isotope Geoscience Section)*, **79**, 155–182.

HURFORD, A. J. 1986. Cooling and uplift patterns in the Lepontine Alps, south-central Switzerland, and an age of vertical movement on the Insubric fault line. *Contributions to Mineralogy and Petrology*, **93**, 413–427.

——— 1990. Standardization of fission track dating calibration: recommendation by the Fission Track Working Group of the I.U.G.S. Subcommission on Geochronology. *Chemical Geology (Isotope Geoscience Section)*, **80**, 171–178.

——— & CARTER, A. 1991. The role of fission track dating in discrimination of provenance. *In:* MORTON, A. C., TODD, S. P. & HOUGHTON, P. D. W. (eds) *Developments in Sedimentary Provenance Studies*. Geological Society, London, Special Publication, **57**, 67–78.

JOHNSTONE, G. S. & MYKURA, W. 1989. *The Northern Highlands of Scotland*, British Regional Geology, HMSO, London.

KONNO, E. & ASAMA, K. 1973. Mesozoic plants from Khorat Thailand. *Geology and Palaeontology of Southeast Asia*, Tokyo University Press, **12**.

LEWIS, C. L. E., GREEN, P. F., CARTER, A. & HURFORD, A. J. 1992. Elevated K/T palaeotemperatures throughout northwest England: three kilometres of Tertiary erosion? *Earth and Planetary Science Letters*, **112**, 131–146.

McINTYRE, D. B., BROWN, W. L., CLARKE, W. J. & MACKENZIE, D. H. 1956. On the conglomerate of supposed Old Red Sandstone age near Tongue, Sutherland. *Transactions of the Geological Society of Glasgow*, **22**, 35–47.

MANGE-RAJETZKY, M. A. 1995. Subsivision and correlation of monotonous sandstone sequences using high-resolution heavy mineral analysis, a case study: the Triassic of the Central Graben. *This volume.*

MARANATE, S. & VELLA, P. 1986. Palaeomagnetism of the Khorat Group, Mesozoic, northeast Thailand. *Journal of Southeast Asian Earth Science*, **1**, 23–36.

MORTON, A. & HURST, A. 1995. Correlation of sandstones using heavy minerals: an example from the Statfjord Formation of the Snorre field, northern North Sea. *This volume.*

MOURET, C., HEGGEMAN, H., GOUADAIN, J. & KRASIDASIMA, S. 1993. Geological history of the siliciclastic Mesozoic strata of the Khorat Group in the Phu Phan Range area, northeastern Thailand. *Proceedings BIOSEA*, Chiang Mai, Thailand, 23–49.

NICOL, J. 1861. On the structure of the North-Western Highlands, and the relations of the Gneiss, Red Sandstone, and Quartzite of Sutherland and Ross-shire. *Quarterly Journal of the Geological Society*, **17**, 85–113.

O'REILLY, K. J. 1983. *Composition and age of the conglomerate outliers around the Kyle of Tongue, North Sutherland.* PhD thesis, University of London, UK.

RACEY, A., GOODALL, J. G. S., LOVE, M. A., POLACHAN, S. & JONES, P. D. 1994. New age data for the Mesozoic Khorat Group of North-west Thailand. *Proceedings of the International Symposium on Stratigraphic Correlation of SE Asia.* IGCP Project 306, Bangkok, Thailand.

SAMBRIDGE, M. S. & COMPSTON, W. 1994. Mixture modelling of multi-component data sets with application to ion-probe zircon ages. *Earth and Planetary Science Letters*, **128**, 373–390.

SATTAYARAK, N., SRIGULWONG, S. & PATARAMETHA, M. 1991. Subsurface stratigraphy of the non-marine Mesozoic Khorat Group, NE Thailand. *Proceedings GEOSEA VII*, Bangkok, 5–8 November, 1991.

SEDGWICK, A. & MURCHISON, R. I. 1827. On the conglomerates and other secondary deposits in the North of Scotland. *Proceedings of the Geological Society*, **1**, 77.

TAGAMI, T., ITO, H. & NISHIMURA, S. 1990. Thermal annealing characteristics of spontaneous fission tracks in zircon. *Chemical Geology (Isotope Geoscience Section)*, **80**, 156–169.

WARD, D. E. & BUNNAG, D. 1964. *Stratigraphy of the Mesozoic Khorat Group in northeastern Thailand; Report of Investigation*, Department of Mineral Resources, Geological Survey of Thailand, **6**, 1–95.

WOODLAND, A. W. 1979. 10" Geological map of the United Kingdom; North. Institute of Geological Sciences, London.

The use of chemical element analyses in the study of biostratigraphically barren sequences: an example from the Triassic of the central North Sea (UKCS)

A. RACEY,[1] M. A. LOVE,[2] R. M. BOBOLECKI[2] & J. N. WALSH[3]

[1]*23 Fernlea, Whitehill, Hampshire, GU35 9QQ, UK*

[2]*Geochem Group Limited, Chester Street, Chester CH4 8RD, UK*

[3]*Department of Geology, Royal Holloway University of London, Egham, Surrey, TW20 0EX, UK*

Abstract: A detailed core sampling programme has been completed on biostratigraphically barren Triassic material from seven wells in the Central Graben of the central North Sea (UKCS). Sedimentological, petrographic and chemical element data from these Skagerrak Formation sequences are presented. The objective of the study was to examine the potential use of chemical stratigraphy in the stratigraphic correlation of sequences and the identification of differences in provenance using inductively coupled plasma–atomic emission spectrometry (ICP–AES). The cores examined were taken mainly from the upper parts of the preserved Skagerrak Formation and ranged in thickness from 21.6–130.5 m. Five major facies associations were identified: (1) meandering channels; (2) braided channels: (3) crevasse splays; (4) bioturbated/pedoturbated floodplain; and (5) desiccated playa/floodplain. Braided channel sandstones comprise c. 66% of the examined sequences. Detailed petrographic examination reveals that most of the sandstones are subfeldspathic, feldspathic or sublithic arenites with rarer quartz of lithic arenites. Detrital minerals comprise quartz, K-feldspar, plagioclase, mica, heavy minerals and rock fragments. Authigenic components include silica, feldspar, carbonates, pyrite, anatase, illite, chlorite, corrensite, kaolinite and illite–smectite.

Within the total sample suite there are a significant number of clay-prone samples in which the clays are both detrital and authigenic in origin. An extensive ICP–AES analysis for 29 elements was performed on 563 samples. The results allow the clay-prone material to be distinguished from the sandstones owing to higher levels of Al_2O_3, MgO, K_2O, TiO_2 and many trace elements (including Li, La and other transition metals) associated with the clays. Since samples were obtained only from cored intervals from the upper parts of the Triassic section, a complete stratigraphic subdivision and correlation for the study area was not attempted.

Bivariate analyses of the ICP–AES data reveal diagnostic correlations between many of the major and trace elements. Moreover, excellent separation between samples from the different wells was seen on a triangular plot of Na_2O, Fe_2O_3 + MgO and K_2O. This separation was even more clearly visible on a triangular plot of Li, Zn and Ni. This appears to be the first use of such a plot for distinguishing Triassic sandstone sequences, and it may yet prove applicable in other intervals.

The chemical data also shows stratigraphic differences within some of the Triassic sequences examined. Two of the more extensively cored wells have clearly identifiable chemical signatures for the upper and lower parts of the sequences examined, which can be attributed to rapidly shifting sources and transport regimes during the deposition of these sediments. The extent of this chemical distinction is currently being investigated. Discriminant analysis of the ICP–AES data from the sand-prone samples permitted an excellent confirmation of the distinction between the chemical signatures for each of the wells studied. Discrimination between facies using data from the clay prone samples was not attempted owing to the limited availability of data. However, sand- and clay-prone facies can be broadly distinguished on the basis of their bulk chemistry.

This study summarizes the results of a two-year investigation into the facies discrimination, zonation and correlation of biostratigraphically barren clastic sequences using chemical elemental analyses. The cored sequences examined comprised mainly fluviatile sandstones from the Triassic Skagerrak Formation from seven central North Sea wells. This formation reservoirs large volumes of gas and condensate but is, invariably, barren of diagnostic microfossils

From Dunay, R. E. & Hailwood, E. A. (eds), 1995, *Non-biostratigraphical Methods of Dating and Correlation*
Geological Society Special Publication No. 89, pp. 69–105

making detailed stratigraphic subdivision difficult. Subdivisions based on lithofacies and electric log characteristics are possible and usually constitute the main basis for the indentification of individual formations and units within the Triassic succession. Further subdivisions on an intraformational scale are difficult, unreliable and localized. It was against this background that the current study into the feasibility of using a chemically based stratigraphic analysis of the Triassic from the Central Graben was developed and conducted. A number of correlation studies using inductively coupled plasma (ICP) on unfossiliferous and fossiliferous sequences from the North Sea (Racey *et al.* 1992), Middle East and Southeast Asia have been undertaken for various oil companies since 1985. Unfortunately, much of this information remains commercially confidential. In order to investigate many of the geological problems currently encountered in UK North Sea Triassic correlations, the authors felt that a detailed sampling program over a well-defined interval, coupled with detailed sedimentological and petrographic data, was necessary. During the course of our investigations over the last several years a number of potential applications of this technique have been identified in petroleum exploration and production which are summarized below.

(1) **Production geology** Individual reservoir sands and shales (potential source rocks or seals) have been 'fingerprinted' and correlated geochemically so that their lateral and vertical extent can be studied. This has proved essential in trying to unravel problems such as reservoir connectivity and partitioning in complex braided and meandering river systems. Within the chalk of the North Sea we have discriminated between allochthonous (better reservoir quality due to sorting during traction) from autochthonous chalk bodies using trace elements and have thus been able to identify better quality reservoir intervals not detectable using electric logs. Prior to this, discrimination was normally based on micropalaeontological analysis with a view to identifying 'transported' i.e. allochthonous, microfaunas (Geochem Group, Internal Reports).

(2) **Diagenetic studies** Major element distributions can be used to map-out compositional changes in different carbonate, evaporite and clay cements, thus assisting in the reservoir modelling of the lateral continuity of such permeability barriers.

The chemistry not only gives an indication of the amount of clay (or carbonate) but also the clay type, once calibrated against selected XRD data.

(3) **E-log ties** Certain element distributions can be tied to specific electric log responses to discriminate between and help correlate various lithological units. This technique, currently under development, could provide an effective downhole geochemical logging tool.

(4) **Basin studies and provenance** Certain changes in chemistry, e.g. levels of Zr, Ti, P, rare earth elements and selected trace elements reflect changes in provenance. These changes can be used in combination with sedimentological and palaeocurrent data to reconstruct basin palaeogeographies through the identification of sediment provenance locations. The technique is ideal for identifying non-sequence boundaries and unconformities within complex or simple sequences, since such boundaries almost always involve a change in provenance and therefore a change in chemistry.

Although the use of inductivity coupled plasma–atomic emission spectrometry (ICP–AES) in geology is not new (Thompson & Walsh 1989), its use in the correlation and subdivision of the central North Sea (UKCS) Triassic is a relatively recent development. The technical analysis itself is quick, inexpensive and can be performed on very small samples ($< 2\,\mathrm{g}$). Consequently, a large number of samples can be analysed quickly and at a reasonable cost. Since only a few grams of rock material are required, cuttings from uncored intervals can be analysed. The potential thus exists for routine, high density, systematic sampling across target intervals at an acceptable cost. In contrast, a comparable sampling programme using more conventional analyses (e.g. thin section, heavy minerals and X-ray diffraction) would be more expensive. The following factors must be considered when applying the ICP–AES technique: (1) the chemical effects of diagenesis and weathering; (2) the relationship between detrital mineralogy and bulk chemistry; (3) the effects of facies on detrital and authigenic mineral composition; (4) limited sample numbers and inadequate sample distribution; (5) statistical validity of the elemental associations and sample data set. The present study considered the above factors by integrating as much relevant geological information as possible. The ICP–AES analyses were thus interpreted in conjunction with additional petrographic and

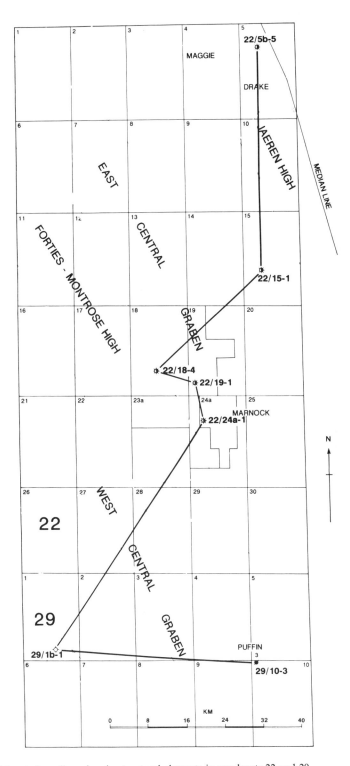

Fig. 1. Location of the study wells and main structural elements in quadrants 22 and 29.

WELL	SEDIMENTOLOGICAL LOGS	PETROGRAPHIC DATA	WIRELINE LOGS	ICP-AES DATA	DEPTH INTERVAL STUDIED (ft)
22/5b-5	✓	✓	✓	✓	10311 - 10498/ 10750 - 10977 (414)
22/15-1	✓	✓	✓	✓	13238 - 13356 (118)
22/18-4	✓	✓	✓	✓	10271 - 10364 (93)
22/19-1	✓	✓	✓	✓	11478 - 11805 (327)
22/24a-1	✓	✓	✓	✓	11640 - 12047 (427)
29/1b-1	✓	✓	X	✓	12690 - 12789 (99)
29/10-3	✓	X	X	✓	14552 - 14797 (245) TOTAL CORE FOOTAGE = 1723

Fig. 2. Database summary for this study.

sedimentological data in order to identify any relationships between these variables. Once identified, the statistical validity of these relationships was tested applying the techniques of Ehrenberg & Siring (1992). This work was undertaken to meet the following broad objectives: (1) examine the relationships between chemical element distribution, lithology and facies types; (2) assess the potential and limitations of chemostratigraphy (using ICP–AES) as a means of subdividing and correlating barren clastic sequences; (3) assess the importance of provenance and the influence of detrital and authigenic mineralogy on the ICP–AES data: (4) identify useful chemical 'signatures' and discriminative elements; (5) objectively assess the strength of any correlations identified via the use of stringent statistical analysis.

Database

The Triassic sequences examined comprised 508.4 m of slabbed core from the Skagerrak Formation from seven wells located in and around the UK Central Graben (Fig. 1). The cores were examined in detail and sampled at the Department of Trade and Industry (DTI) core store in Edinburgh. A total of 563 samples were taken for analysis.

All samples were analysed using ICP–AES and subsamples were selected for thin section, scanning electron microscope (SEM) and X-ray diffraction (XRD) analysis (Fig. 2). Completion logs and core analysis data were also available for six of the seven wells examined.

Analytical Techniques

ICP–AES

ICP–AES is the preferred method of elemental analysis for chemical stratigraphy. The main advantages that it offers are high precision and accuracy, analysis of a wide range of elements, rapid turn-round and, perhaps most importantly, a substantial reduction in cost of analysis. A total of 563 samples were taken at 1 m spacings throughout each of the cored wells studied. This ensured that all lithologies (facies) and grain sizes were examined. Where individual facies units were < 1 m thick additional samples were collected and analysed. Specific attention was paid to any obviously cemented horizons which could represent potential sequence breaks. Sample volumes of 30 g were crushed and analysed using ICP, thus taking account of variations in chemistry due to the presence of large mineral or lithic grains. The maximum grain size encountered was upper–medium sand (i.e. grain diameters of *c.* 0.5 mm), whilst average grain size for the studied sequences was medium to find sand. Consequently, sampling strategies ensured that the samples analysed were truly representative of the studied sequences.

Samples were analysed using the Philips combined simultaneous/sequential optical emission spectrometer (PV8060). The basic analytical protocol follows established procedures (Thompson & Walsh 1989). The samples were first crushed in a Swing Mill grinder – reducing the sample to a fine powder – then weighed (0.15 g) into Teflon crucibles and digested in hydrofluoric and perchloric acids. In this standard preparation method SiO_2 is lost as the sample is evaporated to dryness on a sandbath, and cannot therefore be measured in the analytical programme. However, all the other major element constituents were determined and a reasonable estimate of SiO_2 values made by subtracting the sum of these from 100%. SiO_2 can be determined accurately, if required, using a fusion technique. The remaining sample was dissolved in 10% HCl acid and diluted to 15 ml with distilled water; the overall dilution of 1:100 is optimum for ICP–AES analysis. Thirty elements were measured simultaneously in *c.* 1 min. The high resolution and broad wavelength range allowed most elements of interest to be determined with interferences reduced to a minimum. The elements determined include Al, Fe, Mg, Ca, Na, K, Ti, P and Mn (as weight percent oxide); Ba, Co, Cr, Cu, Li, Nb, Ni, Sc, Sr, V, Y, Zn, La, Ce, Nd, Sm, Eu, Dy, Yb (as p.p.m.). In addition, some other elements could be measured by the related technique of atomic absorption spectrometry (these include Pb, Cd, Rb and possibly Ag). The precision (reproducibility) is comparable to other analytical methods (±2% for the major constituents); accuracy was assessed by analysing a number of international standard reference materials covering the range of compositions found herein and was not significantly worse than the measured reproducibility. Detection limits can be taken as *c.* 5 p.p.m. for most trace elements (although theoretical instrument detection limits are significantly lower). Standards were run every tenth sample so as to correct the machine for instrumental drift.

Zirconium values should be treated with some degree of caution, since zircon (the main source of zirconium), if present in significant amounts, does not fully dissolve quantitatively in the HF/$HClO_4$ attack. The programme analyses rare earth elements (La, Ce, Nd, Sm, Eu, Dy and Yb) to produce an approximate pattern for 'normal' rock compositions. A separate analytical programme has been developed for the accurate determination of all rare earth elements, but was not used in this study.

The elemental compositions recorded by the ICP–AES can be transferred directly to computer. Results are saved on the hard disc of the PC that controls the spectrometer and can be input into a standard spreadsheet, such as Excel. The data can then be plotted out as simple bivariate plots of one element against another to show direct relationships between the major and trace element concentrations. Alternatively, more sophisticated multivariate statistical analysis can be undertaken. Discriminant function analysis and tests of significance were used here to examine elemental relationships within the chemical data (see below).

It is important to appreciate the nature of the samples within the dataset. Major differences in chemistry will arise from changes in lithology, and these can alter the chemistry of the samples fundamentally. Thus, in this study, the clay- and sand-prone samples were treated both together and separately. The elemental analysis provided a simple and very reliable method for discriminating between clay and sand samples, and easily identified sand samples with small amounts of clay present, or vice versa. However, it is preferable to separate the sand and clay samples if possible sources for the detritus are to be sought. In a few cases (< 2%), high carbonate content samples were found (up to 20% carbonate), and these were eliminated from the cross-plots. It may, however, be useful to identify high carbonate samples, and this the ICP dataset will do – from high Ca (calcite), or high Ca and Mg (dolomite) levels. However, this procedure may distort the statistical analysis if small changes in chemistry are sought for different source materials.

Thin-section petrographic analysis

Samples selected from core chips were thin sectioned and stained using: (1) sodium cobaltinitrite to reveal alkali feldspar; and (2) Dickson's combined Alizarin Red-S/potassium ferricyanide to distinguish varieties of calcite and dolomite. Detrital and authigenic constituents were quantified by 200 point modal analysis and the sandstones were classified using the schemes of McBride (1963) and Dott (1964). A total of 127 thin sections were prepared and described from an average sample spacing of 3 m.

Scanning electron microscopy (SEM)

Selected samples were taken as small freshly fractured rock chips and cleaned in xylene for 12 h to remove residual hydrocarbons before being cemented to a Cambridge-type aluminium stub with adhesive. Electrical conductivity was established by applying a thin film of gold or carbon in a Polaron sputter coater and by applying a coating of colloidal carbon dag around the base and sides of the sample. Samples were analysed in a Philips SEM 515 instrument fitted with an EDAX 9100 energy-dispersive X-ray analyser employing an ECON IV light element detector. A total of 50 SEM analyses were undertaken at an average spacing of 6 m.

X-ray diffraction analysis (XRD)

Whole-rock sample preparation
Selected samples were gently disaggregated under alcohol in an agate mortar and the resulting slurry transferred to a micronizing mill and alcohol-ground for a period of 10 min. Samples were dried at 60°C and then back-filled into an aluminium holder to produce a randomly oriented powder mount. (Alcohol was used during preparation to prevent damage to water-sensitive phases.)

Clay fraction (< 2 μm size fraction)
Initial sample disaggregation was conducted under distilled water in an agate mortar. The sample was then transferred to a test tube together with a dispersing agent to prevent flocculation. Liberation of the clay component was achieved by standing the tube in an ultrasonic bath for 30 min. The > 15 μm fraction was removed from suspension by centrifuge and the remaining clay was decanted off. The < 2 μm clay fraction was removed and concentrated using centrification. Known aliquots were magnesium-saturated at this stage if required. The clay suspension was then filtered under vacuum on to a membrane and the resulting clay film transferred to a Pyrex glass mount by membrane filter peel technique. Four traces were run for each sample: (1) after drying the clay mount at room temperature and humidity (angular range: $1.5–30°2\theta$); (2) after ethylene glycol solvation at 25°C for 24 h in order to detect the presence of swelling clays (angular range: $1.5–30°2\theta$); (3) immediately following heating to 375°C for 30 min, causing illite–smectite and discrete illite to coincide at 1 nm (angular range: $8–9.5°2\theta$); (4) after heating to 550°C for 1.5 h, causing the collapse of the brucite layer within chlorite (angular range: $1.5–30°2\theta$).

Instrumentation Philips PW1710 automated powder diffraction system equipped with sample spinner and graphite monochromator. Voltage, 45 kV; current, 44 mA; radiation: $CuK\alpha$; scanning speed, $1.0°2\theta$ min^{-1} for whole-rock and clay fraction. A total of 50 samples was analysed using XRD at an average spacing of 6 m.

Statistical analyses

The numerical ICP–AES data have been statistically evaluated using the PC-based CSS 'Stat-Soft' statistical package. Discriminant function analysis and tests of significance were used to examine the chemical data. Details of these analyses and a discussion of the results are given below.

Stratigraphy and sedimentology

Stratigraphy

The locations of the seven subject wells are shown in Fig. 1 together with the main structural elements which dominate the area of interest. The Triassic Skagerrak Formation intervals from each well, totalling some 488 m were examined in detail.

Deegan & Scull (1977) proposed a simple lithostratigraphy in which sands and conglomerates characterize the Skaggerak Formation in the eastern part of the Central Graben, whilst silty mudstones comprising the Smith Bank Formation dominate in the western part of the graben. Although the Smith Bank and Skagerrak Formations are now established lithostratigraphic units (Lervik *et al.* 1989), relatively little published information is available on their lithofacies and distribution. This factor, combined with the lack of sufficient age-diagnostic microfossils in the Triassic of the Central Graben, has resulted in the use of localized informal terminologies (e.g. Marnock Formation) and ambiguous correlations. An intraformational marker 'shale' horizon has been identified from wireline logs in some wells; it has no formal name, although it has been informally termed the 'Felix' shale by some Shell geologists. This horizon forms the lower 9 m of the cored interval in Well 22/19-1 and was identified using wireline logs in Well 22/24a-1. Similar clay-prone horizons are often interpreted as representing the top of the Smith Bank Formation.

By definition (Deegan & Scull 1977), the Smith Bank Formation 'consists of a monotonous sequence of brick red, somewhat silty, claystones with a few thin sandstone streaks and some anhydrite bands, especially in the lower part'. The Smith Bank Formation has no reservoir (or source) potential and has not, therefore, been cored in any of the wells studied. Apart from Well 22/15-1, which encountered a relatively thick Jurassic succession, the Triassic in the remaining study wells is either overlain by a thin Jurassic sequence or is truncated by the base Cretaceous unconformity.

Facies analysis

During detailed core logging, which included a visual assessment of grain size and sorting, sedimentary structures, lithology and overall sequence relationships, five broad sedimentological facies associations were identified and interpreted as follows: (1) facies 1, meandering channels; (2) facies 2, braided channels; (3) facies 3, crevasse splays; (4) facies 4, bioturbated/pedoturbated floodplain; facies 5, desiccated playa/floodplain. These facies demonstrate that continental clastics dominate the cored Triassic intervals examined. The individual facies associations are discussed in detail below. The percentage in brackets after each heading refers to the proportion of the total core examined represented by each facies.

Facies 1: meandering channels (4.4%)
This facies is not widely distributed in the cores examined and is largely restricted to Well 22/24a-1. Most of its sedimentological characteristics are very similar to those described for the braided channels. The major difference lies in the thicker development and more pronounced form of the fining-upward sequences. The presence of thicker, more bioturbated and pedoturbated abandonment subfacies and lateral accretion surfaces attest to a more stabilized floodplain environment and the existence of fluvial channels possessing a more meandering aspect. The individual sedimentological units (e.g. conglomerate lags, planar cross-beds, ripple trains) are largely the same as those observed and described in facies 2 below.

Facies 2: braided channels (66.8%)
These are the predominant facies observed in the Skagerrak cores and are dominated by cross-bedded fine to medium grained sandstones. Complete channel units occur, comprising fining-upward sequences in the following depositional order: erosionally-based intraformational conglomerates, cross-bedded fine to medium grained sandstones, parallel laminated very fine to fine grained sandstones and ripple cross-laminated very fine grained sandstones. However, the stacking of truncated channel units also results in thicker fluvial sandstone sequences which lack this fining-upward pattern.

A prominent scour surface forms the base of many fluvial channel units and is usually draped by mudstone intraclasts which are crudely stratified. These intraclasts are dominated by compactionally deformed, elongate, green and grey mudstone clasts and pink–brown dolomitized clasts. The conglomeratic channel lags were formed by the basal scour of the fluvial channel by partially lithified clasts derived from the erosion of adjacent floodplains.

Cross-bedded sandstones of both trough and tabular style directly overlie basal conglomerates, or scour surfaces in the absence of a conglomerate lag. Tabular cross-beds are characterized by foresets meeting set bases at an acute angle. They are interpreted as the deposits of linguoid bars developed within shallow braided streams, whereas the trough cross-bedded sets reflect dunes developed within the braid channels between bars.

Parallel laminated and cross-laminated sandstones predominate in the upper parts of the channel units, indicating accretion of planar and ripple bedforms, respectively. The development of plane-beds is favoured by shallow swiftly flowing water, whereas ripples are generated by weak currents.

Isolated tabular cross-beds also occur scattered within sequences of parallel laminated sandstones. These cross-beds are interpreted as the accretionary product of the slip faces of isolated linguoid bars, whereas the parallel laminated sands reflect plane bed conditions developed on top of these bars. Ripple cross-laminated sandstones are commonly interbedded with the parallel-laminated sandstones, and may become dominant towards the top of the depositional unit. These cross-laminated sandstones indicate slackening flow and the replacement of plane-beds by current-rippled units. Commonly, the uppermost beds exhibit climbing ripple trains grading up into laminated micaceous siltstones and mudstones, indicating rapidly waning flow and suspension settling of fines during the final stages of the depositional episode. Mudstone laminae exhibit desiccation features (curled laminae and sand-filled cracks) revealing subaerial exposure and drying of the sediment surface prior to the succeeding depositional episode.

Facies 3: crevasse splays (6.6%)
These units often exhibit an upward progression from fine grained, trough cross-bedded or parallel laminated sandstone to very fine grained, cross-laminated sandstones characterized by abundant mica and climbing ripple cross-lamination.

This depositional pattern indicates an initial erosive event in response to strong traction currents followed by deposition of small dunes, plane beds and current ripples from a decelerating traction current. Climbing ripple cross-lamination indicates a rapid waning of this current with eventual suspension settling of suspended fines from sluggishly moving or standing water. Each sandstone unit therefore represents a single depositional episode and occurs interbedded with floodplain deposits. The sandstones are considered to have accumulated by crevasse splays spilling from adjacent fluvial channels on to the floodplain.

Facies 4: bioturbated/pedoturbated floodplain (13.7%)

This facies association consists predominantly of very fine grained sandstones, micaceous siltstones and silty mudstones, which locally contain calcretized horizons. These deposits are characterized by an abundance of sub-vertical sand-filled burrows assigned to the trace fossil genus *Skolithos*. The degree of bioturbation is often so intense that the sediment is homogenized with the total obliteration of the primary sedimentary fabric. However, in less intensely bioturbated deposits, the sandstones show cross-lamination and are thinly interbedded with laminated siltstones and mudstone beds (centimetre scale). The mudstones exhibit rare small desiccation cracks indicative of periodic exposure and desiccation of the floodplain deposits.

Poorly developed calcretes are common within the bioturbated floodplain deposits and are characterized by irregular white or pink dolomitized concretions. Recent calcretes are found within the surficial sediments of continental areas subjected to semi-arid climatic conditions. Calcretization results from the near-surface precipitation of dissolved carbonates from pore waters (Goudie 1973; Watts 1980). The presence of calcretes within the bioturbated floodplain deposits therefore reveals extended periods of floodplain stability with little or no sedimentation. The conditions favouring their development are consistent with the increased moisture content suggested by bioturbation and periodic aridity indicated by desiccation features.

Facies 5: desiccated playa/floodplain (8.5%)

This facies association consists predominantly of thinly interbedded (centimetre scale) very fine

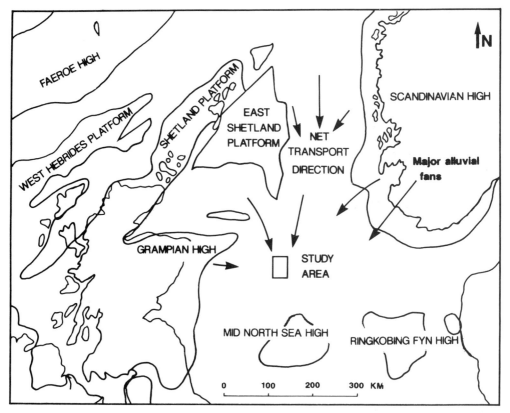

Fig. 3. Structural and depositional setting of the North Sea during Middle and Late Triassic times.

Mineral Phase Well	22/5b-5	22/15-1	22/18-4	22/19-1	22/24a-1	29/1b-1
Mono Quartz	52.2	50.00	42.57	38.26	37.55	42.16
Poly Quartz	9.5	8.50	6.29	5.62	7.47	6.67
Chert	0.09	0.00	0.00	0.00	0.00	0.00
K-Feldspar	6.73	15.09	12.79	14.26	19.24	4.80
Mica	4.79	4.72	7.50	2.84	4.39	4.42
Heavy Minerals	0.21	0.23	0.34	0.09	0.20	0.34
Rock Fragments Nonresolvable	3.74	6.03	7.14	6.74	7.97	8.75
Clay	12.35	1.00	1.93	2.76	4.89	15.20
Plag Feldspar	0.30	2.91	0.72	7.13	2.37	0.37
Illite	2.82	1.16	2.36	1.31	0.84	2.59
Kaolinite	1.80	1.34	0.13	0.51	0.00	0.01
Chlorite	0.45	5.60	2.00	9.08	4.15	0.34
Calcite	4.45	0.00	0.00	2.00	0.50	3.00
Dolomite	4.40	9.54	9.93	6.57	5.63	12.42
Silica	1.56	0.00	1.25	1.70	1.00	0.00
Baryte	1.25	0.00	0.00	0.00	0.00	0.00
Anatase	0.55	0.00	0.50	0.96	0.24	0.26
Feldspar	0.00	0.00	1.00	0.48	0.53	0.00
Opaques	0.92	1.47	6.50	0.21	0.10	0.50
Porosity	6.27	0.44	5.14	4.35	5.16	1.09

Fig. 4. Summary of mean percentage compositional values based on modal analyses of 127 thin sections.

grained sandstones, muddy siltstones and mudstone beds. Generally, these beds are characterized by red–brown haematite staining, unless subjected to diagenetic reduction resulting in a green–black coloration. The sand- and siltstones exhibit parallel-lamination and cross-lamination, and commonly show a rippled upper surface preserved by a mud drape. The mudstones exhibit abundant desiccation features including sand-filled polygonal cracks and curled laminations. Desiccation features also affect underlying sandstone and siltstone beds resulting in soft-sediment deformation structures. Further disruption has been caused by differential compaction between sand and mud beds during burial.

Depositional model

The analysis of facies types and their sequential development in the study wells, combined with limited published information on the Triassic of the central North Sea (Ziegler 1982; Gage & Doré 1986; Cope *et al.* 1992; Smith *et al.* 1993)

has allowed a depositional model to be constructed. The generally thick, widespread nature of Triassic strata encountered in the area between the Forties–Montrose and Jaeren Highs indicates that during Triassic times this area was a low-relief subsiding basin. In addition to this general pattern of subsidence, Fisher & Mudge (1990), Hodgson *et al.* (1992) and Smith *et al.* (1993) present evidence showing that the deposition of Triassic clastics in this area of the North Sea was also influenced by contemporaneous movements of Zechstein salt.

The predominance of fine-grained detritus, intraformational conglomerates and the absence of exotic pebbles indicates that the central North Sea was located far from source areas during Skagerrak Formation times. The great thickness of Triassic clastics indicates that the source areas were undergoing considerable erosion and must have been mountainous regions. Examination of proprietary dipmeter logs and published provenance maps (Ziegler 1982; Cope *et al.* 1991) indicates that net transport was in a southerly direction. The distant mountainous source areas, therefore, must have been located to the north and northwest or northeast of the study area. The heavy mineral suite recognized in thin section samples supports the existence of granitic and metamorphic source terranes, probably associated with the Caledonian and Scandinavian masiffs (Jeans *et al.* 1993). Both Jeans *et al.* (1993) and Mange (1993*a, b*) have attempted, with limited success, to use heavy mineral analyses as a means of subdividing and correlating the Smith Bank and Skagerrak Formations. Their results confirm that basin-wide stratigraphic differences in heavy mineral compositions can be identified.

Detrital mineralogy, sandstone classification and diagenesis

Detrital mineralogy

The detrital mineralogy for all the samples included in the modal analysis from the six wells studied (Fig. 2) have broadly similar compositions and show only relatively minor weathering. However, percentage abundances of each detrital component do vary between the wells studied (Fig. 4). Quartz is the most commonly occurring detrital mineral present in the majority of samples. Feldspars form a high proportion of the detrital component; thus, many samples are classified either as subfeldspathic or feldspathic arenites. Modal analysis data (Fig. 4) show that the dominant feldspar variety is K-feldspar. Detrital mica and rock fragments form subordinate constituents of the overall detrital mineral compositions and together form the lithic component of the sandstone classification diagrams (Fig. 5). The majority of samples contain varying proportions of muscovite, biotite and chlorite which tend to be more abundant in the finer-grained sediments. Rock fragments are dominantly sedimentary in origin, although both igneous and metamorphic grains are also prominent. Sedimentary lithic grains comprise silt and mudstone clasts which, when subjected to strong compaction, have become compressed and distorted. Many grains have lost their original structure and so have formed a type of pseudomatrix. Other lithic grains include those which represent reworked calcretes which have subsequently been replaced by dolomite. Heavy minerals form only a minor constituent of the total detrital component and comprise apatite, zircon, tourmaline, rutile, garnet, rare amphiboles and sphene.

The detrital clay component is somewhat problematic in that it includes part of the non-resolvable clay component. Exact percentages are indeterminable since a substantial proportion is probably formed of authigenic clays. However, SEM analysis has revealed that detrital clays are composed of mixed-layer illite–smectite and illite–chlorite clays which, in the early stages of diagenesis, were reactivated and formed a substrate for the precipitation of corrensite. Detrital clays form an early grain-rimming and pore-occluding phase and are occasionally stained red by the early precipitation of microcrystalline haematite.

Despite the very small differences in detrital mineralogy between the sample suites of each well, there are significant changes in bulk chemistry reflecting both subtle changes in provenance and diagenetic effects. These sediments were mainly derived from a granitic source area with minor metamorphic and intraformational sedimentary components.

Sandstone classification

The classification scheme used is a modified version of the Dott (1964) and McBride (1963) schemes, and is illustrated in Fig. 5. The majority of samples are rich in feldspars, with some rich in lithics, and can therefore be classified as feldspathic arenites with minor subfeldspathic, sublithic and lithic arenites. Secondary pores, present in small quantities in the majority of samples, represent unstable grains which were dissolved during diagenesis. The detrital assemblage therefore originally included slightly more lithic and feldspathic constituent grains (probably ferromagnesian minerals as well as feldspars, micas and other lithics) than are currently present. Samples containing 15% or more detrital clays are termed wackes, whilst authigenic phases which form more than 10% of the modal analysis are included as a qualifier to the lithological name. For example, sample 3593.0 m from Well 22/24a-1 has 28.5% dolomite and was therefore classified as a dolomitic subfeldspathic arenite.

Diagenesis

The Triassic sediments within the wells studied have all undergone a similar diagenetic history. However, certain diagenetic phases are more pronounced in some samples, thus suggesting facies and stratigraphic, depth-related controls. A generalized diagenetic sequence for all the sandstone samples is illustrated in Fig. 6, and has been inferred from the textural relationships between the different mineral phases as observed in optical and SE microscopy in conjunction with XRD analysis. The diagenetic history of

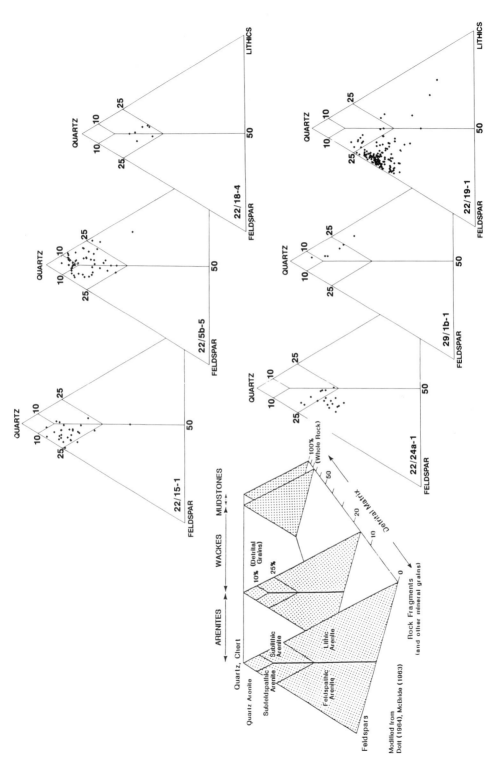

Fig. 5. Sandstone classification of Skagerrak Formation Samples from six wells. Note: This classification considers detrital components only. Lithological terms (as defined above) are prefixed for authigenic phases (including clays and carbonates) where they constitute 10% or more of the whole rock e.g. Calcareous Quartz Arenite.

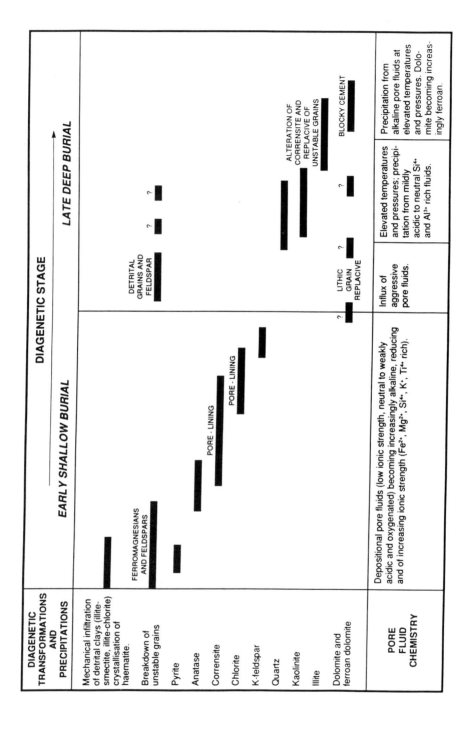

Fig. 6. Generalized diagenetic sequence for the Triassic Skagerrak Formation.

these samples has involved events which can be broadly divided into early (shallow burial) and late (deep burial) diagenetic stages. Early diagenesis includes the mineralogical changes which occurred as a result of the interaction of the original detrital assemblage with the depositional pore waters, followed by reaction with pore waters evolved by bacterial decomposition and fermentation of organic residues during shallow burial. Late diagenesis involves the mineralogical transformations caused by reactions with evolved pore waters at increased temperatures and pressures or with circulating formation waters during deep burial. Dependent upon structural history, meteoric water may be introduced as a late 'flush' at this stage. The diagenetic sequence and the authigenic mineralogy of the Skagerrak Formation samples are described below.

Early (shallow burial) diagenesis

Shallow burial diagenesis began soon after sediment deposition. Detrital clay complexes in the sediment may have suffered minor redistribution within the pore network as depositional fluids migrated through the sediment column. Thin-section analysis suggested they are optically non-resolvable but XRD and SEM analyses have revealed these clays to be mixed layer, Fe-rich, illite–smectite and illite–chlorite species, including corrensite. The precipitation of haematite also occurred at this time and causes the red coloration of the clay within some of the samples examined. The formation of microcrystalline haematite is favoured by neutral to mildly acidic, oxygenated pore waters. Under these conditions haematite could have formed from iron liberated during the decomposition of biotite, other ferromagnesian minerals, and also from the recrystallization of amorphous iron hydroxides, transported and deposited with the detrital clays. Such clays form grain-tangential rims to framework grains and occasionally infill intergranular pores.

The presence of apparently undeformed mudclasts within the detrital assemblages suggests that these clasts were lithified prior to deposition. Some show evidence of dissolution, with the subsequent formation of secondary porosity, whereas others show extensive replacement by dolomite. Both these features suggest former calcite cementation of the clasts, with subsequent dissolution or dolomite replacement of the calcite. Some of these clasts are therefore considered to have been reworked from contemporaneous calcretes. Calcretization is caused by near-surface precipitation and pedogenic

fixing of calcite from alkaline pore waters enriched by evaporation in Ca^{2+}, Mg^{2+}, Na^+ and K^+, partly derived from the breakdown of unstable grains and clays. With the onset of bacterial reduction of sulphate ions, pyrite was formed. It occurs in some of the samples as a minor grain replacive phase. Small crystals of anatase are found in minor amounts in many samples. They occur as euhedral to subhedral aggregates located in dissolution voids which possibly represent the remains of dissolved, unstable, ferromagnesian grains. These grains could have provided a rich source of Ti which would have been utilized for the *in situ* precipitation of anatase.

Chloritic clays are abundant in many of the samples examined and appear to have developed after the initial phase of grain dissolution and alteration. XRD and SEM analyses have allowed the identification of two types of clays. These can be divided into: (1) an early mixed-layer species, with higher Al, distinctive K and/or Ca content which has been identified as corrensite; and (2) a later-formed Fe-rich 'true' chlorite. Corrensite occurs as large, irregular plates aligned perpendicular to grain surfaces and forms isopachous fringes which line intergranular pores and early-formed secondary pores. It is apparent from SEM analysis that they formed due to the reactivation of detrital clays, which also acted as a substrate for their formation and precipitation. Discrete or 'true' chlorite formed subsequently, but still during early diagenesis. However, the presence of detrital clays was not a prerequisite for its formation as in the case of corrensite. True chlorite crystals are observed to be essentially more euhedral, crystalline and smaller than corrensite and tend to exhibit the characteristic rosette-type habit of chlorite. They too form grain-tangential rims to framework grains and are observed to infill pores. The chloritic clays would have precipitated from reducing alkaline pore waters enriched in cations liberated by the dissolution and alteration of unstable silicate grains during the preceding stages. The dominance of chloritic clays indicates that Fe and Mg cations were abundant, implying the breakdown of ferromagnesian grains. Dissolution of amphiboles, micas and feldspars would have released significant amounts of magnesium, iron, potassium, calcium, aluminium and silica into solution. The preferred precipitation of either the corrensite or discrete chlorite most probably reflects slightly different physico-chemical conditions of the pore waters and/or mineralogy of dissolved grains.

Authigenic feldspar forms a minor component.

It occurs locally as rhombic crystals growing through chloritic and detrital clay coatings on feldspar grains, and continued growth has resulted in the formation of partial, and less commonly complete, epitaxial overgrowths. Authigenic K-feldspar crystal faces are frequently fresh but occasionally appear etched owing to a later phase of dissolution. The precipitation of K-feldspar requires essentially similar Eh and pH conditions, where the precipitation of chlorite leads to a likely change in cationic composition of the pore waters such that Mg^{2+} is depleted and K^+ is relatively enriched. This results in the passage from the chlorite to the K-feldspar stability field and hence its precipitation.

Late (deep burial) diagenesis

Deep burial diagenesis involves mineralogical changes caused by reactions at increased temperature and pressure utilizing evolved pore waters developed during early diagenesis or introduced fluids from adjacent formations. This stage begins with the influx of acidic fluids and the dissolution of feldspar and other unstable framework grains and cements. The influx of such fluids may relate to leaching below the base-Cretaceous unconformity and may have been generated during deep burial either from the thermal decarboxylation of organic matter (Schmidt & McDonald 1979) or from the release of organic carboxylic acids from maturing organic matter, derived from connate fluids released from adjacent mudstone formations (Surdam *et al.* 1984). Acid pore fluids are also considered to cause the corrosive margins shown by many dolomite crystals. Some renewed, partial kaolinitization of micas may have occurred at this time with the formation of vermicular stacks of kaolinite, as observed in a few samples from Wells 22/18-4 and 22/5b-5, where they form a pore-occluding cement. The enrichment of pore water in Si^{4+} encouraged the precipitation of authigenic quartz, present within the majority of samples examined.

Authigenic illite precipitation is developed to varying degrees within the sediments examined, with illite generally occurring as fibres and laths growing from detrital lithic grains, detrital clays, micas and authigenic chlorites. It is best observed as fibrous laths on corrensite crystals which line pores and in certain samples extensively bridge pore throats. Authigenic illite can form over a wide range of pH conditions, depending primarily upon K^+ concentration and the temperature of the pore water. The preceding dissolution of K-feldspar would have released a high concentration of K^+, Al^{3+} and Si^{4+} into solution providing the necessary cations, while deep burial has provided high temperatures favouring illite precipitation.

The final stages of diagenesis are marked by a return to neutral to alkaline conditions and the precipitation of minor dolomite together with more extensive ferroan dolomite. Dolomites occur as rhombic crystals, either replacing lithic grains and interstitial clays and cements, or scattered within intergranular and secondary pores. Crystals commonly exhibit outer zones of ferroan dolomite which have compositionally evolved due to the eventual uptake of ferrous iron. Dolomite is primarily associated with lithic clasts and probably reflects substitution, under elevated temperatures, of Mg^{2+} and Fe^{2+} into the calcite crystal lattice of calcrete fragments.

SEM and XRD analyses of clays

SEM analysis has revealed four major interstitial clay types, described below, which are present in all samples, although their proportions vary considerably.

Detrital mixed-layer substrates

Detrital clays consist of thickly aggregated, grain-tangential plates which line large proportions of the intergranular pore networks and provide a nucleation substrate for subsequent clay authigenesis. In composition, EDS analysis suggests a Fe-rich mixed-layer phase, but with a significant Mg, Ca and K content. In the X-ray diffraction data, this clay type is included within the illite–smectite portion.

Authigenic corrensite or 'swelling' chlorite

Corrensite or 'swelling chlorite' is a platy, authigenic clay, usually developed on detrital clay substrates. The plate morphology is irregular, often with well defined 'spine' or 'fibrous' margins, and plate diameters are relatively large (10–30 μm). Individual plates aggregate to form a distinct, 30 μm (maximum) pore lining fringe, and plates occasionally show a well-developed boxwork structure. Illite fibres and ribbons are occasionally recorded on plate margins. In composition this authigenic clay is characterized by a distinct K and Ca content in addition to Mg, Fe, Si and Al. Corrensite has a high surface area, is H_2O sensitive and susceptible to swelling. Occasional larger (> 30 μm) plates are recorded, suggesting that this corrensite formation has occurred either under reservoir conditions or during recovery/sampling.

Authigenic chlorite

Highly crystalline chlorite is distinctive as a major pore-lining phase in Well 22/24a-1. In form, the plates are more crystalline, uniformly smaller (< 5 μm) than corrensite, forming rigid grain-coats where extensive, or irregular, pore filling pseudorosettes. This microcrystalline chlorite, although often nucleating upon detrital clays, does not require a clay substrate as a prerequisite for development. Thus, chlorite is primarily developed in intergranular pores, but is also recorded in secondary dissolution voids, suggesting a later phase or precipitation under longer residence times relative to corrensite. Where corrensite and chlorite intermix the latter often appears to be later, developed upon the more irregular platy surfaces. In composition, chlorite is Fe-rich with reduced Ca and K.

Authigenic illite

Illitic clays are generally subordinate to the Fe-rich phases, and several different mechanisms for illitization are recorded by SEM. (1) Fibrous and platy replacement of expanded detrital mica flakes; (2) fibrous/ribbon development after K-feldspar dissolution; (3) reactivation of grain-tangential detrital clay plates to elongate fibres and filaments; (4) reactivation of authigenic platy corrensite to short fibres and ribbons. Mechanisms (1) and (2) are sporadically recorded in each of the wells studied. Mechanisms (3) and (4) are more commonly observed (especially (4)).

The sample database suggests that individual wells show either dominant corrensite distribution (Well 22/18-4) or chlorite distribution (Well 22/24a-1), but with no significant transition with burial depth. In the chlorite dominant lithologies the XRD data suggest that illite > illite–smectite. In corrensite-dominated wells the opposite is true. It should be noted that the proportion of interlayered smectite in mixed-layer species is low ($< 10\%$) in all lithologies. Examination of samples from Well 22/19-1 suggests that there may be downwell variations in major chlorite clay type which could be facies controlled. The relative difference in chlorite or corrensite distribution reflects varying physico-chemical fluid conditions.

ICP–AES analysis

The distributions and concentrations of the elements from the ICP–AES analysis can be examined for correlations with facies type, geographical location and stratigraphic position. The analytical data have been subdivided into two major subgroups – the sand- and clay-prone samples. This procedure separates the samples that are most useful in determining provenance (i.e. the sandstones) from samples that are less likely to be useful in assessing provenance (i.e. the clay-prone samples). The elemental analytical data can be used to provide a sensitive test for the introduction of small amounts of clay material into a predominantly sand sample (with increases in diagnostic element levels). However, large differences in sand and clay contents of samples (where sandstones are interbedded with clay bands) will distort any chemical distribution pattern. Similarly, two samples in this dataset contained very high levels of carbonates (presumably of calcrete origin) and these were also eliminated from the statistical analysis.

Aluminium (Al_2O_3)

The range of values for both sand- and clay-prone samples is large. In the sands the Al_2O_3 percentage reflects feldspar (and mica) contents and in the clays it is a direct reflection of clay mineral content. In the clay samples good correlations exist between Al_2O_3 and Fe_2O_3 and Cr.

Iron (Fe_2O_3)

The wide range of iron values found in the samples can be attributed to the increased concentrations of haematite, pyrite, biotite and rock fragments in some samples. Iron values show good correlations with other transition metals – Ni, V and Cr, for example. The plot of Fe_2O_3 v. Li (Fig. 9) shows the clear discrimination between wells; this is especially marked for the discrimination between Wells 22/5b-5 and 29/1b-1, and may indicate differences in provenance for their Triassic sections. Although the sequences studied are often referred to as 'red beds', with the red coloration due to iron remobilization staining the rock, only small amounts (tens of parts per million) of Fe are required to stain a sequence red. As iron is present at levels of 1–14% it can be reasonably assumed that much of the iron is locked in authigenic or detrital mineral phases.

Magnesium (MgO)

A wide range of MgO levels has also been found. This probably reflects both variations in concentrations of chlorite and, for the highest values, dolomite in the samples. However, the chemical data show that for the great majority of the samples dolomitization has not occurred on any significant scale.

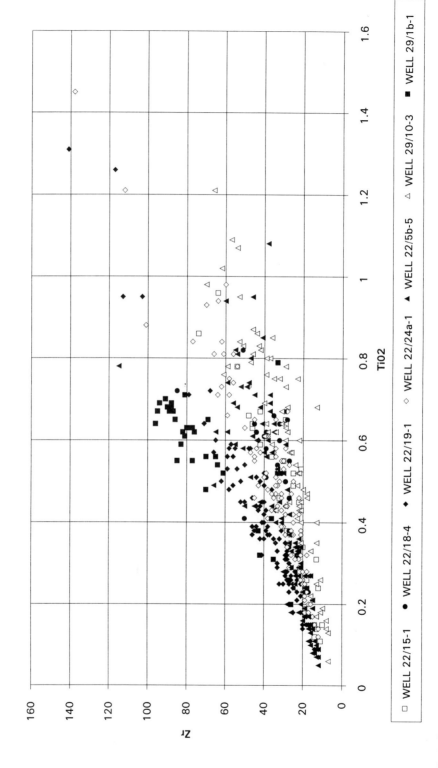

Fig. 7. Crossplot of Ti and Zr for each well. These correlate expectedly with the concentration of detrital zircon and rutile. Higher concentrations of Zr are observed in the most westerly well.

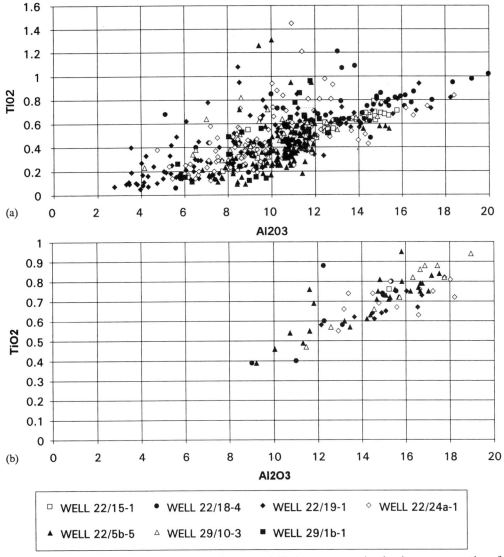

Fig. 8. Crossplots of Al and Ti for both (**a**) sand-prone and (**b**) clay-prone samples showing poor separation of individual wells. No obvious separation by well for sand- or clay-prone samples is seen.

Potassium (K₂O)

Alkali feldspar, mica and clay minerals (especially illite) all contain potassium. The presence of variable (and often high) levels of these minerals is reflected in the range of K_2O values. The original detrital K-feldspar (4.80–19.24%) has undergone little or no weathering, i.e. the feldspars appear unaltered in thin section. Feldspar overgrowths and secondary porosity due to feldspar dissolution are minimal (maximum 1%). Owing to the susceptibility of alkali feldspar to chemical degradation it is reasonable to conclude from the K_2O data that the sand-prone samples are relatively immature sands

that have undergone only moderate chemical weathering. Immature sediments offer a better possibility in trying to establish provenance and correlation using chemical data in that the likelihood of remobilization of chemical elements will be less. Although muscovite (2.8–7.5%) and illite (0.8–2.8%) are also present, K-feldspar is the dominant potassium-bearing phase recorded.

Titanium (TiO₂)

This element is often associated with heavy minerals (i.e. rutile, sphene and anatase). Figure 7 shows that a broad correlation exists between

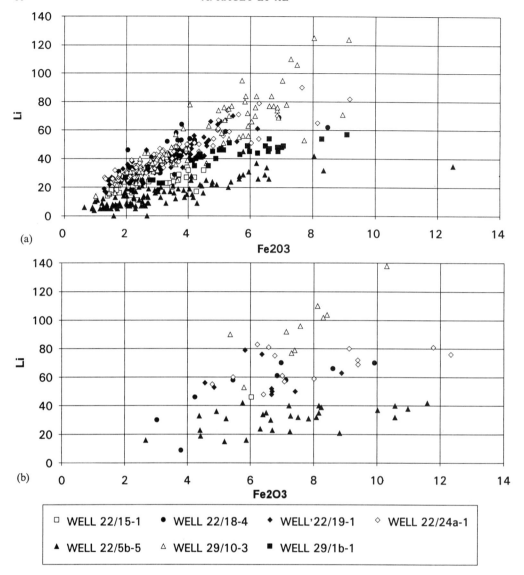

Fig. 9. Crossplots of Fe and Li for both (**a**) sand-prone and (**b**) clay-prone samples showing increasing discrimination with increasing inter-well distance.

TiO_2 and Zr (which would normally give a direct measure of the zircon content of the rock). There is a considerable 'scatter' on this plot, but the correlation is clear and discrimination between some wells can be seen. Samples from Wells 29/10-3 and 29/1b-1 are clearly segregated from the other samples.

Cross-plots of sand- and clay-prone samples

The bulk of the samples in the dataset are sand-prone, and the comments above on the relation-ships of some of the major elements and trace elements apply mainly to the sand-prone samples. The clay-prone samples make up some 12% of the database and have been plotted separately to allow elemental concentrations to be studied for this group. This segregation also allows the role of detrital and authigenic clays in controlling some of the element correlations to be examined. Selected cross-plots for both sand- and clay-prone samples have been produced for comparison and these are discussed below. The reduced number of samples for clay-prone material, and their uneven distribution with

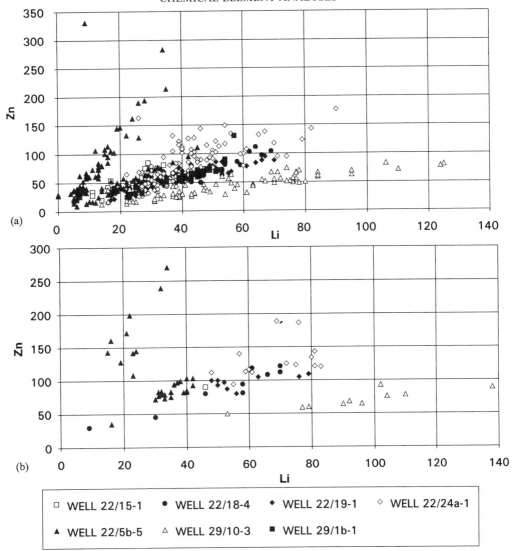

Fig. 10. Crossplots of Li and Zn for both (**a**) sand-prone and (**b**) clay-prone samples showing a fairly good separation by well. Well 22/5b-5 shows distinct separation into two units for both clay-prone and sand-prone lithologies.

respect to each well, have biased the clay-prone dataset; detailed statistical analysis of this dataset has not been possible.

Cross-plots for Al_2O_3 v. TiO_2, Al_2O_3 v. Fe_2O_3 and Fe_2O_3 v. Ni show broad correlations between these four groups of elements. Figure 8 may be taken as a representative example. There is a reasonably good correlation for the clay-prone samples, but well discrimination is poor. The plot of Fe_2O_3 v. Li (Fig. 9) does show that chemical discrimination between wells is possible for these elements. Wells 29/10-3 and 22/5b-5 can be separated both for sand- and clay-prone samples. Cross-plots of Ni v. Zn, Ni

v. Li and Li v. Zn demonstrate discrimination between wells. The cross-plot of Li v. Zn shows this most clearly (Fig. 10), where discrimination is seen in both sand- and clay-prone material. The most marked separation is again between Wells 22/5b-5 and 29/10-3, i.e. two of the most widely separated wells geographically. Most of the well samples plot into reasonably coherent linear trends.

Facies discrimination using ICP–AES data

The samples in this Triassic dataset were divided into subgroups based on facies type. For the five

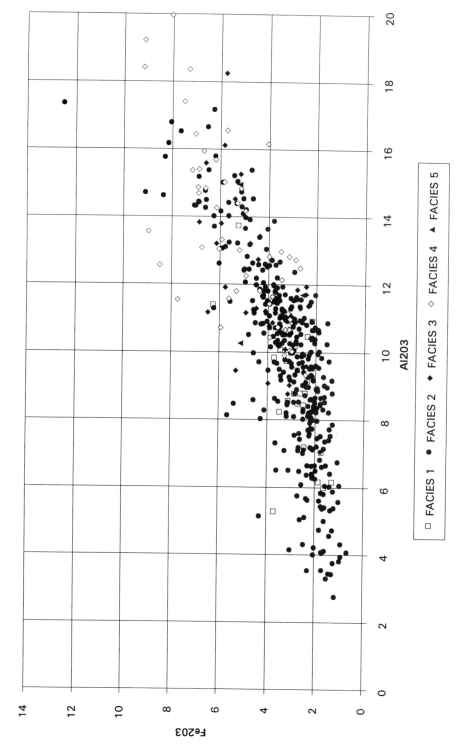

Fig. 11. Crossplot of Al and Fe discriminated by facies type. Facies can be broadly discriminated as either sand-prone (Facies 1 and 2) or clay-prone (Facies 3–5).

□ FACIES 1 • FACIES 2 ♦ FACIES 3 ◇ FACIES 4 ▲ FACIES 5

different facies identified, samples were coded 1–5 accordingly.

Several bivariate plots were prepared to test whether facies types could be identified chemically. Because the dataset is dominated by the braided channel facies type (66.8% of the samples) it is not surprising that there is only partial separation of the facies types. However, the plots do show the separation of the low and high energy facies types present, as seen in Fig. 11, where facies 1 and 2 (meandering and braided channel sands) separate from facies 4 and 5 (floodplain silts and mudstones). The plots of Fe_2O_3 v. Li and MgO v. Ni also show identifiable separation of facies types.

Cross-plots of trace elements for sand-prone samples

Ni v. V
The large spread of Ni and V values is probably associated with changes in clay mineral composition rather than relative proportions of detrital grains. The overall correlation is considered predictable, i.e. relatively linear (Fig. 12).

Ni v. Zn
Although a wide scatter of results is exhibited (Fig. 13), a weak correlation is demonstrated by some wells (e.g. 29/10-3). A limited degree of well separation is achieved and it appears that the datapoints for Well 22/5b-5 can be separated into two populations (equating with an upper and lower sequence).

Li v Ni
Well discrimination is clearly evident in the cross-plot (Fig. 14) for all the wells studied. Well 22/15-1 tends to cluster near the origin for all subject wells.

Ni v. rare earth elements
No obvious correlations were anticipated or seen in the various cross-plots generated (e.g. Fig. 15). However, all the rare earth elements analysed showed a similar distribution pattern with respect to Ni.

Li v. Zn
This combination of elements has been highly successful in discriminating between the study wells (Fig. 16). Wells 22/5b-5 and 29/10-3 exhibit virtually no overlap of datapoints. Remaining wells exhibit increased levels of overlap but can still be differentiated using these two elements. The differentiation probably arises from differences in regional clay mineral and mica compo-

sitions. Well 22/5b-5 also exhibits two clearly distinct sample populations arising from markedly different Zn concentrations.

Cross-plots of clay-prone samples

Clay-prone samples comprise just over 12% of the database and have been plotted separately in order to examine the relationships between this group and the observed elemental correlations. Furthermore, by doing this, the role of detrital and authigenic clays in controlling some of the correlations described can be examined. Selected cross-plots for both sand- and clay-prone samples (Figs 8–10) have been produced for comparison and these are discussed below.

Discrimination between wells

A triangular plot of Na_2O, $MgO + Fe_2O_3$ and K_2O (Fig. 17) reveals that most samples can be successfully grouped into their parent wells. Although Wells 22/5b-5 and 29/10-3 exhibit a large degree of overlap, the remaining wells are clearly distinguishable using this particular combination of elements. An even better discrimination was achieved using combinations of Li, Zn and Ni, as shown in Fig. 18. This plot, in contrast to Fig. 17, successfully discriminates Wells 29/10-3 and 22/5b-5, but reveals an increased degree of overlap between Wells 22/15-1 and 22/5b-5. Furthermore, the latter well exhibits two distinct subpopulations which are thought to indicate differences in provenance.

The discriminative elements Li, Zn and Ni used in Fig. 18 are most likely to be concentrated in the clay minerals of both detrital and authigenic origin. Although not discussed in this study, the relative importance of these two origins certainly warrants further investigation.

Owing to the restricted stratigraphic distribution of sample material, mainly constrained by sample availability, no ICP–AES data were available across large intervals of the Triassic succession in the subject wells. It was not, therefore, possible to assess the full potential of using ICP–AES data in the stratigraphic subdivision of the Triassic. Chemostratigraphic units, therefore, would have to be restricted to the available cored intervals. Unfortunately, much of this part of our investigation remains commercially confidential, although it can be said, that in the thickest sequence studied, 14 chemostratigraphic units were identified.

In order to examine any possible stratigraphic resolution using the current ICP–AES database three wells were chosen for closer examination

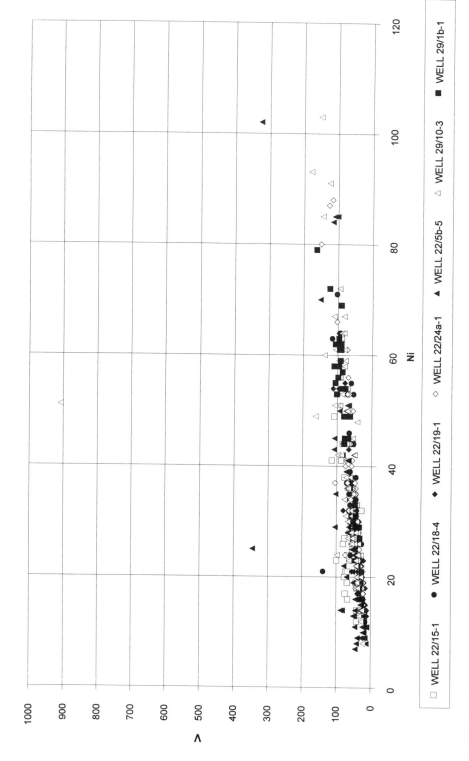

Fig. 12. Crossplot of Ni and V for each well. Values of Ni/V show little variation within and between wells. The linear trend for these elements suggests a strong interdependency.

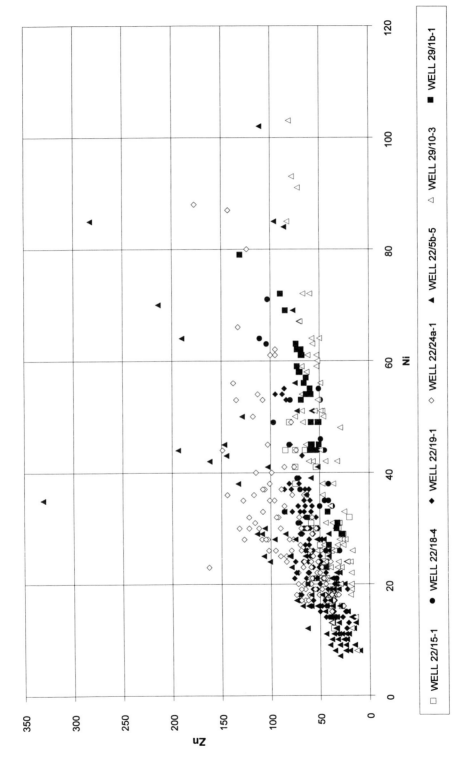

Fig. 13. Crossplot of Ni and Zn for each well showing relatively poor separation by well although a generally higher Ni content is present in the more southerly wells.

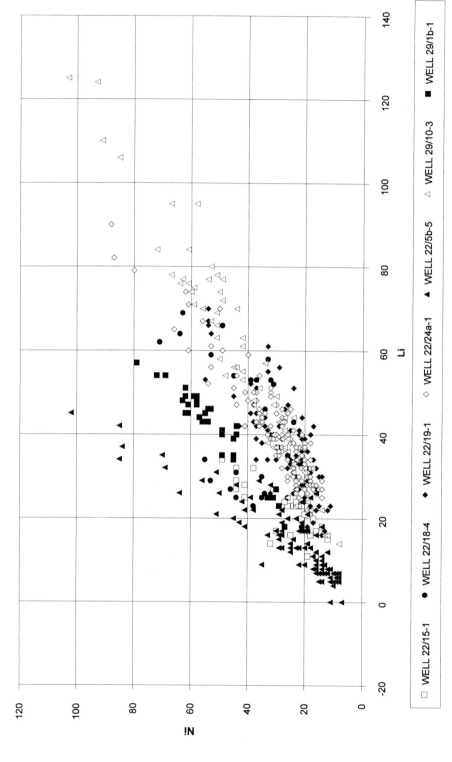

Fig. 14. Crossplot of Ni and Li for each well showing a reasonable separation by well. The most geographically separated wells show greatest compositional differences.

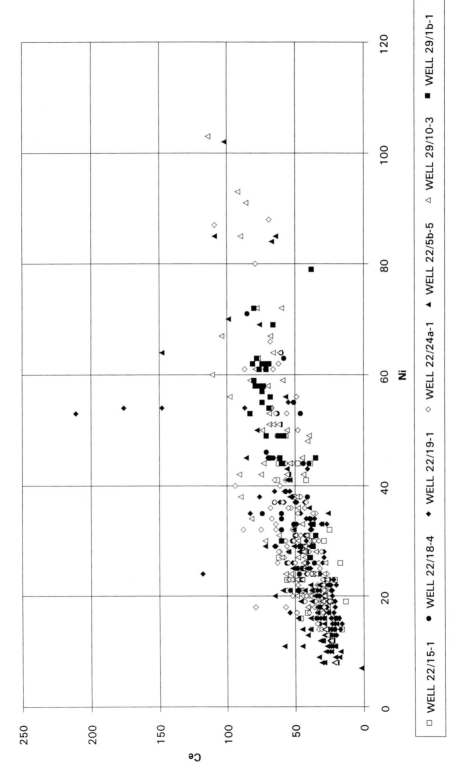

Fig. 15. Crossplot of Ni and Ce for each well showing poor discrimination of individual wells. Overall, however, this shows a fairly tightly constrained co-variation for the Skagerrak samples.

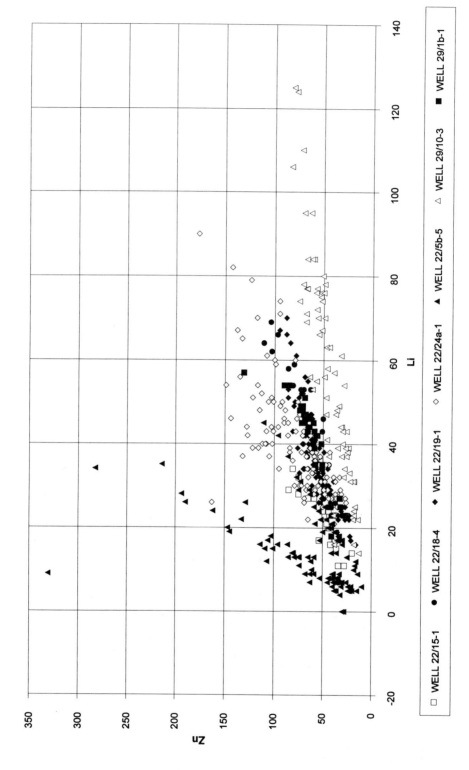

Fig. 16. Crossplot of Li and Zn for each well showing good well separation. The most geographically separated wells show the greatest chemical differences, i.e. 29/10-3 and 22/5b-5.

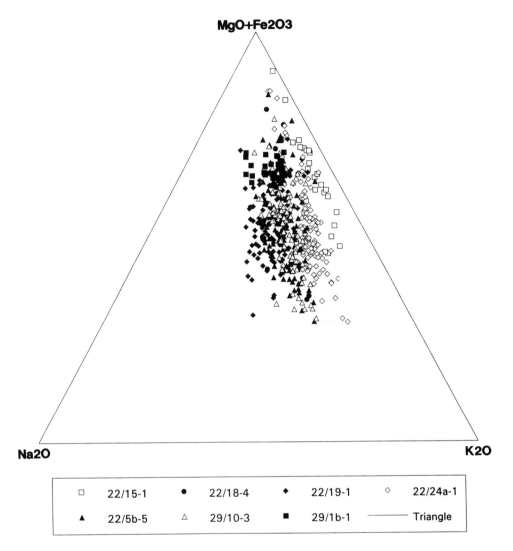

Fig. 17. Triangular plot of Na, Mg + Fe and K for each well showing some differentiation by well but no clear geographical separation. The compositional field for well 29/1b-1 is tightly constrained.

(Wells 22-5b-5, 22/24a-1 and 29/10-3). Well 22/5b-5 possesses the greatest stratigraphic core coverage and its samples can be easily discriminated from those of Well 22/24a-1.

Some results are shown in Figs 19 & 20. ICP–AES data for each of these wells were grouped into five depth intervals (dcat 1–5) in increasing depth order. Each quintile is coded on each plot and the discriminative elements chosen for the example illustrated were Li and Zn. In well 22/5b-5 (Fig. 19) a depth discrimination between the two deepest subgroups (dcat 4 and 5) and the two shallowest subgroups (dcat1 and 2) is clearly

demonstrated. The middle subgroup (dcat3) is missing owing to a large core gap of 76 m in this well. This bimodal subgrouping is also apparent (Figs 10 & 18), It is through the integration of data from a number of such plots for various elements that a 'chemical zonation' can be established for the studied sequence.

A similar presentation for Well 22/24a-1 is given in Fig. 20 where the separation of the shallowest (dcat 5) quintiles is more obvious, but this plot excludes the intermediate depth categories. In both these wells the observed sub-populations may represent differences in prove-

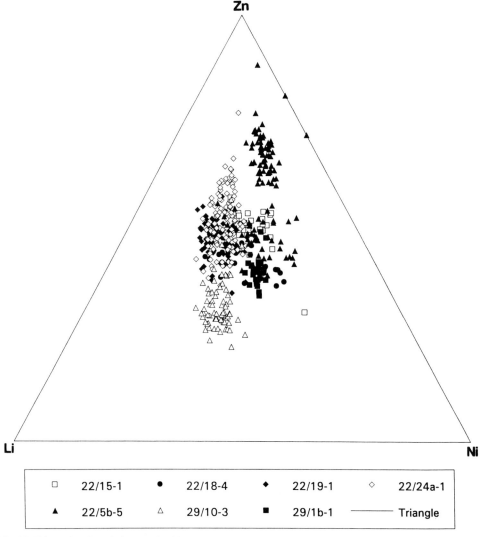

Fig. 18. Triangular plot of Li, Zn and Ni for each well showing that differentiation is almost directly proportional to the degree of geographical separation. The compositional field for well 29/1b-1 is tightly constrained.

nance, the occurrence of hiatuses and/or differences in their diagenetic histories. A difference in provenance is the most likely explanation based on our petrographic observations.

Statistical analysis of ICP–AES data

Discriminant function analysis

In this study discriminant function analysis has been used as a tool in performing exploratory data analysis (Ehrenberg & Siring 1992). Initially, it was considered that the chemical data generated from analysis of the 491 sandstone

samples might provide diagnostic information on facies type in barren sequences. However, when the data were analysed in scatter plots and using discriminant function analysis (DFA) there was limited evidence for discrimination between facies using the values of elements and oxides obtained by ICP–AES spectrometry. The data set was further examined and good discrimination between wells was obtained using DFA.

Prior to undertaking DFA, all the elemental bivariate cross-plots were studied to investigate relationships between variables and between categories. This was done by colour coding co-

Well 22/5b-5

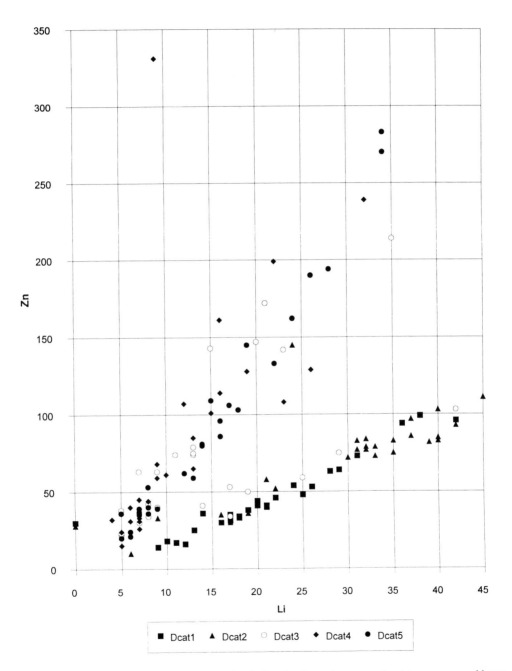

Fig. 19. Crossplot of Li and Zn grouped into depth quintiles showing a clear separation into an upper and lower unit. Through the integration of a number of such plots for various element pairs (shown to be statistically significant) for each well, it is possible to build up a chemical zonation.

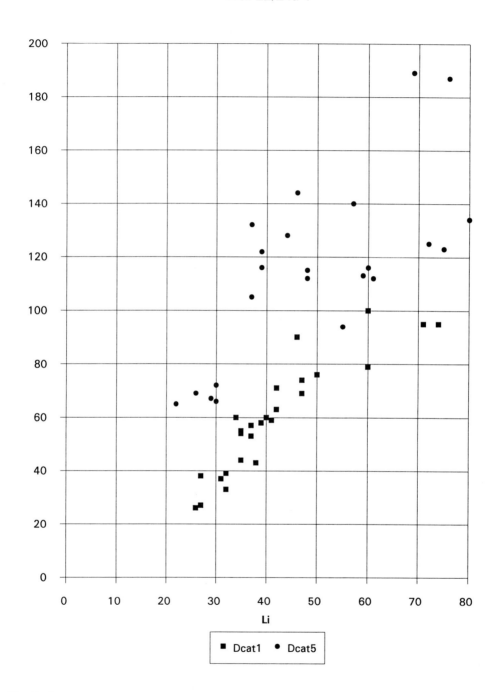

Fig. 20. Crossplot of Li and Zn revealing only shallowest and deepest quintiles in order to demonstrate more clearly the stratigraphic separation of the data.

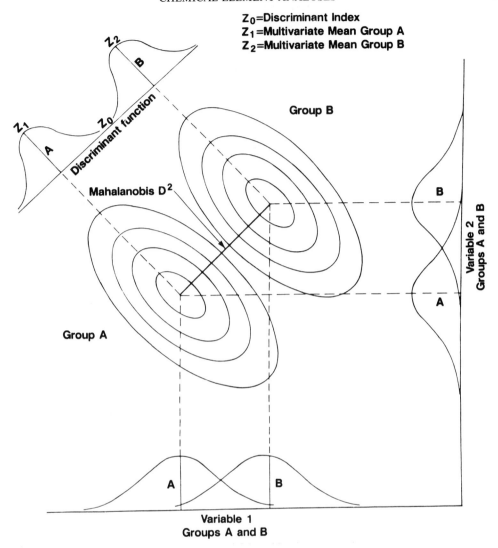

Z_0=Discriminant Index
Z_1=Multivariate Mean Group A
Z_2=Multivariate Mean Group B

Fig. 21. Graphical illustration of discriminant analysis. Plot of two bivariate distributions, showing overlap between groups A and B along both variables 1 and 2. Groups can be distinguished by projecting members of the two groups onto the discriminant function line (after Davis 1986).

ordinate points so that the categorical data for each facies and well were also displayed. The clearest relationships identified are presented in the scatter plot of Zn v. Li and the triangular plots of Li, Zn, Ni and Na_2O, $MgO + Fe_2O_3$, K_2O (Figs 10, 17 & 18). Because these variables were observed to provide some discrimination between wells in bi- and trivariate space, we had an a priori reason for attempting to build a model for discriminating between wells in multivariate space using these variables. If these variables were also found to be efficient discriminators between wells in multivariate space

we would not have the task of interpreting the results of a stepwise DFA using all 29 chemical species contained in the dataset, most of which had failed to provide any insight into associations with facies or well in bivariate plots.

The purpose of deriving the discriminant functions (also known as canonical roots) is to transform the univariate data (i.e. wt% oxides and p.p.m. of elements) to multivariate data, where the ICP–AES data for each analysed rock sample are reduced to one number termed the score. Each score occupies a position on the vector of the discriminant function. In doing this

Variable Enter/Remove	Step	F to entr/rem	df 1	df 2	p-level
Li	1	110.7512	6	483	0.000000
Ni	2	77.7131	6	482	0.000000
Zn	3	69.8270	6	481	0.000000
K_2O	4	31.3087	6	480	0.000000
$MgO + Fe_2O_3$	5	4.2595	6	479	0.000341

Fig. 22. Summary of stepwise analysis.

Variable	DF1	DF2	DF3	DF4	DF5
Li	1.61480	-0.46263	-0.672509	-0.293397	-0.320372
Ni	-1.02351	1.98723	-0.061362	0.653495	-0.523393
Zn	-0.77900	-1.28569	-0.371736	0.632784	-0.124981
K_2O	0.60389	-0.12336	1.130836	0.210494	0.509142
$MgO + Fe_2O_3$	0.09586	-0.24564	-0.193052	-0.323493	1.247815
Eigenval	3.71567	1.01137	0.215987	0.084733	0.024479
Cumul. %	0.73545	0.93563	0.978384	0.995155	1.000000

Fig. 23. Standardized coefficient for discriminant functions.

DFA maximizes the separation between the average value scores for each of the seven wells (group centroids) whilst minimizing the inflation of each group. A graphical example of the separation obtained from a two variable, two group discriminant function is presented in Fig. 21.

The best model for discrimination between facies using all variables, in a stepwise discriminant analysis, achieved a Wilk's Lambda value of 0.48 and provided poor discrimination between facies. The more successful model reported here in greater detail for discrimination between wells has a Wilk's Lambda value of 0.08. In terms of efficiency of discrimination a value of 0 for this statistic represents a model with perfect discriminatory power and a value of 1 indicates a model with no discriminatory power whatsoever. After considering and rejecting experimental and sampling artifacts, it became evident from these analyses that some natural mechanism had caused the unexpected result of being able to discriminate between wells on the basis of chemistry.

Data conditioning

All the raw data generated for the variables in the ICP–AES study were reported in units of either p.p.m. or wt% oxide. Inspection of these data did not reveal any serious outliers in the variables used to discriminate by well, and no

Group (Well)	DF1	DF2	DF3	DF4	DF5
29/1B-1	-0.37001	1.85420	-0.882472	0.486489	-0.013873
29/10-3	2.43912	1.38497	0.431246	-0.220828	0.129681
22/5B-5	-3.30327	0.06412	0.112116	-0.122769	0.101908
22/24A-1	0.96399	-1.11691	0.170141	0.320316	0.077228
22/19-1	0.83843	-0.72091	-0.438146	-0.352008	-0.105751
22/18-4	0.18867	0.88613	-0.620917	0.263470	-0.141974
22/15-1	-0.93882	0.52462	1.023338	0.159409	-0.503224

Fig. 24. Group means on discriminant functions.

Functions Removed	Eigen-value	Wilks' Lambda	Chi-Sqr	df	p-level
0	3.715674	0.078021	1232.027	30	0.000000
1	1.011370	0.367920	482.947	20	0.000000
2	0.215987	0.740023	145.419	12	0.000000
3	0.084733	0.899859	50.965	6	0.000000
4	0.024479	0.976106	11.681	2	0.002911

Fig. 25. Chi-square tests with successive functions removed.

cases were removed. For DFA it is desirable that the distribution of each variable in each group (well) approximates to a normal distribution. To this end a comparison was made between the raw distributions and the $\log_{10}$ (log normal) transformation of the raw data. A test for normality in each case was made using the Kolmogorov–Smirnov statistic. The results showed that for 20 variables from the seven groups both raw data and $\log_{10}$ transforms approximated to a normal distribution. Eleven variables were normally distributed only after the log transform, and in the case of four variables the raw data only approximated to a normal distribution. In the light of these results the logarithmic transform was applied. It was considered that although four variables from two wells would not have the desired distribution, any ill effects of the DFA might not be too serious since the $\log_{10}$ transform would reduce

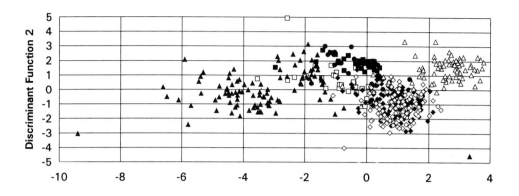

Discriminant Function 1

□ WELL 22/15-1	● WELL 22/18-4	◆ WELL 22/19-1	◇ WELL 22/24a-1
▲ WELL 22/5b-5	△ WELL 29/10-3	■ WELL 29/1b-1	

Fig. 26. Discriminant function plot of sandstone samples, showing good separation by well.

any problems with long tails in these distributions.

Forward stepwise discriminant analysis

A model for the discrimination between wells was generated and is presented below. This analysis allows discriminant functions to be determined by selecting variables (chemical species) which are significant in discriminating between wells. Solutions are found for the unknown weights for the variables which are used to build linear equations of the type shown below:

. . . discriminant function . . .

Z (discriminant score) $= b0 + b1X1 + b2X2 + \ldots bnXn,$

where $b0$ is a constant, and the other b values are the weights (discriminant coefficients) applied to the variables (chemical species) $X1$–Xn.

With the conditions of significance for entry of variables to the model set, the stepwise analysis proceeds to add variables to the model in order of decreasing statistical significance. In

the case of the five variables selected by us, all were found to be significant in discriminating between wells. The number of discriminant functions in the model in this instance is equal to the number of variables in the model, which is five.

The summary of the stepwise analysis is presented in Fig. 22, and standardized coefficients for the discriminant functions are shown in Fig. 23. The presentation here of the standardized coefficients allows interpretation of the effect of each variable in contributing to discrimination since the data are still in different units. The effect of each variable on the score is directly proportional to the magnitude of its coefficient. The eigenvalues for each discriminant function and their cumulative percentages show the amount of variance in the data accounted for by each function. It can be seen that the first two functions account for nearly 94% of variance in the data. Figure 24 presents the mean values of the scores of the seven groups (wells) along the vectors of the five discriminant functions. Figure 25 presents the results of tests of significance of the discriminant functions. All functions have discriminatory power well

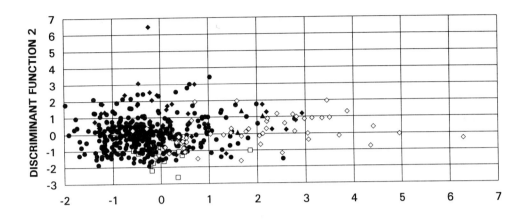

DISCRIMINANT FUNCTION 1

□ FACIES 1 • FACIES 2 ◆ FACIES 3 ◇ FACIES 4 ▲ FACIES 5

Fig. 27. Discriminant function plot of sandstone samples, showing good separation by facies.

beyond the 0.05 significance level chosen at the outset of the study, as the acceptable risk of falsely rejecting a correct null hypothesis. However, only the first two roots are useful at a practical level.

Figure 26 presents a cross plot of sample scores on discriminant function 1. v. discriminant function 2. The power of the model is evident, with good separation achieved in most cases. Since the mineralogy (authigenic and detrital) is virtually identical (including clay type and amount) and the sedimentary facies fairly uniform from one sequence to another, it is concluded that provenance changes are the most likely cause of these element distributions. In the case of Wells 22/19-1 and 22/24A-1 discrimination is poor along both roots. It is considered that this may be due to the close proximity of the wells to each other, in which case similar sediment sourcing may have imprinted a similar chemistry on both wells, such that discrimination cannot be achieved using the five variables selected.

Figure 27 presents a cross-plot of the sample scores on the two most significant discriminant functions obtained in attempting to discriminate between facies using a stepwise procedure on all 29 variables. The same approach was taken in building this model as already described. The model provides poor discrimination. However, this result is consistent with the idea that the intervals sampled are variably sourced. In this case good discrimination between facies could not be expected.

Conclusions

The sequences studied are very similar in terms of facies and detrital mineralogy. They are mineralogically immature and have undergone limited chemical weathering. Diagenetically all had very similar histories, though certain diagenetic phases are more pronounced in certain samples. Consequently the elemental separations seen in this study are valid and cannot be the result of diagenetic processes, since all the wells studied had similar diagenetic histories. Moreover, both clay- and sand-prone samples often show similar elemental discriminations (Fig. 10).

The variables Li, Ni, Zn, K_2O and the sum of $MgO + Fe_2O_3$ showed some discriminatory power between wells in the cross-plots and provided us with an a priori hypothesis as to which chemical species might be entered into a multivariate model with which to discriminate between wells. All the selected variables were significant when discriminant functions were derived, although only the first two functions were of practical use; we therefore felt justified in using those variables. Models with more power for discriminating between wells were generated using the data from all 29 chemical species analysed in a stepwise DFA. However, some of the most significant species in multivariate space did not appear to be useful in bivariate space, and the models were therefore more difficult to interpret and open to the criticism of stepwise DFA capitalizing on chance.

The interpretation of the results of discrimination by well is that the wells may be sourced variably, and that within the 'system' studied, Wells 22/5b-5 and 29/10-3 represent end members, whereas Wells 22/19-1 and 22/24A-1 show least separation because they have been similarly sourced.

The writers wish to acknowledge Amerada Hess, Amoco, British Gas and Enterprise for their full support during the course of this project. Thanks are due to the Geochem Group for providing technical assistance in the production of this paper. Don Cameron (BGS) and another anonymous referee are also thanked for their constructive comments on an earlier draft of this paper. The editors and publisher are gratefully acknowledged for their patience.

References

COPE, J. C. W., INGHAM, J. K. & DAWSON, P. F. (eds) 1992. *Atlas of Palaeogeography and Lithofacies*. The Geological Society, London.

DAVIS, J. C. 1986. *Statistics and Data Analysis in Geology*. 2nd ed. John Wiley & Sons, Inc., New York.

DEEGAN, C. E. & SCULL, B. J. 1977. *A standard lithostratigraphic nomenclature for the central and northern North Sea*. Institute of Geological Science, Report 77/25. HMSO, London.

DOTT, R. L. 1964. Wacke, greywacke and matrix – what approach to immature sandstone classification. *Journal of Sedimentary Petrology*, **34**, 625–632.

EHRENBERG, S. N. & SIRING, E. 1992. Use of bulk chemical analyses in stratigraphic correlation of sandstones: an example from the Statfjord North Field, Norwegian Continental Shelf. *Journal of Sedimentary Petrology*, **62**, 318–330.

FISHER, M. J. & MUDGE, D. C. 1990. Triassic. *In:* GLENNIE, K. W. (ed.) *Introduction to the Petroleum Geology of the North Sea*. Blackwell Scientific Publications, Oxford, 191–218.

GAGE, M. S. & DORÉ, A. B. 1986. A regional perspective of the Norwegian offshore exploration provinces. *In:* SPENCER, A. M. (ed.) *Habitat of Hydrocarbons on the Norwegian Continental Shelf*. Norwegian Petroleum Society, Graham and Trotman, London, 31–38.

GOUDIE, A. 1973. *Duricrusts in Tropical and Subtropical Landscapes*. Oxford University Press, Oxford.

HODGSON, N. A., FARNSWORTH, J. & FRASER, A. J. 1992. Salt-related tectonics, sedimentation and hydrocarbon plays in the Central Graben, North Sea. *In:* HARDMAN, R. F. P. (ed.) *Exploration Britain: Geological Insights for the Next Decade*. Geological Society, London, Special Publication, **67**, 31–63.

JEANS, C. V., REED, S. J. B. & XING, M. 1993. Heavy mineral stratigraphy in the UK Trias: Western Approaches, onshore England and the Central North Sea. *In:* PARKER, J. R. (ed.) *Petroleum Geology of Northwest Europe: Proceedings of the 4th Conference*. The Geological Society, London, 609–624.

LERVIK, K. S., SPENCER, A. M. & WARRINGTON, G. 1989. Outline of Triassic Stratigraphy and Structure in the Central and Northern North Sea. *In: Correlation in Hydrocarbon Exploration*. Norwegian Petroleum Society, Graham & Trotman, London, 173–189.

MCBRIDE, E. F. 1963. A classification of common sandstones. *Journal of Sedimentary Petrology*, **33**, 664–669.

MANGE, M. A. 1993a. Subdivisions and correlation of monotonous sandstone sequences using high resolution heavy mineral analyses: Examples from North Sea reservoirs. Abstract in: *Dating and Correlating Biostratigraphically Barren Strata*. The Geological Society of London Meeting, May 1993.

—— 1993b. Zonation and correlation of the Smith Bank and Skagerrak Formations, Central Graben. Abstract in: *Dating and Correlating Biostratigraphically Barren Strata*. The Geological Society of London, Meeting, May 1993.

RACEY, A., WALSH, J. N. & LOVE, M. A. 1992. Applications of chemical stratigraphy in the Jurassic and Triassic of the UK: Examples from the North Sea and Cheshire Basin. *EAPG, Paris*, April 1992.

SCHMIDT, V. & MCDONALD, D. A. 1979. Texture and recognition of secondary porosity in sandstones. *In:* SCHOLLE, P. A. & SCHLUGER, P. R. (eds) *Aspects of Diagenesis*. SEPM Special Publication, **26**, 209–225.

SMITH, R. I., HODGSON, N. & FULTON, M. 1993. Salt control on Triassic reservoir distribution, UKCS Central North Sea. *In:* PARKER, J. R. (ed.) *Petroleum Geology of Northwest Europe: Proceed-

ings of the 4th Conference. The Geological Society, London, 547–557.

SURDAM, R. C., BOESE, S. W. & COSSLEY, L. T. 1984. The chemistry of secondary porosity. *In:* MC-DONALD, D. A. & SURDAM, R. C. (eds) *Clastic Diagenesis.* American Association of Petroleum Geologists Memoir, **37**, 127–149.

THOMPSON, M. & WALSH, J. N. 1989. *Handbook of ICP Spectrometry.* Blackie and Co, Glasgow.

WATTS, N. L. 1980. Quaternary pedogenic calcretes from the Kalahari (southern Africa), mineralogy, genesis and diagenesis. *Sedimentology,* **27**, 661–686.

ZIEGLER, P. A. 1982. *Geological Atlas of Western and Central Europe.* SIPM, Elsevier, Amsterdam.

High-resolution chemostratigraphy of Quaternary distal turbidites: a case study of new methods for the analysis and correlation of barren sequences

TIMOTHY J. PEARCE[1] & IAN JARVIS[2]

[1]*Chemostrat Consultants, 20 Maer-yr-Eglwys, Llansantffraid-ym-Mechain, Powys, SY22 6BE, UK*

[2]*School of Geological Sciences, Kingston University, Penrhyn Road, Kingston upon Thames, Surrey, KT2 1EE, UK*

Abstract: The late Quaternary distal turbidites of the NE Atlantic Madeira Abyssal Plain (MAP), represent one of the most intensely studied sequences in the modern deep ocean. More than 160 sediment cores have been recovered from the $68 \times 10^3 \, km^2$ of the plain, and existing litho- and biostratigraphies, when integrated with oxygen stable-isotope data for other areas of the NE Atlantic, provide a tight spatial and temporal framework. We have used the MAP as a test-bed to assess the applications of inorganic geochemistry as a sedimentological and stratigraphic tool. Data for 22 major and trace elements were obtained in over 500 samples taken from representative cores. Chemostratigraphic sequences are highly uniform across the plain, and several turbidites have unique geochemical signatures, providing a means of establishing bed-by-bed correlations over distances of $> 500 \, km$. Cluster and principal component analysis confirm the statistical validity of empirically derived correlations and groupings. Vertical and lateral geochemical trends within beds document the sedimentological evolution of flows, and may be used to establish palaeotransport pathways. Four compositional groups are interpreted as representing derivations from different source areas: (1) the margins of the NW Africa off Morocco; (2) off Western Sahara; (3) the Canary Islands; (4) the Cruiser–Great Meteor East seamount chains. The study amply demonstrates the viability of chemostratigraphy for the correlation and analysis of sedimentary sequences, providing a stratigraphic resolution and reliability which matches or exceeds that obtainable by other techniques.

A significant proportion of the World's hydrocarbon resources occur in sequences which are barren of biostratigraphically significant index fossils. A number of methods are currently being developed to address the stratigraphic problems associated with such successions; one of these new techniques, known as 'chemostratigraphy', involves the use of inorganic geochemical data to characterize and correlate strata. Sequences are subdivided into units with diagnostic geochemical signatures, which in most siliciclastic sediments are controlled largely by the composition of their source regions. These signatures, or 'fingerprints' potentially provide a means of defining and correlating sedimentary units over wide areas. This paper presents results of a detailed test of this technique on sequences with an existing well-defined stratigraphic framework.

Geochemical data have been used to characterize sediment sequences for several years (e.g. Bhatia & Taylor 1981; Spears & Amin 1981; Bhatia 1983; Bhatia & Crook 1986; Floyd & Leveridge 1987; Wronkiewicz & Condie 1987). However, many previous workers have concentrated on the application of major-element concentrations and ratios. Bhatia (1983), for example, working on Palaeozoic mudstones from a number of basins in eastern Australia, attributed stratigraphic variations in Al_2O_3/SiO_2, K_2O/Na_2O and other ratios to a change in source rock-type from andesite/dacite to granite/gneiss and sedimentary rocks. Geochemical data have also been used successfully to predict the tectonic setting of ancient sequences (Bhatia 1983, 1985; Bhatia & Crook 1986; Floyd & Leveridge 1987). Recent research has emphasized the importance of trace elements (e.g. Floyd *et al.* 1991; Humphreys *et al.* 1991; McCann 1991; Pearce 1991), particularly those located predominantly in refractory heavy minerals, for the interpretation of sediment provenance.

The idea that geochemical heterogeneities can be employed to establish a stratigraphic framework is not new, but the interest shown by the

From Dunay, R. E. & Hailwood, E. A. (eds), 1995, *Non-biostratigraphical Methods of Dating and Correlation* Geological Society Special Publication No. 89, pp. 107–143

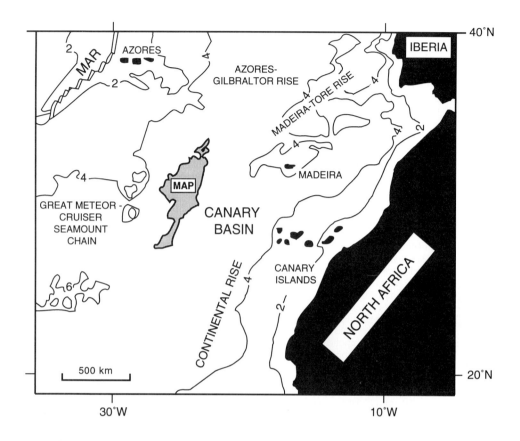

Fig. 1. Regional setting of the Madeira Abyssal Plain (MAP). Water depths are indicated by the 2 and 4 km isobaths.

petroleum industry has been limited, principally because of the high cost and slow turnround associated with obtaining large high-quality multi-element data sets. Jørgensen (1986), however, successfully correlated six wells in a Chalk reservoir from the Danish sector of the North Sea, despite being hampered by the use of laborious analytical procedures involving single-element determinations by atomic absorption spectrometry (AAS). Only five elements were investigated and correlations were based solely on changes in broad vertical geochemical trends within wells, rather than a higher resolution marker-bed stratigraphy. Further research on chemostratigraphy has been limited, due, at least in part, to a rather sceptical response to the early work. Data had been obtained primarily for major-elements (e.g. Ca, Mg and K), constituents which are commonly involved in diagenetic processes, and as a consequence, are inherently limited in their potential for correlating ancient

depositional sequences.

During the 1980s, significant advances in analytical geochemistry, particularly improvements in inductively coupled plasma–atomic emission spectrometry (ICP–AES) and the advent of ICP–mass spectrometry (ICP–MS), for the first time enabled the rapid and cost-effective acquisition of high-quality data for a comprehensive suite of major, minor and trace elements (see Jarvis & Jarvis, 1992a, b for recent reviews). Up to 50 elements could now be determined in a matter of minutes for solutions prepared from < 1 g of material, thereby overcoming the tedium of single-element determination by AAS. When combined with mineralogical information, these data enabled detrital geochemical 'fingerprints' to be identified with great confidence. This approach has been employed by a number of workers (de Lange et al. 1987; Wray 1990; Pearce, 1991; Pearce & Jarvis, 1992a; Wray & Gale 1993).

Wray (1990) and Wray & Gale (1993), for example, undertook an ICP–AES-based study of marl bands from Upper Cretaceous Chalk outcrops in the Anglo–Paris Basin, and developed a high-resolution correlation based on geochemically distinct marker beds.

It was considered timely, therefore, to reassess the potential of chemostratigraphy for more routine use in the correlation of biostratigraphically barren sequences. Three requirements were identified for the study: (1) to test the principle that sequences may be characterized by their detrital geochemical signature, the succession analysed should not have been modified significantly by diagenesis; (2) to confirm the validity of chemostratigraphic correlations, it would be necessary to choose a study area with an existing well-constrained stratigraphic framework; (3) to determine whether the method was sufficiently robust to be applied successfully to other sequences, it would be necessary to confirm its validity using statistical methods. It was decided that a modern deep-ocean turbidite sequence would provide the ideal starting point to test the technique. The area chosen was the Madeira Abyssal Plain (MAP), an area of intensive research since the early 1980s.

Madeira Abyssal Plain (MAP): a case study

The MAP lies in the deepest part of the Canary Basin (Fig. 1) at water depths of c. 5400 m. It consists of three sub-basins which combined occupy an area of c. 68 000 km². The morphology of the MAP is virtually flat, interrupted only by small abyssal hills that rise a few hundred metres above the plain. The late Quaternary sediments from the MAP have been studied in great detail over the last 10 years. The initial impetus was provided by the nuclear industry (Nuclear Energy Agency 1984) who were searching for potential sites for the disposal of high-level radioactive waste. However, following the end of that programme, research has been continued, particularly by workers at, or in collaboration with, the Institute of Oceanographic Sciences Deacon Laboratory (Colley et al. 1984; Colley & Thomson 1985; Kuijpers & Weaver 1985; Wilson et al. 1985, 1986; Thomson et al. 1986, 1987; Weaver et al. 1986, 1989a, b, 1992, 1994; de Lange et al. 1987, 1989; Jarvis & Higgs 1987; Kidd et al. 1987, Searle 1987, Weaver & Rothwell 1987; Pearce 1991; Jones et al. 1992; Pearce & Jarvis 1992a, b; Rothwell et al. 1992; Weaver 1993; Weaver & Thomson, 1993). As a consequence, virtually all aspects of the late Quaternary sedimentary record have been studied.

Sediments have been collected primarily by piston corers recovering to a depth of up to 35 m. To date, more than 160 cores have been collected (Fig. 2), sequences typically comprising metre-thick fine-grained distal turbidite muds separated by centi- to decimetre intervals of pelagic ooze, marl or clay (Weaver et al. 1986; Weaver & Rothwell 1987). The thicker turbidites commonly have a coarser basal facies comprising laminated thin sands and silts (Units T_{bcd} of Bouma 1962), but these units progressively disappear distally, leaving only wispy laminated silts and muds at the base (Units T_{3-5} of Stow & Shanmugam 1980). Throughout the area turbidites are composed dominantly of T_e (T_{6-8}) ungraded muds.

Turbidite sequences are constrained by the nannofossil biostratigraphy of the intervening pelagic intervals (Kuijpers & Weaver 1985; Weaver et al. 1986, 1989a, b, 1992, 1994; Weaver 1993). These may be correlated with entirely pelagic sequences on the adjacent abyssal hills where climatic cycles are recorded (Kuijpers & Weaver 1985) as regular alternations of clays and marly clays (glacial intervals) with marls and oozes (interglacials). These alternations may be related in turn, to oxygen isotope stages, some of which are identifiable from their distinctive nannofossil contents; the intervening sediments may then be assigned to their corresponding isotope stages based on lithological criteria. Using this approach the chronostratigraphy developed for the abyssal hill successions can be traced back into the turbiditic sequences on the abyssal plain (Fig. 3), providing a time framework for turbidite emplacement over the last 750 ka. The turbidites themselves exhibit distinctive colour and thickness variations and yield unique mixed coccolith assemblages (Fig. 3) derived from their sediment source areas (Weaver et al. 1986; Weaver 1993; Weaver & Thomson, 1993). By combining these data, individual turbidites may be identified and correlated across the area. Each turbidite has been assigned a letter code, labelled alphabetically from the top downwards, with thin laterally inextensive units being assigned numbered suffixes (e.g. a, a_1, b).

Despite their young age and minimal burial, MAP turbidites display evidence of modification by post-depositional oxidation (Colley et al. 1984; Colley & Thomson 1985; Wilson et al. 1985, 1986; Thomson et al. 1986, 1987; Jarvis & Higgs 1987; Weaver et al. 1989a). The uppermost decimetre of each bed is commonly 'bleached', reflecting the progressive downward post-depositional migration of an oxidation front while the top of the turbidite remained

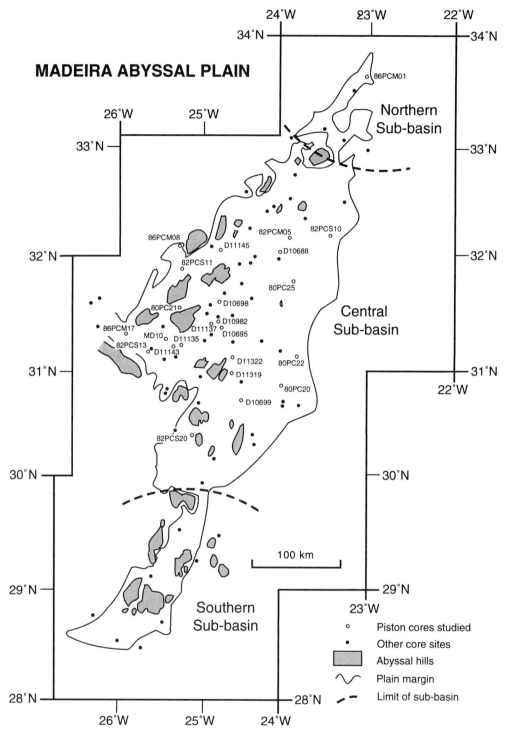

Fig. 2. Map showing the limits of MAP sub-basins and the positions of major piston core sites. Core numbers that commence with 'D' were collected by RRS Discovery and are stored at IOSDL; those starting with a number are Dutch Geological Survey cores; MD10 was collected on the ESOPE cruise (Auffret *et al.* 1989).

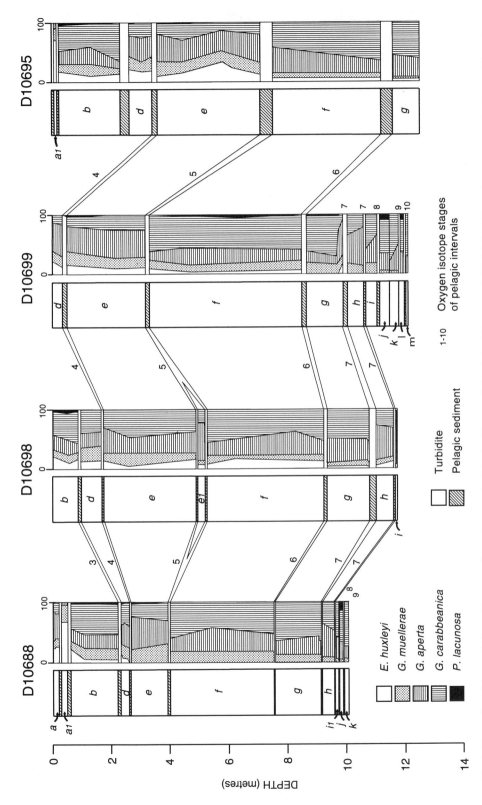

Fig. 3. Biostratigraphic correlation of four cores from the central sub-basin. The correlation (after Weaver *et al.* 1986) is based on calcareous nannofossil assemblages from the pelagic intervals, which were then assigned to their corresponding oxygen isotope stages; note that individual turbidites also have distinct mixed coccolith assemblages.

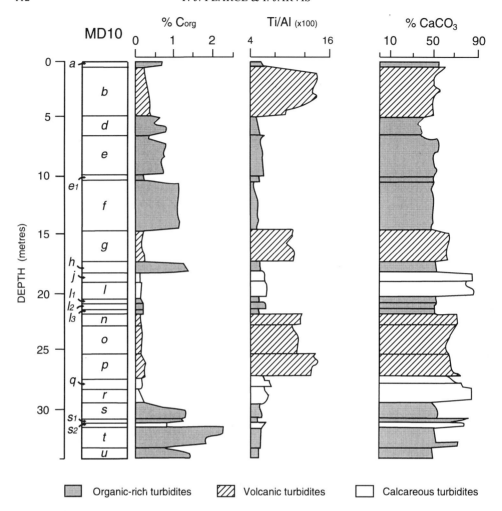

Fig. 4. Stratigraphic variation of C_{org}, $CaCO_3$ (%) and Ti/Al ratios in core MD10 turbidites. These and other data were used by de Lange *et al.* (1987) to define their three compositional groups. Turbidite lettering scheme after Weaver & Kuijpers (1983) and Weaver *et al.* (1989*a, b*); solid black lines separating turbidites are pelagic intervals.

exposed on the seafloor. This 'bleaching' effect, caused by the destruction of organic matter and the oxidation of reduced transition metal species, is particularly apparent in organic-rich (> 0.3% organic carbon) turbidites, which display a pronounced two-tone coloration. A number of elements, particularly Co, Cu, Fe, Mn, Ni, V, U and Zn (Jarvis & Higgs 1987), have been mobilized and relocated in response to changing redox conditions. However, it is important to note that many others, including Si, Ti, Al, Mg and Zr, appear to be unaffected by these early diagenetic processes (Jarvis & Higgs 1987; Jarvis *et al.* 1988).

The bulk geochemistries of turbidites from two cores were studied by de Lange *et al.* (1987,

1989), who recognized three broad compositional groups (Fig. 4): (1) organic-rich turbidites with 0.3–2.5% organic carbon (C_{org}); (2) 'volcanic' turbidites containing volcaniclastic debris and high concentrations of elements such as Ti; (3) calcareous turbidites with high $CaCO_3$ (> 75%) contents. The geochemical compositions of individual turbidites were shown to be remarkably consistent between two cores (D10688 and MD10) situated in the NE and SW of the central sub-basin (Figs 2 & 4), and it was proposed that these signatures might remain similar for each turbidite over the whole of the plain. The material appeared to be ideal, therefore, to test the potential of chemostratigraphy as an independent means of correlation.

Table 1. *Details of MAP piston cores sampled for geochemical analysis*

Core site	Position		Water* depth	Length[†]	Turbidites[‡]
	Lat. (°N)	Long. (°W)	(m)	(m)	
D10688	32°03.0′	24°12.1′	5428	10.0	$a, a_1, b, b_1, d, e, f, g, h, i_1, j, k$
D10695	31°23.7′	24°46.3′	5433	11.5	b, b_1, d, e, f, g
D10698	31°38.2′	24°49.8′	5433	12.8	$a, a_1, b, b_1, d, e, f, g$
D10699	30°43.8′	24°29.0′	5431	12.2	$a_1, b, d, e, f, g, h, i, j, k, l, m$
D10982	31°27.0′	24°48.3′	5434	17.7	$a, b, b_1, d, e, f, g, h, j, l$
D11135	31°15.9′	25°11.1′	5442	8.4	a, b, b_1, d, e
D11137	31°23.5′	24°46.0′	5441	10.8	a, a_1, b, b_1, d, e
D11143	31°11.9′	25°11.3′	5444	11.8	a, b, b_1, d, e
D11145	32°06.4′	24°45.5′	5433	12.0	$a, a_1, b, b_1, d, d, e, f$
D11319	30°56.5′	24°36.0′	5436	9.0	a_1, b, b_1, d, e, f
D11322	31°10.9′	24°35.0′	5435	4.5	b, b_1, d
MD10[§]	31°16.7′	25°23.3′	5439	33.8	$a, b, d, e, e_1, f, g, h, j, l, l_1, l_2, l_3,$ $n, o, p, q, r, s, s_1, s_2, t, u$
80PC20	30°56.7′	24°06.0′	5439	11.7	$a, b, b_1, d, e, f, g, h, i$
80PC21	31°35.5′	25°11.8′	5436	14.8	b, b_1, d, e, e_1, f, g
80PC22	31°13.9′	23°43.4′	5406	1.2	b_1
80PC25	31°52.4′	23°55.5′	5432	4.5	b, d, e, e_1, f, g
82PCM05	32°11.7′	24°03.4′	5426	10.6	a, a_1, b, d, d_1, e, f
82PCS10	32°09.1′	23°27.5′	5410	8.8	b, b_1, d, d_1, e, f, g
82PCS11	31°56.5′	25°12.7′	5433	17.8	$a, b, b_1, d, d_1, e, e_1, f$
82PCS13	31°09.5′	25°36.2′	5433	21.6	$a, b, c, d, e, e_1, e_2, f, g$
82PCS20	30°22.8′	25°02.9′	5433	15.2	$a_1, b, b_1, c, d, e, f, g, h, i$
86PCM01	33°51.2′	22°58.0′	5431	10.9	b, d, e, f, g, g_1, o, p
86PCM08	32°09.2′	25°13.8′	5431	16.3	d, e, f
86PCM17	31°17.5′	25°48.7′	5422	11.9	$b, b_1, c, d, e, e_1, e_2, f, g$

*Corrected depths after Rothwell *et al.* (1992); [†]excluding disturbed sections; [‡]turbidites logged by Pearce (1991); [§] data from de Lange *et al.* (1987).

Material and methods

Samples were collected from 23 cores (Table 1) from the central sub-basin and one (86PCM01) from the northern sub-basin (Fig. 2). The existing litho- and biostratigraphy was used to constrain the collection of samples; a total of 528 were collected from turbidites $a, a_1, b, b_1, d, d_1, e, e_1, f, g, g_1, h, i, j, k, l, m, o$ and p (representing deposition over the last 500 ka). Each sample consisted of a 1 cm thick, 3 cm diameter plug taken from the middle of each split core. Sampling intervals were generally c. 30 cm but ranged from 1–80 cm, the frequency being decreased in thick monotonous turbidites and increased in the vicinity of turbidite bases. In addition, sampling of core material was controlled by the necessity to gain an extensive geographical coverage, although few cores were available from the northern and southern sub-basins at the time of this work.

This geochemical investigation involved the determination of major, minor and selected trace elements in all samples by ICP–AES. In addition, 100 of these were analysed by ICP–MS, providing values for an additional 22 constituents; these data are discussed by Pearce (1991) and will not be considered further here.

Sample preparation

Sample preparation methods are described elsewhere (Jarvis 1992; Totland *et al.* 1992). Briefly, unwashed samples were dried at 105°C for 24 h, ground by hand in an agate mortar and pestle, and homogenized. Samples were prepared for analysis using two techniques: (1) 0.500 g subsamples were digested in open PTFE beakers using hydrofluoric (HF) and perchloric ($HClO_4$) acids, and evaporated to near dryness; final solutions were made up in 50 ml 1 M nitric acid (HNO_3); (2) 0.250 g subsamples were fused with 1.250 g of lithium metaborate ($LiBO_2$) at 1050°C, and the melts dissolved in dilute HNO_3; final solutions were prepared in 250 ml 0.5 M HNO_3.

Table 2. *Data obtained for US Geological Survey rock reference materials*

Material	Element	Mean	SD	Reference
SCo-1 Cody shale				
Majors (%)				
	SiO_2	62.3	1.1	62.78
	TiO_2	0.590	0.020	0.628
	Al_2O_3	13.6	0.1	13.67
	$Fe_2O_3{}^T$	5.08	0.11	5.14
	MnO	0.050	0.004	0.053
	MgO	2.75	0.03	2.72
	CaO	2.55	0.08	2.62
	Na_2O	1.07	0.05	0.899
	K_2O	2.69	0.07	2.77
	P_2O_5	0.211	0.010	0.206
AGV-1 andesite				
Majors (%)				
	SiO_2	58.1	1.0	59.25
	TiO_2	1.04	0.02	1.06
	Al_2O_3	16.7	0.1	17.15
	$Fe_2O_3{}^T$	6.57	0.02	6.76
	MnO	0.100	0.002	0.096
	MgO	1.55	0.03	1.53
	CaO	4.94	0.09	4.94
	Na_2O	4.31	0.10	4.25
	K_2O	2.89	0.09	2.90
	P_2O_5	0.470	0.010	0.48
Traces ($\mu g\ g^{-1}$)				
	Ba	1230	10	1221
	Co	16.3	0.2	15.1
	Cr	14.3	4.0	12
	Cu	62.0	2.1	60
	Li	13.6	0.5	12
	Ni	18.1	0.3	17
	Sc	13.5	0.5	12.1
	Sr	652	12	662
	V	118	3	123
	Y	22.4	0.4	21
	Zn	86.0	4.3	88
	Zr	229	6	225

Reference values from Potts *et al.* (1992); number of determinations: SCo-1, 21; AGV-1, 25; SD, standard deviation (σn); total iron expressed as $Fe_2O_3{}^T$.

ICP–AES analysis

Geochemical data were obtained using a Jobin Yvon JY70 Plus ICP–AES at Kingston University. Analytical procedures and operating conditions are listed in Pearce (1991) and Jarvis & Jarvis (1992*b*). In this study 22 elements were determined – Si, Ti, Al, Fe, Mn, Mg, Ca, Na, K, P, Ba, Co, Cr, Cu, Li, Ni, Sc, Sr, V, Y, Zn and Zr. Calibration of the ICP–AES was achieved using artificial composite matrix-matched stan-

dards for solutions prepared by acid digestion, and using rock reference materials (RRMs) for fusion solutions. Reported values for Ti, Fe, Mg, Mn, Na and K represent mean data derived from solutions prepared by both acid digestion and alkali fusion. Calcium, P, Ba, Co, Cu, Li, Ni, Sc, Sr, V, Y and Zn data were obtained from samples prepared by acid digestion; Si, Al, Cr and Zr data were generated from fusion preparations. Data are reported as wt% oxides for Si, Ti, Al, Fe, Mg, Na, K and P, and as $\mu g\ g^{-1}$ (p.p.m.) for Ba, Co, Cr, Cu, Li, Ni, Sc, Sr, V, Y, Zn and Zr. Calcium data are also presented as $CaCO_3$, since the bulk of the Ca present occurs in the carbonate fraction.

Analytical precision and reproducibility were determined by replicate analyses of multiple digestions of RRMs and in-house reference materials analysed on a routine basis with each batch of unknowns. Assessments (Table 2) were based on preparations and determinations made on a number of different days, over a period of more than one year. Long-term reproducibility for the major elements was generally better than 2% (with a short-term precision of 0.5–1%), and for the trace elements was *c.* 3%. With reference to international RRMs (Table 2), absolute accuracy is considered generally to lie within the range of the reproducibility.

Data were interpreted as absolute concentrations but also as values recalculated on a carbonate-free basis (CFB) in order to exclude the masking effects of high but variable $CaCO_3$ contents (cf. Jarvis 1985).

Compositional groups and geochemical correlation

As a first step, geochemical data for each core were plotted stratigraphically to assess the variation in each sequence. Subsequent analysis utilized binary and ternary plots to further elucidate stratigraphic and inter-element relationship in the dataset.

Core D10982

As an example, geochemical data for core D10982 from the middle of the central sub-basin (Fig. 2) are plotted stratigraphically in Fig. 5. The mean compositions of the turbidite muds from this and other representative cores are presented in Table 3. The profiles demonstrate the high degree of compositional homogeneity but different geochemical signatures displayed by individual turbidite muds. In contrast, basal sands and silts are marked by spikes and troughs relating to higher and variable concentrations of coarser material.

Differences in bulk geochemistry relate predominantly to mineralogical variation between beds. Carbonate (calcite, with very small amounts of dolomite occurring in some organic-rich turbidites) provides the only discernible control on the distributions of Ca and Sr. It is notable that Mg and Mn, the other two elements which commonly occur at significant levels in calcites, do not follow similar trends and must be controlled largely by other mineral phases. However, carbonate variation (41–82% $CaCO_3$) does markedly affect the geochemical trends of other elements (Fig. 5) due to dilution effects. Recalculating the data on a CFB (Table 3) enables differences in the composition of the non-carbonate fraction to be assessed more easily.

High concentrations of Si and Al, which broadly co-vary with K, Li and Sc, are indicative of quartz, feldspar and clay minerals (illite, smectite, chlorite, kaolinite) dominating the non-carbonate fraction of the sediments (cf. Pearce & Jarvis 1992a,b). Elements such as Ti, Fe, Mg and Zr, on the other hand, are associated for the most part with lithic grains (basalt, pumice, trachyte, volcanic glass) and heavy minerals (amphibole, olivine, pyroxene, rutile, sphene, zircon). All elements show the greatest variation in the basal silts and sands which contain highly variable proportions of carbonate, quartz, feldspar, mica, lithic and heavy mineral grains.

The three turbidite compositional groups (Fig. 4) of de Lange et al. (1987, 1989) are easily identified in the D10982 profiles. Organic-rich turbidites (a, d, e, f, h) are also distinguished by high Si, Al and K values, volcanic turbidites (b, g) are particularly enriched in Ti, Fe, Mg, Zr; and calcareous turbidites (j, l) display high Ca and Sr. It is also notable that the basal sands of the organic-rich turbidites, which contain an assemblage enriched in quartz, orthoclase feldspar, rutile and zircon (Pearce & Jarvis 1992b), display peaks in Si, Ti, K and Zr but depletion of most other elements, while the volcaniclastic sands and silts of the volcanic turbidite group are more enriched in Al, Ti, Fe, Mg, V, Zr and many other trace elements, reflecting their more mafic grain assemblage.

Correlation

The geochemical characteristics displayed by core D10982 are repeated in all of the other cores studied from the MAP. Correlation of representative sequences taken on N–S (D11145, D11137, D10982, D10699) and E–W (82PCS20, 86PCM17) transects (Fig. 2), demonstrates that the lateral geochemical variation for individual beds (Figs 6–8), is considerably less than the differences in geochemistry that characterize each bed. The geochemical trends and compositional groups defined in core D10982 are clearly identifiable in the other cores, enabling each turbidite to be correlated over distances of > 500 km totally independently of any other stratigraphic information. Throughout the area Al, Ti and Sr (Figs 6–8), for example, provide excellent discriminators of the organic-rich, volcanic and calcareous turbidite groups, respectively.

Turbidite groups

The three turbidite groups, and to a lesser extent individual beds, may be readily identified by plotting all of the turbidite mud data (441 samples) from the MAP on a TiO_2–Al_2O_3–$CaCO_3$ ternary diagram (Fig. 9).

Organic-rich turbidites (a, a₁, d, e, e₁, f, h, i, k)
These define an array (218 samples) trending away from their most Al_2O_3-rich end member (a_1) towards the $CaCO_3$ apex (Fig. 9). Individual turbidites form well-defined clusters on this carbonate mixing line, although there are varying degrees of overlap and some beds (d, i and e, e_1, f) cannot be distinguished individually. The scatter of a few anomalous compositions around the main a_1 cluster is indicative of the unavoidable inclusion of some pelagic sediment introduced by bioturbation in these samples.

The organic-rich turbidites, therefore, display a wide range of $CaCO_3$ contents, ranging from c. 18% in turbidite a_1 up to 57% in a (Table 3). They are also generally enriched (Figs 5 & 6) in Si (averaging c. 54% SiO_{2CFB}), Al (typically 17–20% Al_2O_{3CFB}) and Li (c. 90 $\mu g\,g^{-1}$, on a CFB). It is noteworthy that the distributions of many elements (e.g. Ni, V, Zn) are rather variable (Fig. 5), reflecting their sensitivity to redox-mediated diagenesis. Compositional differences in the non-carbonate fraction are masked by this large variation in carbonate content. Examination of carbonate-free data (Table 3), however, suggests that two compositional subgroups may be recognized, as exemplified by an Al_2O_3–SiO_2–K_2O diagram (Fig. 10) which separates a high-K subgroup 1 (a, d, e, e_1) from a less potassic association (subgroup 2) of a_1, f, h, i and k.

Subgroup 1 (a, d, e, e₁) turbidites (146 samples) have K_2O_{CFB} contents of > 2.9%, with average concentrations of c. 3.3%. They are

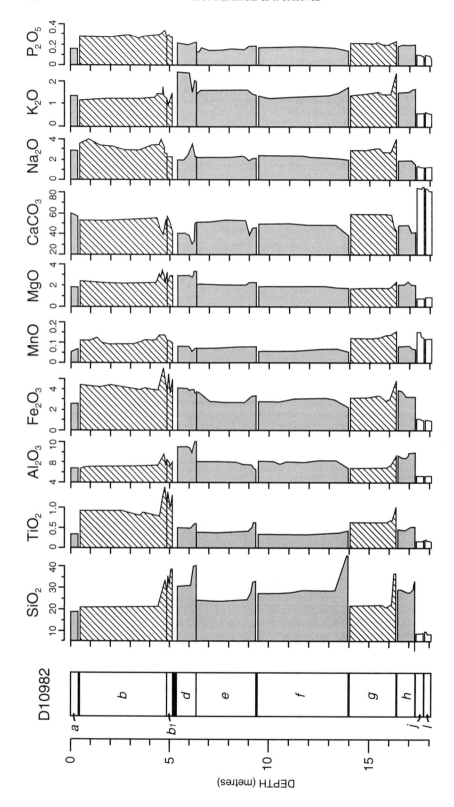

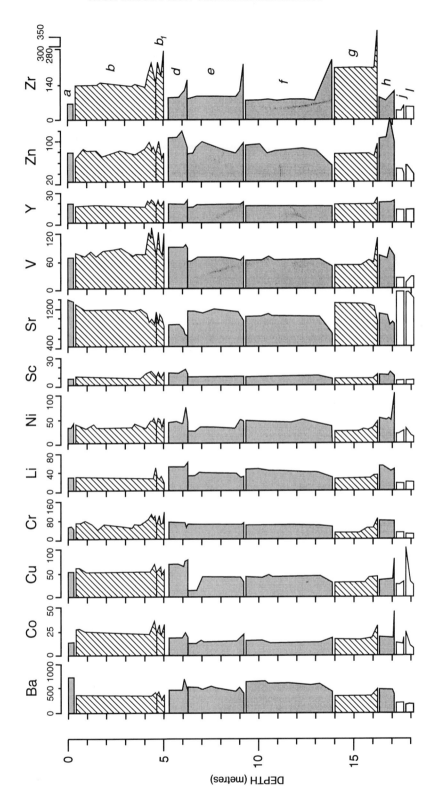

Fig. 5. Bulk geochemical profiles for turbidites in central-basin core D10982. The compositional group of each turbidite is indicated; data for the pelagic intervals are not shown.

Table 3. *Representative mean compositions of MAP turbidite muds expressed on a carbonatefree basis*

| Turbidite | Grp* | Core | Major elements (%) | | | | | | | | | |
			SiO_2	TiO_2	Al_2O_3	$Fe_2O_3^T$	MnO	MgO	$(CaCO_3)$	Na_2O	K_2O	P_2O_5
a	O1	D10982	47.6	0.816	16.2	6.09	0.132	4.16	56.8	6.49	2.94	0.357
a_1	O2	D10699	53.8	0.913	23.0	8.53	0.069	3.04	18.2	3.71	2.18	0.412
b	V1	D10982	47.5	1.85	15.9	8.81	0.229	4.81	54.7	6.86	2.40	0.581
b	V1	86PCM17	46.1	1.66	15.8	8.58	0.256	4.55	55.9	6.15	2.71	0.509
b_1	V1	D10982	50.0	1.69	16.3	8.43	0.292	4.65	54.4	5.85	2.39	0.547
d	O1	D10982	52.0	0.865	19.1	6.90	0.124	4.82	40.9	3.71	3.97	0.314
d	O1	D10699	53.9	0.890	19.7	7.66	0.183	4.96	40.6	3.22	3.72	0.443
d_1	V1	D11145	48.2	1.46	16.6	8.10	0.241	4.50	56.9	6.46	3.50	0.548
e	O1	D10982	51.2	0.904	17.4	6.14	0.150	4.51	53.7	4.71	3.40	0.420
e	O1	D10699	55.1	0.987	18.4	5.24	0.180	4.75	55.6	5.04	3.25	0.472
f	O2	D10982	53.3	0.711	16.0	5.88	0.110	3.57	49.0	4.18	2.55	0.346
f	O2	D10699	59.4	0.770	16.9	6.08	0.133	3.83	51.5	4.50	2.84	0.379
g	V2	D10982	55.4	1.56	17.3	8.07	0.341	4.30	56.2	5.77	2.57	0.545
g	V2	86PCM17	48.4	1.49	16.8	7.63	0.258	4.04	62.6	6.48	3.29	0.466
h	O2	D10982	50.9	0.816	16.7	6.21	0.139	3.95	47.7	3.23	2.59	0.381
h	O2	D10699	53.0	0.811	16.8	5.62	0.166	3.82	48.8	3.39	2.81	0.389
i	O2	D10699	48.2	0.916	18.9	6.52	0.154	3.86	41.5	3.53	2.87	0.399
j	C	D10982	45.1	0.771	15.8	5.64	0.756	3.75	83.1	6.48	2.55	0.591
j	C	D10699	47.5	0.792	16.7	5.36	0.602	3.93	82.3	5.34	2.34	0.609
k	O2	D10699	49.5	0.739	20.2	7.14	0.031	3.19	45.5	3.46	2.63	0.412
l	C	D10699	47.2	0.865	15.6	6.36	0.736	3.90	82.1	6.42	2.95	0.602
l	C	D10982	45.0	0.819	15.0	5.79	0.653	3.84	81.8	6.40	2.67	0.584
m	C	D10699	49.7	0.929	16.6	6.81	0.553	4.67	73.5	5.64	3.18	0.566
o	V2	86PCM01	49.4	1.61	16.7	8.27	0.344	4.23	60.8	6.60	3.21	0.414
p	V1	86PCM01	45.8	1.83	15.3	9.36	0.251	4.79	61.8	6.88	2.81	0.445

Table 3. (continued)

	Grp	Sample	Ba	Co	Cr	Cu	Li	Ni	Sc	(Sr)	V	Y	Zn	Zr
a	O1	D10982	1620	29.5	118	117	69.9	80.6	19.3	1320	146	40.8	171	144
a_1	O2	D10699	1120	34.4	76.1	188	119	112	22.2	491	152	27.1	111	169
b	V1	D10982	749	51.3	123	137	67.1	79.5	19.2	1160	169	37.2	174	278
b_1	V1	86PCM17	757	54.4	86.1	130	65.9	103	20.9	1160	171	38.6	128	256
b_1	V1	D10982	734	50.6	113	92.4	67.3	81.9	19.1	1100	155	36.9	185	280
d	O1	D10982	773	30.7	62.9	122	87.2	70.7	19.5	806	146	27.2	186	169
d	O1	D10699	825	36.1	130	116	103	55.5	21.0	789	160	30.6	116	163
d_1	V1	D11145	1220	44.1	103	93.7	83.3	75.7	22.4	1240	181	40.6	131	232
e	O1	D10982	1050	30.7	85.0	120	84.0	69.4	18.8	1140	142	35.0	182	155
e	O1	D10699	1140	36.1	95.0	136	98.6	83.8	21.2	1160	171	38.8	133	212
f	O2	D10982	1160	26.8	82.9	116	79.1	80.4	17.9	1020	117	30.7	158	121
f	V2	D10699	1240	27.8	92.9	150	94.8	81.9	19.5	1060	142	33.3	118	200
g	V2	D10982	2090	41.3	85.3	130	83.3	68.3	17.5	1260	143	43.4	129	434
g	V2	86PCM17	815	44.4	124	106	49.0	79.7	16.0	1350	135	41.4	122	385
h	O2	D10982	855	33.6	70.0	134	83.8	89.4	19.0	949	133	30.5	188	115
h	O2	D10699	887	32.4	102	140	92.6	84.8	19.3	1010	144	32.8	111	212
i	O2	D10699	914	40.5	135	106	84.0	105	21.3	900	151	35.6	110	224
j	C	D10982	1220	78.1	164	107	89.7	124	22.8	1520	111	74.4	236	148
j	C	D10699	774	34.5	84.9	116	75.9	63.2	19.9	1360	94.0	93.3	96.6	256
k	O2	D10699	1120	38.0	82.8	104	115	108	23.7	992	153	31.9	120	149
l	C	D10699	1020	128	379	91.4	102	220	25.2	1510	144	80.4	142	243
l	C	D10982	1130	55.6	215	87.6	110	95.3	27.1	1480	128	83.2	209	178
m	C	D10699	1460	53.9	82.9	122	107	102	24.7	1540	154	57.0	142	222
o	V2	86PCM01	1040	58.9	123	125	51.0	86.2	21.1	1370	157	44.4	136	368
p	V1	86PCM01	1020	52.7	135	137	51.2	96.7	24.3	1380	166	39.7	129	256

Trace elements (μg g^{-1})

*Grp = compositional group; O1, O2, organic-rich subgroups 1 and 2; V1, V2, volcanic subgroups 1 and 2; C, calcareous turbidites; total iron expressed as $Fe_2O_3^T$; all data reported on a carbonate-free basis except $CaCO_3$ and Sr (shown in parentheses).

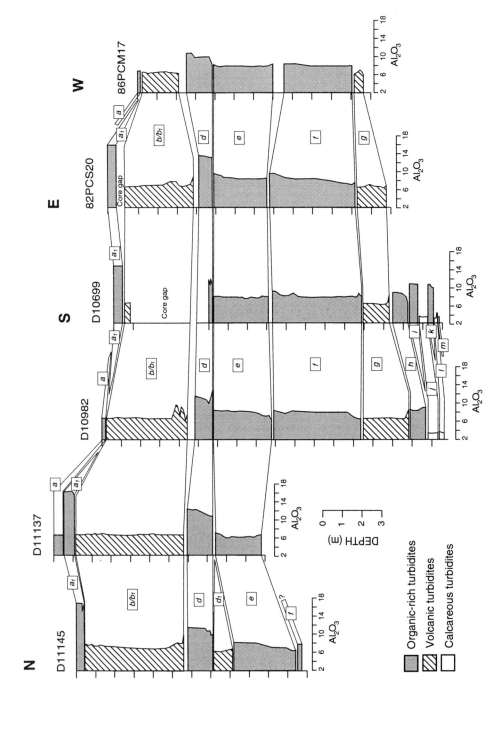

Fig. 6. Chemostratigraphic correlation based on Al$_2$O$_3$ for six cores from the central sub-basin. Individual beds have relatively uniform but distinct compositions across the entire area; note that Al contents are consistently highest in the organic-rich turbidite group.

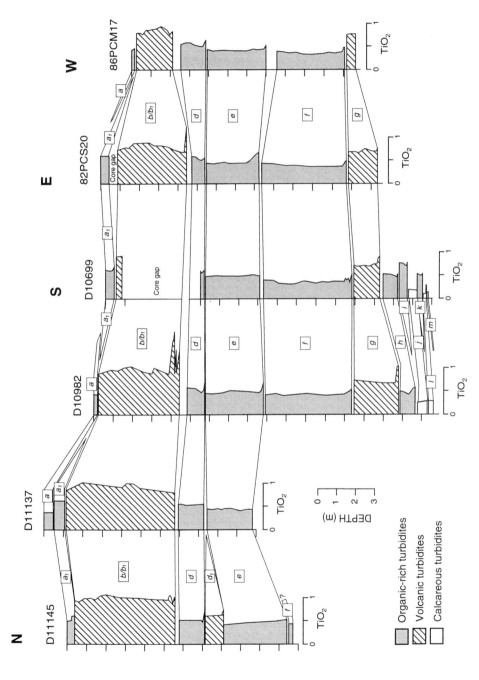

Fig. 7. Chemostratigraphic correlation based on TiO$_2$ for six cores from the central sub-basin. Values for Ti are consistently highest in the volcanic turbidite group.

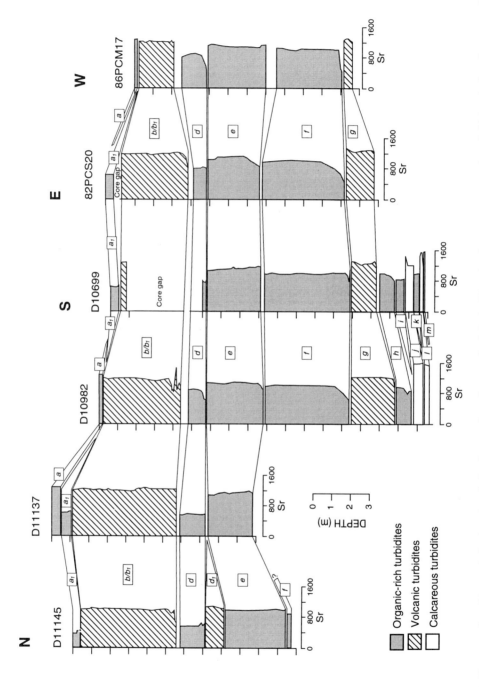

Fig. 8. Chemostratigraphic correlation based on Sr for six cores from the central sub-basin. Note that Sr contents are consistently highest in the calcareous turbidite group.

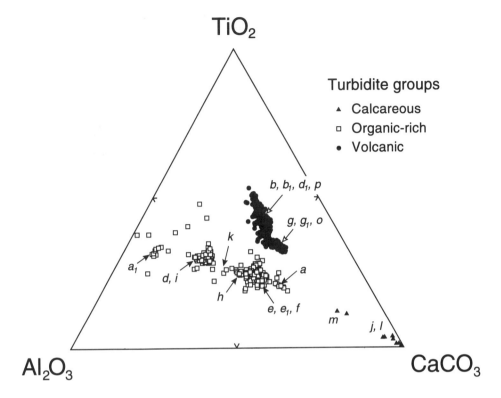

Fig. 9. TiO_2–Al_2O_3–$CaCO_3$ ternary plot of MAP turbidite muds (441 samples). The three main compositional groups and a number of well-defined turbidite clusters are clearly identifiable. Data plotted using the method of de Lange *et al.* (1987).

also characterized by higher Mg values (Table 3) than subgroup 2; MgO_{CFB} contents are > 4.0% with typical levels of *c.* 4.5%. Elevated K and Mg contents are consistent with the higher concentrations of both illite and chlorite observed in these sediments (Auffret *et al.* 1989; Pearce & Jarvis 1992*b*). Iron contents are variable, reflecting its susceptibility to diagenetic remobilization, and cannot be related clearly to differences in chlorite contents. No quantitative data are available for dolomite, the presence of which might also increase the Mg content of these sediments. All members of this subgroup exhibit lower C_{org} (<1.3% on a CFB), and commonly have lower Si and Al values than subgroup 2 turbidites. Carbonate contents are variable ($CaCO_3$ *c.* 38–57%) but serve to distinguish individual beds (Fig. 9).

Subgroup 2 (*a₁*, *f*, *h*, *i*, *k*) turbidites (72 samples) typically exhibit K_2O_{CFB} contents of 2.6–2.8% and lower Mg values (< 4% MgO_{CFB}); Si and Al contents are commonly higher than those of subgroup 1. Turbidite *a₁* is assigned to this subgroup, although samples

form a rather poorly defined cluster offset towards the Al_2O_3 apex on the Al_2O_3–SiO_2–K_2O diagram (Fig. 10). This turbidite, therefore, exhibits more Al-rich compositions which point to a kaolinite-rich, quartz and orthoclase feldspar-depleted composition (cf. Pearce & Jarvis 1992*b*). Significantly, all members of the subgroup (including *a₁*) contain higher C_{org} contents of > 1.3% on a CFB (Fig. 4; de Lange *et al.* 1987, 1989). $CaCO_3$ contents are very variable (18–52%) and again serve as one of the best means of distinguishing individual turbidites.

Volcanic turbidites (b/b₁, d₁, g/g₁, o, p)
Muds from this group (213 samples) are easily distinguished by their high proportion of Ti on the TiO_2–Al_2O_3–$CaCO_3$ ternary diagram (Fig. 9), where they form two very tight, but slightly overlapping clusters constituting high-Ti (*b/b₁*, *d₁*, *p*) and low-Ti (*g/g₁*, *o*) subgroups. In general, volcanic turbidites are characterized by the highest Ti, Fe, Mg, Zr (TiO_{2CFB} *c.* 1.5–2.4%, Fe_2O_{3CFB} *c.* 7.5–10.5%, MgO_{CFB} *c.* 4.5%,

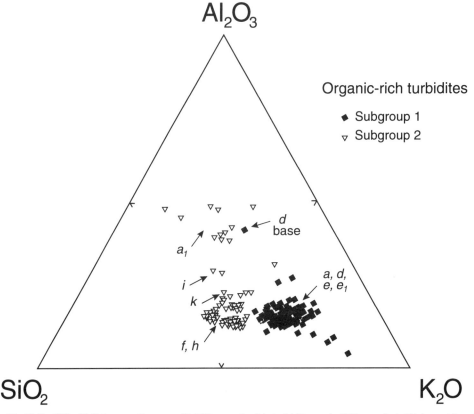

Fig. 10. Al_2O_3–SiO_2–K_2O ternary diagram of MAP organic-rich turbidite muds (218 samples). High- and low-K subgroups 1 and 2 are effectively separated but individual beds are generally not distinguishable. Data plotted using the method of de Lange *et al.* (1987).

Zr_{CFB} *c.* 250–460 $\mu g\,g^{-1}$; Table 3) and lowest Li concentrations (Li_{CFB} *c.* 50–80 $\mu g\,g^{-1}$). They also display relatively high $CaCO_3$ contents (50–60%), but their SiO_{2CFB} contents are low compared to the organic-rich turbidites (typically < 50%); they are characterized by low C_{org} contents of < 0.3%.

The two subgroups are best separated on a Zr–Sc–TiO_2 diagram (Fig. 11). In this case, turbidite b/b_1 samples define a tightly clustered array of Zr-depleted compositions (subgroup 1) that also encompass d_1 and p. Turbidite g/g_1 defines a similarly tight but totally separate and more steeply inclined Zr-rich array (subgroup 2) which incorporates turbidite o. The general characteristics of these subgroups may be summarized as: (1) *Subgroup 1* (b/b_1, d_1, p) turbidites (174 samples) are characterized by very high TiO_{2CFB} ($\geq 1.7\%$, but levels commonly attain 1.9%), Fe_2O_{3CFB} (> 8.5%) and MgO_{CFB} concentrations (> 4.5%, typically 4.8–5.0%). They are also enriched in Sc_{CFB} (19–

25 $\mu g\,g^{-1}$) and Ni_{CFB} (*c.* 100 $\mu g\,g^{-1}$), although Ni concentrations are variable as this element is highly susceptible to diagenetic relocation. The Ti content of d_1 is lower than is typical of the subgroup, but samples consistently lie close to the end of the appropriate compositional array on a range of discrimination diagrams (e.g. Fig. 11). (2) *Subgroup 2* (g/g_1, o) turbidites (39 samples) contain the highest Zr_{CFB} concentrations (> 350 $\mu g\,g^{-1}$, commonly over 400 $\mu g\,g^{-1}$); Al_2O_{3CFB} and K_2O_{CFB} contents are also high, with values generally *c.* 17% (as compared to 16% for subgroup 1) for the former, and $\geq 2.6\%$ (but commonly exceeding 3.0%) for the latter.

The distinctive geochemical signature of the volcanic turbidites essentially reflects the presence of variable amounts of volcaniclastic debris (Weaver & Rothwell 1987, Pearce & Jarvis 1992a, b), including abundant fragments of volcanic glass, basaltic to trachytic lithic clasts, together with grains of diopside, feldspar (ranging from plagioclase to sanadine), horn-

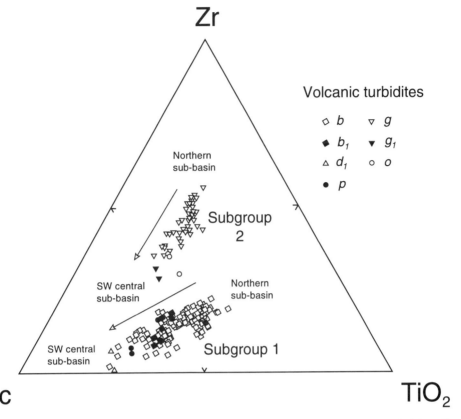

Fig. 11. Zr–Sc–TiO$_2$ ternary plot of MAP volcanic turbidite muds (213 samples). The low and high-Zr linear arrays of subgroups 1 and 2 are defined by variation in turbidites *b* and *g*, respectively. The location of samples within these arrays relates directly to their geographical location on the plain and defines proximal (N) to distal (SW) variation within these beds. Data plotted using the method of de Lange *et al.* (1987).

blende, kaersutite, nepheline, olivine, opaques and titanaugite. This grain assemblage is most abundant in the basal sands and silts where it is commonly interlaminated with foraminiferal sands, producing very irregular geochemical profiles (e.g. Fig. 5 turbidites *b*/*b$_1$*). However, its geochemical signature persists throughout the entire bed due to the incorporation of silt- and clay-grade volcaniclastic debris in the mud fraction.

Calcareous turbidites (j, l, m)
Turbidites *j* and *l* (eight samples) define the CaCO$_3$ apex on the TiO$_2$–Al$_2$O$_3$–CaCO$_3$ ternary diagram (Fig. 9), while *m* (two samples) has a more marly composition and lies closer to the other turbidite groups. All three are characterized by high CaCO$_3$ contents in excess of 70% (Table 3, Fig. 5), and they also exhibit the highest levels of Sr (*c.* 1500 μg g^{-1}; Figs 5 & 8). Concentrations of all other elements are low.

Viewed on a CFB (Table 3), the composition of the non-carbonate fraction lies closest to that of the organic-rich group. Anomalously high concentrations of Mn and Y in the CFB data (MnO$_{CFB}$ > 0.5%; Y$_{CFB}$ > 55 μg g^{-1}) are due, at least in part, to the occurrence of a portion of these elements in the carbonate fraction. Turbidite bases comprise laminated foraminiferal sands and silts, but these have geochemical compositions that are broadly similar to the remainder of the bed.

Lateral variation in turbidite geochemistry

Correlation of geochemical profiles (Figs 6–8) and discrimination diagrams (Figs 9–11) demonstrate that most turbidites display little compositional variation throughout the area. Exceptions are provided by the two main volcanic turbidites sampled, *b* and *g*. These define very tight linear arrays on the TiO$_2$–Al$_2$O$_3$–CaCO$_3$ (Fig. 9) and

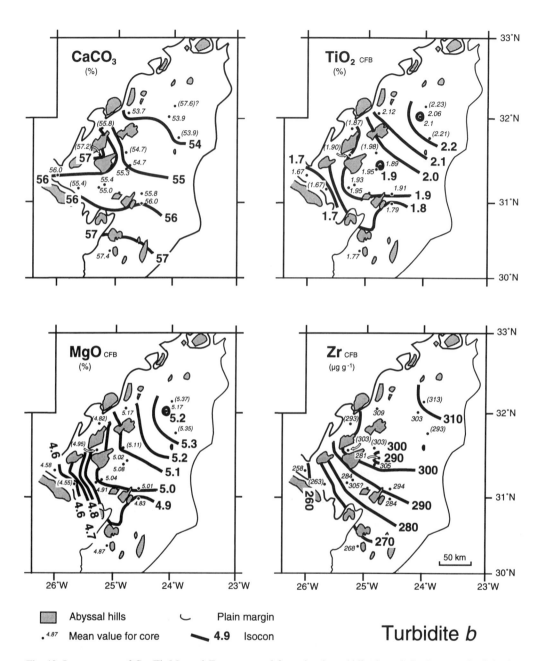

Fig. 12. Isocon maps of Ca, Ti, Mg and Zr constructed for volcanic turbidite *b* muds in the central sub-basin. Mean Ca data are presented as carbonate (% CaCO₃); TiO₂, MgO (%) and Zr (μg g⁻¹) values are reported on a carbonate-free basis (CFB); isocon values (large numerals) are shown in conjunction with mean values (small numerals) at each core location; values in brackets represent data for incomplete vertical profiles resulting due to sampling difficulties or poor core recovery; values with question marks are unresolved anomalies. See Pearce & Jarvis (1992a) for further details.

Zr–Sc–TiO$_2$ (Fig. 11) plots, within which the positions of samples relate closely to their geographical location. Samples from the northern and northernmost central sub-basins contain the highest proportions of Zr and Ti (even on a CFB; Fig. 11), while those taken towards the SW become progressively more depleted in these elements but enriched in CaCO$_3$. Similar, but slightly less well-defined trends occur in ternary plots containing Fe or Mg in place of Zr or Ti or Al in place of Sc.

Lateral variation in turbidite *b* is best illustrated using isocon maps (Fig. 12) which show graphically the geographical controls on geochemical composition. Pearce & Jarvis (1992*a*) attributed progressive decreases in Ti, Mg and Zr from the NE to the SW across the central sub-basin to the downstream 'fall out' of heavy minerals and mafic clasts within the turbidity flow, leaving the more distal (SW) portions enriched in carbonate and clay minerals, reflected by higher Ca and Al contents. Similar trends are displayed by turbidite *g*. Organic-rich turbidites, on the other hand, are generally finer grained and lack an abundant and diverse heavy mineral assemblage, so although subtle vertical geochemical trends may be observed in more marginal cores, systematic geographical variation cannot be demonstrated using isocon maps. No significant vertical or lateral geochemical variation has been detected in the calcareous turbidites.

Statistical analysis of turbidite geochemistry

It has been demonstrated empirically that individual turbidites retain their distinct geochemical characteristics throughout the area, enabling the accurate geochemical correlation of strata across the entire MAP. However, it is apparent that individual turbidites and turbidite groups are characterized by several constituents which together constitute a 'fingerprint'; some elements are clearly more diagnostic than others for this purpose. A statistical approach was employed, therefore, to assess more rigorously which elements might be most suitable for chemostratigraphic purposes and to develop a quantitative correlation model based on the complete geochemical dataset.

Bulk geochemical data from a single core (MD10; Fig. 2) were analysed previously by Middelburg & de Lange (1988) using Fuzzy C-means clustering. This analysis distinguished three compositional groups broadly consistent with those reported by de Lange *et al.* (1987). The effects of diagenesis on the bulk composi-

tion of the turbidite were shown to be minimal, despite the distributions of some individual components (such as C$_{org}$) being strongly influenced by oxidation. However, the diagenetic overprint result in organic-rich turbidites being subdivided into unaltered and oxidized subgroups, with thin oxidized units such as turbidite *d* being classified with the oxidized tops of the thicker organic-rich turbidites.

Statistical analysis of our more extensive geochemical dataset was undertaken using cluster and principal components methods to: (1) determine whether individual turbidites have unique geochemical 'fingerprints' that might enable them to be used as chemostratigraphic marker beds; (2) further test the validity of the compositional groups and subgroups; (3) better assess lateral variation in turbidite geochemistry.

Cluster analysis

The cluster analysis computer program (Clustan) employed uses the Ward's hierarchical grouping method. Cluster analysis is a classification technique which involves the assigning of observations into distinct groups. Hierarchical clustering joins the most similar observations, then successively connects the next most similar observations (Davis 1986). Firstly, $n \times n$ matrix of similarities between all pairs of the observations is calculated. Those pairs having the highest similarities are then merged and the matrix is recomputed. This is done by averaging the similarities that the combined observations have with other observations. This process is repeated until the similarity matrix is reduced to 2×2. The levels of similarity at which observations are merged is used to construct a dendrogram.

A preliminary study employing cluster analysis was undertaken on the muds from core D10982. The turbidites analysed included *a*, *b/b$_1$*, *d*, *e*, *f*, *g*, *h*, *j*, *l* (plus turbidite *a$_1$* from neighbouring core D11137), and only ten elements (Si, Ti, Al, Fe, Mg, Ca, Na, K, P and Zr) were considered initially. This exercise was followed by cluster and principal component analyses of the complete dataset for all turbidites, firstly using only values for turbidite muds, and secondly expanded to include basal sands and silts. Finally, turbidite mud data recalculated on a CFB were also tested. Results were analysed in the form of dendrograms and tables that define the numerical characteristics of the clusters, and finally as principal component graphs.

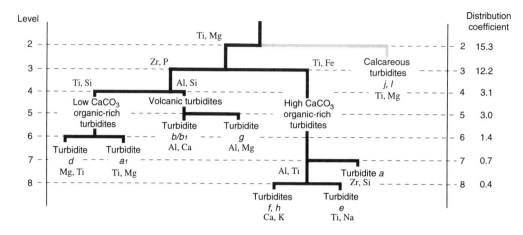

Fig. 13. Cluster analysis dendrogram of turbidite mud geochemistry (41 samples) from mid-central basin core D10982. The first two diagnostic elements at each separation are shown; see Fig. 5 for geochemical profiles and Table 4 for numerical data.

Table 4. *Main clusters defined for turbidite muds from piston core D10982*

Cluster	Turbidites	n	Variable	Mean	SD
Level 4					
1	a, e, f, h	17	TiO_2	0.39	0.04
			Fe_2O_3	2.96	0.23
2	a_1, d	5	TiO_2	0.58	0.06
			SiO_2	33.2	2.3
3	$b/b_1, g$	14	Al_2O_3	7.13	0.18
			SiO_2	21.1	0.8
4	j, l	5	TiO_2	0.13	0.01
			MgO	0.64	0.02
Level 8					
1	a	3	Zr	62.4	0.3
			SiO_2	20.0	0.1
2	a_1	3	TiO_2	0.62	0.01
			MgO	2.20	0.01
3	b/b_1	10	Al_2O_3	7.22	0.11
			$CaCO_3$	54.3	0.7
4	d	2	MgO	2.85	0.01
			TiO_2	0.51	0.01
5	e	5	TiO_2	0.42	0.01
			Na_2O	2.14	0.05
6	f, h	9	$CaCO_3$	48.6	1.0
			K_2O	1.32	0.03
7	g	4	Al_2O_3	6.90	0.12
			MgO	1.63	0.02
8	j, l	5	TiO_2	0.13	0.01
			MgO	0.64	0.02

Major elements quoted as wt%, Zr as $\mu g\ g^{-1}$; n, number of samples; SD, standard deviation; see Fig. 13 for dendrogram.

Core D10982

Results for the cluster analysis of bulk geochemical data from core D10982 are summarized in Fig. 13 and Table 4. The data may be considered at different levels of correlation, but the dendrogram essentially shows four compositional groups: (1) high-$CaCO_3$ organic-rich turbidites a, e, f, h; (2) low-$CaCO_3$ organic-rich turbidites

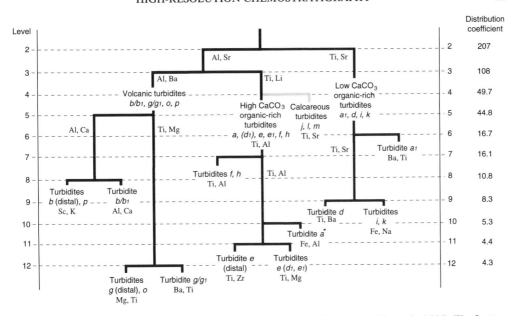

Fig. 14. Cluster analysis dendrogram of turbidite mud geochemistry (441 samples) from the MAP. The first two diagnostic elements at each separation are shown; see Tables 5 & 6 for numerical data.

Table 5. *Main clusters defined for MAP turbidite muds at level 4*

Cluster	Turbidites	n	Variable	Mean	SD
1	a, d_1, e, e_1, f, h	148	TiO_2	0.41	0.04
			Al_2O_3	8.09	0.47
2	a_1, d, i, k	73	TiO_2	0.62	0.01
			Sr	791	91
3	$b/b_1, g/g_1, o, p$	210	Al_2O_3	7.14	0.4
			Ba	347	22
4	j, l, m	10	TiO_2	0.16	0.04
			Sr	1510	40

Major elements quoted as wt% oxides, trace elements as $\mu g\ g^{-1}$; n, number of samples; SD, standard deviation; see Fig. 14 for dendrogram.

a_1, d; (3) volcanic turbidites $b/b_1, g$; (4) calcareous turbidites j, l. Clusters are defined on the basis of variables with the lowest degree of variance and the classification is obviously controlled strongly by $CaCO_3$ contents and their associated dilution effects.

It is notable that, as with our qualitative analysis, Ti is commonly the diagnostic chemostratigraphic index element at all levels of correlation (Table 4); an exception is the volcanic turbidite group which is subdivided principally on Al contents. At the lower levels of correlation individual turbidites are distinguished, except f and h, and j and l, which cannot be separated without subdividing data

from other individual beds. Significantly, not one of the 41 samples incorporated in the test clustered with the wrong turbidite.

Turbidite muds
Statistical evaluation of turbidite mud geochemistry was extended to include data from all of the study cores. At this stage, only samples of the mud portions below visible bioturbation and above the silty/sandy bases of the turbidites were included, since the basal facies is characterized by extreme geochemical compositions which might hamper the definition of turbidite fingerprints.

Table 6. *Geochemical compositions of individual clusters defined from the MAP turbidite mud dataset*

Cluster	1			2			3			4			5			6			7			8			9		
Turbidites	a			a_1			$b/b_1, p$			d			d_1, e, e_1			f, h			$g/g_1, o$			i, k			j, l, m		
Group	Organic			Organic			Volcanic			Organic			Organic			Organic			Volcanic			Organic			Calcareous		
n	7			9			170			59			83			58			40			5			10		
	Order	Mean	SD	Order	Mean	SD	Order	Mean	SD	Order	Mean	SD	Order	Mean	SD	Order	Mean	SD	Order	Mean	SD	Order	Mean	SD	Order	Mean	SD
Major elements (%)																											
SiO$_2$	5	20.3	0.4	15	34.4	5.9	8	21.3	1.2	6	31.1	1.5	9	23.7	1.4	6	27.2	1.2	7	19.9	1.2	13	27.6	5.3	9	9.03	2.35
TiO$_2$	3	0.36	0.02	2	**0.62**	**0.05**	12	0.85	0.10	1	**0.53**	**0.02**	1	**0.43**	**0.03**	1	**0.38**	**0.03**	1	**0.61**	**0.03**	9	0.50	0.07	1	**0.16**	**0.05**
Al$_2$O$_3$	2	**6.87**	**0.08**	9	15.8	1.2	1	**7.23**	**0.33**	8	11.3	0.8	3	8.17	0.46	2	**8.14**	**0.27**	5	6.76	0.41	5	11.1	0.5	3	3.01	0.77
Fe$_2$O$_3$	1	**2.63**	**0.02**	12	5.56	0.56	9	4.08	0.30	9	4.12	0.36	13	2.95	0.51	3	2.96	0.12	4	3.14	0.15	1	**3.98**	**0.11**	6	1.19	0.34
MgO	8	1.82	0.06	3	2.16	0.10	10	2.20	0.16	4	2.82	0.11	2	**2.07**	**0.08**	5	1.82	0.10	2	**1.65**	**0.06**	11	2.01	0.22	12	0.78	0.25
CaCO$_3$	12	58.7	2.4	8	25.2	4.5	2	**55.7**	**1.6**	11	41.7	4.1	8	54.7	2.3	4	50.3	1.5	9	60.3	3.3	8	42.7	2.7	13	81.1	4.3
Na$_2$O	14	2.95	0.25	10	2.83	0.37	15	3.04	0.39	7	1.94	0.22	12	2.15	0.26	7	2.09	0.18	12	1.32	0.15	2	**1.84**	**0.10**	11	1.09	0.39
K$_2$O	7	1.25	0.06	7	1.56	0.16	5	1.19	0.09	13	2.30	0.25	6	1.63	0.11	9	1.35	0.15	11	1.32	0.15	12	1.43	0.37	4	0.55	0.16
P$_2$O$_5$	10	0.16	0.01	14	0.31	0.04	13	0.25	0.03	14	0.21	0.03	15	0.17	0.04	14	0.17	0.03	15	0.20	0.36	3	0.22	0.01	7	0.11	0.02
Trace elements ($\mu g\,g^{-1}$)																											
Ba	11	677	27	1	**776**	**17**	3	349	22	2	**451**	**21**	5	484	26	12	573	51	3	337	18	7	569	34	14	231	82
Li	6	31.7	1.7	5	81.8	4.4	7	29.0	2.8	5	55.9	3.8	11	41.1	4.5	11	41.6	4.8	10	25.7	4.4	6	62.1	3.7	5	19.0	5.2
Sc	4	8.3	0.2	11	16.8	1.2	4	9.1	0.4	10	12.4	0.9	10	9.2	0.5	10	9.3	0.7	8	6.8	0.5	4	12.5	0.4	10	4.7	1.1
Sr	9	1300	30	6	591	65	6	1210	40	3	810	36	4	1150	50	8	1040	60	6	1300	40	10	929	67	2	**1510**	**40**
Y	15	16.9	0.7	13	19.5	0.8	14	17.3	0.6	15	17.2	0.8	14	17.2	0.7	15	16.2	0.7	13	16.9	0.7	15	19.8	2.0	15	13.6	1.2
Zr	13	73.2	10.4	4	125	8	11	128	12	12	103	16	7	93.6	9.6	13	89.8	15.7	14	172	19	14	118	35	8	38.6	13.6

n, Number of samples in cluster; SD, standard deviation; main discriminating elements shown in bold type; see Fig. 14 for dendrogram.

Table 7. *Geochemical compositions of basal facies clusters*

Cluster	10			11		
Turbidite bases	f, h			$b/b_1, g/g_1$		
Group	Organic			Volcanic		
n	21			26		
	Order	Mean	SD	Order	Mean	SD
Major elements (%)						
SiO_2	14	38.2	8.3	9	31.3	5.6
TiO_2	**2**	**0.46**	**0.09**	11	1.07	0.32
Al_2O_3	4	7.55	0.62	4	8.96	1.56
Fe_2O_3	5	2.66	0.34	7	4.59	0.88
MgO	7	1.81	0.22	5	2.39	0.45
$CaCO_3$	11	43.2	6.31	8	45.1	7.9
Na_2O	**1**	**1.97**	**0.18**	15	3.44	0.59
K_2O	3	1.52	0.12	12	1.77	0.55
P_2O_5	8	0.15	0.03	10	0.29	0.06
Trace elements (μg g^{-1})						
Ba	10	456	64	**1**	**344**	**37**
Li	9	30.0	6.6	**2**	**28.4**	**5.8**
Sc	6	7.9	0.8	6	10.5	1.9
Sr	13	816	163	3	919	125
Y	12	16.3	1.57	13	22.7	2.9
Zr	15	179	85	14	215	78

n, number of samples in cluster; SD, standard deviation; main discriminating elements shown in bold type.

Data from the oxic portions of the turbidites (below the bioturbation zone) are also subtly different from the remainder of their beds (Jarvis & Higgs 1987, Middelburg & de Lange 1988). However, samples from the oxic portions or those in the vicinity of redox fronts, could not be excluded because thinner turbidites (particularly organic-rich examples) are commonly either completely oxidized or totally pervaded by redox-controlled element enrichments. As a consequence, in order to minimize the effects of oxidation on the bulk geochemical signature, data for C_{org} and other redox-sensitive constituents (Co, Cr, Cu, Mn, Ni, V, Zn) were omitted, leaving a dataset comprising 15 elements (Si, Ti, Al, Fe, Mg, Ca, Na, K, P, Ba, Li, Sc, Sr, Y and Zr) for a total of 441 samples from turbidites $a, a_1, b/b_1, d, d_1, e, e_1, f, g/g_1, h, i, j, k, l, m, o$ and p.

Results of the cluster analysis for the complete bulk sediment geochemical dataset are summarized in Fig. 14 and Tables 5 and 6. Out of total of 441 samples, only five samples plotted anomalously outside their expected turbidite cluster, confirming the homogeneity of individual turbidite compositions across the plain. The test subdivided the dataset (Table 5) into essentially the same four categories defined in the D10982 dataset at level 4 (Table 4): (1) high-$CaCO_3$ organic-rich turbidites a, d_1, e, e_1, f, h;

(2) low-$CaCO_3$ organic-rich turbidites a_1, d, i, k; (3) volcanic turbidites $b/b_1, g/g_1, o, p$; and (4) calcareous turbidites j, l, m. Turbidites d_1, e_1, i, k, o and p, which were not represented in core D10982, therefore fell into their expected turbidite groups, except 'volcanic' turbidite d_1 which clustered with the high-$CaCO_3$ organic-rich category.

The four clusters could be subdivided into a number of subgroups (Fig. 14, Table 6) composed of single turbidites (a, a_1, d), but other clusters consisted of two or three turbidites of indistinguishable compositions. The calcareous turbidites (j, l, m) could not be separated (level 4) and similarly, turbidites f and h (level 7), b/b_1 and p (level 8), i and k (level 9), d_1, e & e_1 (level 11) and g/g_1 and o (level 12) clustered together. Furthermore, in order to subdivide the organic-rich turbidites, it was necessary to split the b/b_1 cluster at level 8 into proximal and distal portions. At lower levels of confidence the proximal and distal portions of turbidites e and g/g_1 were also distinguished. Titanium was again highlighted as the most important chemostratigraphic index-element defining clusters (Fig. 14). In detail, diagnostic elements (Table 6) vary between the compositional groups, but combinations of Ti with Al, Fe, Mg and Ba are the commonest discriminators. It is noteworthy that groups are generally defined by elements

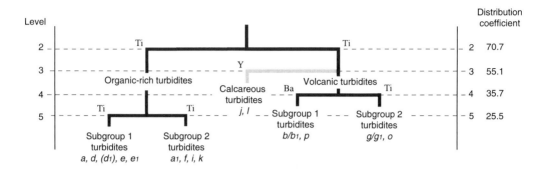

Fig. 15. Cluster analysis dendrogram of turbidite mud geochemistry recalculated on a CFB (346 samples) from the MAP. The main element at each separation is shown; see Table 8 for numerical information.

Table 8. *Geochemical clusters defined from the turbidite mud CFB dataset*

Cluster	Grp	Turbidites	n	Anomalies	Variable	Mean	SD
1	O1	a, d, d_1, e, e_1	121	9	TiO_2	0.95	0.14
2	O2	a_1, f, h, i, k	76	13	TiO_2	0.80	0.11
3	V1	$b/b_1, p$	96	0	Ba	782	53
4	V2	$g/g_1, o$	44	3	TiO_2	1.53	0.09
5	C	j, l	9	0	TiO_2	0.84	0.07

Grp, compositional group: O1, O2, organic-rich subgroups; V1, V2 volcanic subgroups; C, calcareous turbidites; TiO_2 values are Wt% oxide, Ba is $\mu g\ g^{-1}$; n, number of samples; SD, standard deviation; see Fig. 15 for dendrogram.

occurring at intermediate levels in the cluster, rather than those exhibiting extreme concentrations; this contrasts to our empirical analysis which gave more weight to elements displaying the greatest variance.

It can be concluded that it is only possible to define a unique geochemical fingerprint for turbidites a, a_1 and d. However, it is highly likely that terminal clusters composed of two statistically indistinguishable turbidites were derived from very similar source sediments. It is important to note that none of these composite clusters conflicts with the groupings based on sedimentological or other geochemical criteria, except the two mud samples of 'volcanic' turbidite d_1 which consistently fall in the high-$CaCO_3$ organic-rich turbidite cluster. The subdivision of turbidites b/b_1 into proximal and distal portions at a higher statistical level than the distinction between some organic-rich turbidites, confirms a higher degree of lateral heterogeneity in b/b_1 than in most of the organic-rich units.

Turbidite muds, silts and sands
This investigation involved the entire sample set, combining basal sediments with those of the

muds described above, to yield a total of 528 samples and the same 15 variables. The inclusion of the basal facies had little effect on the turbidite clusters defined using the mud dataset, but the number of anomalously plotting samples rose to 15, reflecting the more variable composition of the sands and silts. Basal facies samples clustered into two main groups (Table 7): (1) organic-rich turbidite bases f and h; (2) volcanic turbidite bases b/b_1 and g/g_1; these could be distinguished at a lower statistical level. The basal compositions of turbidite d vary considerably, and silt samples plotted in both basal groups. However, the silty mud bases of distal cores correlated with the main turbidite d mud cluster. Sands and silts of turbidite e in proximal cores plotted with muds from volcanic turbidites g/g_1, while the silty mud bases from more distal cores were grouped with turbidite e muds.

It was also observed that in proximal cores where turbidites b and b_1 could be clearly distinguished, the coarsest-grained basal samples from b correlated with those of b_1. The similarity in composition of coarse-grained sediments from turbidites b/b_1 and g/g_1 indicates their derivation from a common source, whereas clear differences between the finer-grained fractions high-

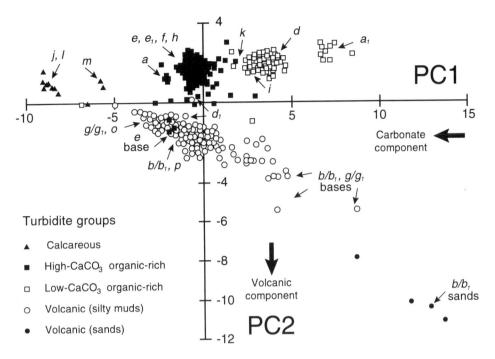

Fig. 16. Scatter plot illustrating the spatial relationship between the first two principal geochemical components (PC1, PC2) for MAP turbidites. Many beds form unique compositional clusters; note the distinction of four compositional groups.

light the geochemical differences which define the individual subgroups.

Complete geochemical dataset
The complete dataset, including the redox-sensitive elements, was also analysed statistically. Results were broadly the same as for the selected data, indicating that despite the diagenetic relocation of certain trace elements, provenance remains the primary control on turbidite geochemistry. However, the number of anomalous samples increased to 21. These samples were primarily from organic-rich turbidites and originated from the vicinity (< 20 cm) of relic redox fronts; two anomalous samples from calcareous turbidites also came from levels displaying metal-enriched diagenetic laminations.

Carbonate-free data
The CFB values of the mud dataset were investigated in an attempt to remove the dilution influence of $CaCO_3$. It was found to be necessary to exclude samples from above or in the vicinity of redox fronts because carbonate dissolution effects (cf. Jarvis & Higgs, 1987) introduced significant bias into the recalculated data. The

test was applied to 346 samples and successfully categorized the volcanic, calcareous and organic-rich turbidites (Fig. 15, Table 8), although a larger number of anomalous samples contaminated the clusters, particularly those of the organic-rich turbidites. These anomalies are attributed to the slightly poorer precision of the Ca data which carried forward into the calculation of the carbonate-free values. However, it is considered that these data provide a better guide to the compositions of the non-carbonate fractions. Furthermore, the test also clearly distinguished the subgroups within the organic-rich and volcanic turbidites; significantly, the organic-rich group was subdivided into two provenance subgroups (Fig. 15) which were identical to those defined using other geochemical, mineralogical and sedimentological criteria. Interestingly, 'volcanic' turbidite d_1 again plotted out of place, with organic turbidite subgroup 1, suggesting a mixed provenance for its sediment.

Comparison with published data
To test the validity of the cluster groupings, and to see whether turbidite identity could be distinguished from the published datasets of

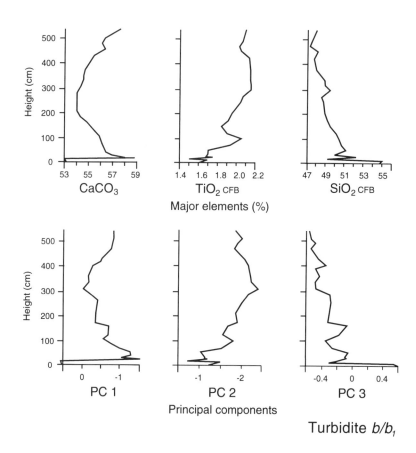

Fig. 17. Vertical geochemical profiles for $CaCO_3$, TiO_{2CFB} and SiO_{2CFB} compared with profiles of the first three principal component scores (PC1–3) through volcanic turbidites b/b_1 in west central sub-basin core D11143. A close correspondence between the PC scores (below) and trends in their corresponding index elements (above) is evident.

Jarvis & Higgs (1987) and de Lange *et al.* (1987), the results of these authors for turbidites a, a_1, b/b_1, d, d_1, e, f, g/g_1, i, j and k (22 variables) were incorporated into our total dataset. The test proved successful in categorizing the data into the correct turbidite compositional groups. In addition, all samples, except one from turbidites a_1, b/b_1, d, g/g_1, and j, fell into their correct turbidite clusters. However, published data for turbidites e, f, d_1, h and k plotted as an undifferentiated group between the clusters of turbidites e and f. It is likely that the classification of these turbidites was hampered by analytical differences between the three datasets, particularly Al and Zr data which were obtained from acid digestions (which in our experience may lead to low values) in the earlier studies.

Furthermore, Jarvis & Higgs' (1987) sampling was deliberately concentrated around redox fronts, leading to the inclusion of a large proportion of highly anomalous compositions in their dataset.

Principal component analysis

Principal components are nothing more than the eigenvectors of a variance–covariance or a correlation matrix (Davis 1986). Principal components analysis (PCA) was applied to the total (muds and basal facies) and muds-only geochemical datasets excluding the redox-sensitive elements. This test was undertaken to further investigate the validity of turbidite groupings and to study geochemical variation within beds.

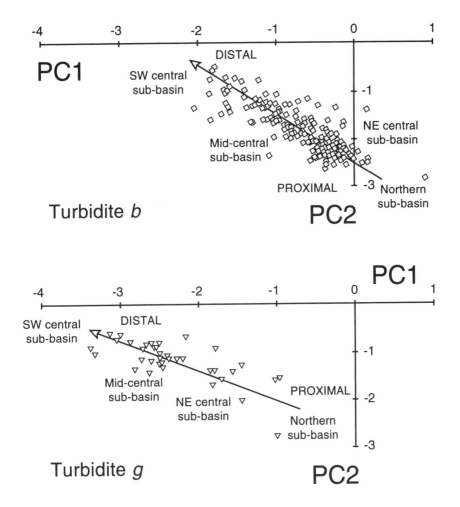

Fig. 18. Scatter plots of PC1 v. PC2 scores for volcanic turbidites *b* (above) and *g* (below) muds. The positions of samples in each linear array relates to their geographical location on the MAP; samples with high PC1 and low PC2 scores correspond to the most proximal muds, recovered from the northern part of the plain.

As a first step, principal component (PC) scores were plotted on *XY* scatterplots to provide a visual expression of the turbidite groupings. The first three principal components account for 78% of the variation: PC1, 41.2%; PC2, 28.3%; PC3, 8.1%; Fig. 16 illustrates the relationship between the first two principal components. It is apparent that the PC scores very effectively distinguish between the main compositional clusters (volcanic, calcareous, high- and low-CaCO$_3$ organic turbidites).

As with elemental plots, most turbidites form distinct and tightly defined clusters, but the volcanic turbidites b/b_1 and g/g_1 again define elongate linear arrays (Fig. 16). Volcaniclastic

sands from turbidite *b* and b_1 bases fall on the same trend but beyond the limits of the main b/b_1 mud array. By comparison with the primary geochemical data, it is clear that PC1 relates predominantly to carbonate contents and their associated dilution effects. The most calcareous samples (*j, l*) display high negative values of *c*. −8 and the low-carbonate volcaniclastic sands of b/b_1 having high positive scores of up to +14. PC2 effectively describes variation in the volcaniclastic component, ranging from −11 for b/b_1 sands, to +3 for turbidite a_1 muds. It is noteworthy that basal samples of organic-rich turbidite *e* lie in the middle of the volcaniclastic turbidite cluster. This is consistent with the

presence of a significant proportion of volcaniclastic debris in the basal sediments, as noted
previously for organic-rich turbidites *d* and *e* by
Pearce & Jarvis (1992*b*).

Geochemical controls on the PC scores are
confirmed by comparisons between stratigraphic
plots of the elemental and PC data. As an
example, the first three PC scores for samples
taken through volcanic turbidites b/b_1 in SW
central sub-basin core D11143 (Fig. 2) are
plotted along with $CaCO_3$, TiO_{2CFB} and
SiO_{2CFB} data for the same material (Fig. 17).
The markedly curved vertical structure of the
elemental plots for distal cores of turbidites b/b_1
was discussed in detail by Pearce & Jarvis
(1992*a*) who related it to mineralogical variation. Particularly in distal cores, the preferential
settling of sediment components with larger
grain sizes and higher grain densities leads to a
progressive sequence of volcaniclastic basal
muddy silts, overlain by carbonate-rich muddy
silts, then volcaniclastic silty muds and finally,
towards the top of the bed, calcareous nanno-
fossil-rich muds. The close similarity of PC1 to
$CaCO_3$, PC2 to TiO_{2CFB}, and PC3 to SiO_{2CFB}
confirms that these may be regarded as largely
carbonate, volcaniclastic and aluminosilicate
components, respectively.

In detail, elongation of the compositional
arrays displayed by the volcanic turbidites on
the PC1 v. PC2 diagram (Fig. 16), may again be
attributed to geographical variation within each
bed (Fig. 18). It is clear that in both cases there is
a systematic decrease in PC1 and increase in PC2
scores from the northern sub-basin towards the
southwest of the central sub-basin, reflecting
increasing carbonate and decreasing volcaniclastic material in that direction. Geographical
variations in PC2 scores across the central sub-
basin may be contoured (Fig. 19) in the same
way as the elemental isocon maps (Fig. 12).
However, clear trends in the elemental data were
only obtained by plotting the mean concentrations of samples taken from the central portion
(Unit 2 of Pearce & Jarvis 1992*a*) of turbidite *b*
in each core. In our case, PC2 scores were derived from the entire mud dataset, but they also
vary systematically across the plain, displaying
the same trends but in an opposite sense to the
Ti, Mg and Zr isocon maps. It is concluded that
the PC2 scores provide a very robust proximality
indicator, high negative scores (-3.5) indicating
the most proximal (volcaniclastic-rich) and low
negative scores (> -1.0) the most distal (heavy
mineral-depleted) sediments.

Geochemical fingerprints and chemostratigraphy

Statistical analysis confirms that the geochemical
composition of individual turbidites is relatively
uniform across the MAP. It has also been
demonstrated that a geochemical fingerprint
can be identified for a number of individual
turbidites (*a*, a_1, *d*). These geochemical fingerprints are consistent across the central sub-
basin. As a consequence, these turbidites can be
regarded as geochemical marker beds and can be
used for stratigraphic correlation between cores.
Certain turbidites have statistically indistinguishable compositions (Fig. 14), but if the
geochemical data are placed in stratigraphic
order it is possible to identify virtually all
turbidites individually. For example, the muds
of turbidites *b* and *p* are very similar (Table 3),
preventing them from being regarded as unique
marker beds, but by considering the stratigraphic and geochemical characteristics of the
intervening turbidites, it is easy to distinguish
between them. Strong similarities exist between
all of the core geochemical profiles from the
central sub-basin. This combination of geochemical marker horizons and distinct chemostratigraphic sequences enables detailed bed-by-bed
correlation (Figs 6–8) across the central sub-
basin.

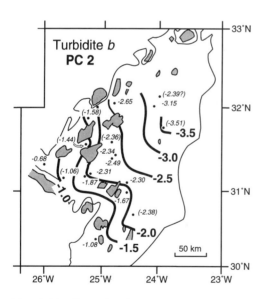

Fig. 19. Map illustrating the geographical distribution
of PC2 mean scores for volcanic turbidite *b* muds
across the central sub-basin. A close similarity to the
isocon distributions of Ti, Mg and Zr is apparent
(Fig. 12).

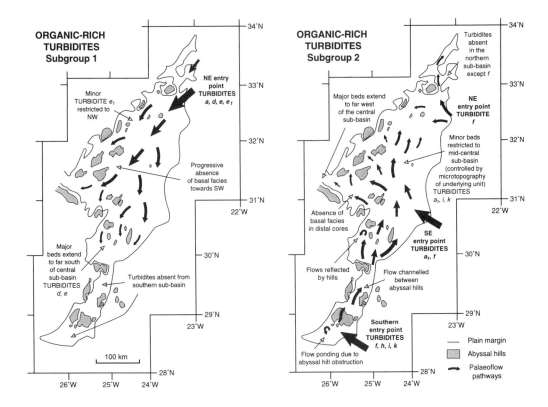

Fig. 20. Summary map illustrating the postulated palaeoflow characteristics of organic-rich turbidites. The contrasting emplacement pathways of the two geochemical subgroups is clearly displayed.

The chemostratigraphy developed for the MAP is entirely consistent with the established basin-wide litho- and biostratigraphy. The success of the technique is highly dependent on the geographical homogeneity in the composition of the beds but, as has been demonstrated, the MAP sequences are relatively uniform in composition and grain-size across the basin. Significantly, although MAP turbidites have been subjected to early diagenetic alteration, their distinct geochemical provenance signatures have been maintained by non-redox sensitive elements. MAP sediments have presented an ideal testing ground for chemostratigraphic techniques, and it is now considered proven that turbidite geochemistry can be used as a highly sensitive and selective correlation tool.

Turbidite provenance

Geochemical data may be integrated with sedimentological (Jones 1988; McCave & Jones 1988; Pearce 1991; Pearce & Jarvis 1992a; Rothwell et al. 1992), mineralogical (modal analysis and grain chemistry; Weaver & Rothwell 1987; Pearce 1991; Pearce & Jarvis 1992b) and geological (Weaver et al. 1989a, b, 1992, 1994) data to provide very tight constraints on the emplacement (Figs 20 & 21) and provenance (Fig. 22) of the MAP turbidites.

Organic-rich turbidites

Organic-rich turbidites a, a_1, d, e, e_1, f, h, i and k generally display a distinctive two-tone olive and pale grey coloration (totally pale grey if completely oxidized). They are characterized by high C_{org}, Si, Al, K and Li contents and occur in all three sub-basins (Rothwell et al. 1992). Two geochemically distinct subgroups have been recognized.

Subgroup 1 turbidites (a, d, e and e₁)

These contain higher amounts of K and Mg but are relatively depleted in C_{org}, and commonly Si and Al. Their geochemical characteristics correspond to enhanced illite and chlorite, and lower quartz and kaolinite contents. The subgroup is

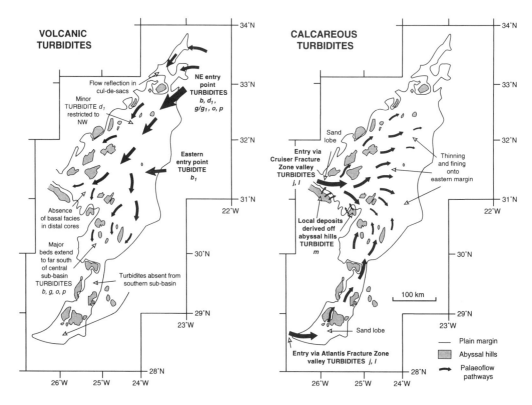

Fig. 21. Summary map showing the suggested palaeoflow characteristics of volcanic (left) and calcareous (right) turbidites. Volcanic turbidites display emplacement patterns similar to those of organic-rich subgroup 1 turbidites (Fig. 20).

best exemplified by turbidites *d* and *e*. Basal sands and silts are composed of continental-derived quartz and alkali feldspar, but additionally contain a significant amount of debris from trachytic and altered basaltic volcanic sources. The presence of altered lithic clasts and basic minerals indicate a high degree of submarine exposure and alteration of this volcaniclastic fraction.

Subgroup 1 turbidites are absent from the southern sub-basin (Fig. 20) and the basal facies coarsens and thickens northeastwards across the central and into the northern sub-basin, indicating that the turbidity currents entered from the NE. Minor organic-rich turbidite e_1 is restricted to the NW of the central sub-basin. Geochemical variation confirms that the muds of subgroup 1 turbidites become more distal southwestwards across the MAP, and grain-size data for the fine-grained portion of turbidite *d* (Jones *et al.* 1992) are also consistent with a NE source.

Relatively enhanced illite and chlorite contents in subgroup 1 turbidites, and their low C_{org}

and Al values are consistent with a continental slope source north of the Canaries (Chamley *et al.* 1977), probably off the Moroccan coast and possibly transported via the Agadir Canyon (Fig. 22). This hypothesis is supported by the discovery of turbidite *a* on a turbidity current pathway between the Canaries and Madeira (Dutch Geological Survey, unpublished data). The volcanic component was most likely incorporated during transportation past the Canary Islands and Madeira. However, green turbidites have been discovered on the Seine and Horseshoe Abyssal Plains (Weaver, pers. comm.) to the north, and these may represent the proximal portions of subgroup 1 turbidites. If this proves to be the case, some organic-rich turbidites may have originated from further north (Fig. 22), off the southern Iberian margin.

Subgroup 2 turbidites a_1, f, h, i and k
These are characterized by lower K and Mg, higher C_{org} and generally higher Si and Al contents, reflecting enhanced proportions of

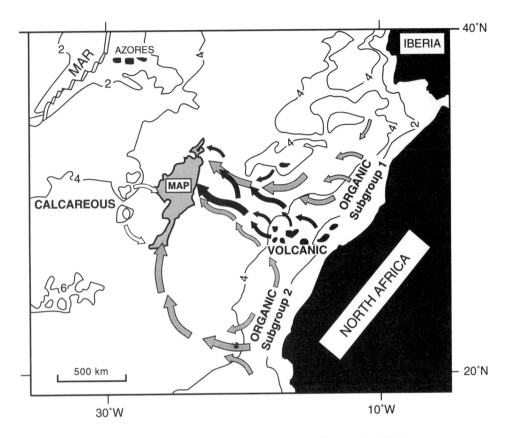

Fig. 22. Map illustrating the postulated source areas for MAP turbidites and their likely transport pathways. Stippled arrows are organic-rich, black arrows volcanic and open arrows calcareous turbidite pathways.

quartz, kaolinite and organic matter. Their basal sands and silts are dominated by quartz and K-feldspar with only minimal amounts of volcanigenic debris. These turbidites, with the exception of f, are restricted to the southern and central sub-basins (Fig. 20). The distribution of basal sands and silts, isopach data and geochemical trends in the central sub-basin muds, all indicate that subgroup 2 turbidites were emplaced primarily via the southern sub-basin, and then transported downslope and northwards along the axis of the MAP into the central sub-basin. This points to a provenance south of the Canaries on the continental slope off West Africa. Continental slope sediments in the vicinity of 24–20°N have high C_{org} contents, and are enriched in kaolinite, quartz and feldspar (Hartmann *et al.* 1976), supplied primarily by the Saharan dust plume. Kidd *et al.* (1987) detected turbidity current pathways that flowed northwestwards from this region, although their GLORIA coverage did not extend on to the MAP. It is postulated that subgroup 2 turbidites followed this route.

The distribution of f is atypical of the subgroup. Turbidite f represents by far the thickest (up to 5 m) and most extensive bed on the MAP. Sedimentological data (Rothwell *et al.* 1992) demonstrate that, in addition to being emplacement via the southern sub-basin, it was also introduced directly into the central sub-basin (Fig. 20), from where it flowed into the northern sub-basin. This presents difficulties in defining flow pathways for the turbidite, as no compositional differences are apparent between the different lobes. The massive volume (190 km³) of turbidite f must have been generated by an enormous slide which affected a huge geographical area. It is possible that the resultant turbidity currents divided on the continental rise because the direct route between their source area on the continental slope and the MAP was obstructed by the Saharan seamount and other topographic highs. Consequently, flows were directed both to the north-west and to the west (Fig. 22), and entered the MAP concurrently from the east and south.

Volcanic turbidites

Volcanic turbidites b/b_1, d_1, g/g_1, o and p are generally pale brown in colour, have low C_{org} and Li contents, and are characterized by high Ti, Fe, Mg and Zr concentrations. All volcanic turbidites are restricted to the northern and central sub-basins (Fig. 21). The majority have basal facies that coarsen and thicken northeast-wards across the central and into the northern sub-basin (Rothwell et al. 1992), indicating that turbidity currents entered from the NE. Geo-chemical proximality indicators clearly demon-strate that these turbidites become more distal southwestwards across the MAP. An exception to this pattern is turbidite b_1, which thickens eastwards across the central sub-basin and appears to have been introduced directly into the sub-basin from the east. Two compositional subgroups have been defined on the basis of our geochemical and mineralogical data.

Subgroup 1 turbidites b/b_1, d_1 and p

These are characterized by higher proportions of Ti, Fe, Mg and Sc, but less Zr. In addition, they contain more volcanic debris (including basalt clasts, volcanic glass, olivine, titanaugite and titanomagnetite) which has been derived pri-marily from an alkali basalt source (Pearce & Jarvis, 1992b). The most detailed information is available for turbidites b/b_1.

Turbidite b_1 is dominated by alkali basalt volcaniclastic sands. Volcanic glass shards and mineral assemblages point to an origin on the western Canary Islands, such as Hierro, La Palma and Gomera, or possibly the basaltic complexes on Tenerife. Recent TOBI surveys have identified a large submarine slide on the slopes of Hierro (Weaver, pers. comm.), which is now considered to be the origin of this turbidite. It is postulated that the b/b_1 doublet originated from a single event and that b_1 sands were deposited rapidly before being overrun, prob-ably within a few hours, by the main portion of turbidite b. This may have been achieved by the transportation of coarse-grained sands of b_1 along a direct channelized pathway on to the MAP (Fig. 22). The mud-dominated turbidite b, on the other hand, travelled in the wake of the first phase of flow, following a less direct pathway, which entered the MAP from the NE (Fig. 21). Turbidity current b may have been forced to flow further to the north due to the 'blockage' of the direct pathway by b_1.

Subgroup 2 turbidites g/g_1 and o

These exhibit higher Al, K and Zr concentra-tions. The characteristics of the subgroup are best exemplified by turbidites g/g_1. These contain a higher proportion of grains derived from trachytic/syenitic provenance (including alkali feldspars, trachyte clasts and amphiboles; Pearce & Jarvis 1992b) than subgroup 1 turbidites, which is consistent with a more fractionated volcanic source. In addition, turbidites g/g_1 contain K-feldspars and quartz of probable continental origin, phases which are virtually absent from turbidites b/b_1. Furthermore, the occurrence of spilitized feldspars, altered vol-canic glass and basalt clasts, and higher smectite contents (in the absence of comparable enrich-ments in kaolinite) indicate a more highly altered volcanic component. Like most volcanic turbidites, g/g_1 have low C_{org} contents.

Sedimentological, isopach and mud geochem-ical data all indicate that turbidites g/g_1 were emplaced on the MAP from the northeast (cf. Rothwell et al. 1992). Similar palaeoflow directions are envisaged for turbidite o (Fig. 21). It is suggested, therefore, that these turbidites were supplied from a source to the north of the Canaries. The dominance of a fractionated volcanic provenance may indicate an origin on the northern flanks of the central and eastern Canary Islands, which is characterized by more acidic volcanic complexes. These islands also lie in the pathway of strong Saharan-derived winds, and it is postulated that continentally derived quartz and K-feldspar were introduced to the source areas by aeolian processes.

Calcareous turbidites

Calcareous turbidites j, l and m are white to very pale brown in colour, and are strongly enriched in $CaCO_3$ and Sr, a consequence of the high proportion of pelagic biogenic carbonate (for-aminifera and coccoliths). This indicates that their source area was isolated from coarse-grained detrital input, although siliciclastic material (principally clay minerals) constitutes up to 30% of their total composition. The clay mineral component, dominantly smectites, was probably introduced by the Saharan dust plume, or possibly from the submarine weathering of basalts.

Calcareous turbidites j and l thicken and coarsen into the fracture zone valleys that dissect their proposed source area, the Great Meteor–Cruiser Seamount Chain which flanks the western margin of the basin (Figs 21 & 22). Two distinct lobes have been identified in turbidite j (Rothwell et al. 1992); one emplaced via the Cruiser Fracture Zone valley into the central sub-basin, and the second deposited in the southern and central sub-basin, being introduced along a trough that follows the trace of the Atlantis Fracture Zone. The presence of two lobes demonstrates that either the initial turbidity current separated into two subflows, or that separate local flows were initiated from along the margins of the seamount chain. No geochemical differences were detected between these lobes, although this is to be expected, given the probable homogeneity of the source area.

Turbidite m has a lower $CaCO_3$ content, reflecting its more marly composition. This turbidite is restricted to the vicinity of the abyssal hills which occur on the western margin of the plain (Fig. 21). The lower carbonate content of m is attributed to a source area lying within or below the lysocline. Turbidites j and l were probably derived from shallower water seamount-edge sources.

Conclusions

Our MAP case study has demonstrated the scientific validity of chemostratigraphic methods for correlating and interpreting distal turbidite sequences. MAP turbidites display distinctive geochemical compositions and chemostratigraphic sequences that enable them to be identified and correlated over distances of > 500 km. Turbidites may be classified into three compositional groupings and several subgroups which are consistent across the entire MAP. Cluster and principal components analysis confirm and further elucidate interpretations based on empirical analysis of the data. When combined with sedimentological information, geochemical data may be used successfully to constrain depositional processes and palaeoflow pathways, and to define turbidite source areas.

Clearly, the commercial potential of chemostratigraphy is greatest in successions devoid of biostratigraphic markers or in lithologies dominated by long-ranging fossils. In sequences of monotonous barren sediments, subtle differences in geochemical composition may provide the only means of refining crude lithostratigraphic divisions and constitute an independent means of correlation. The success of the present study has provided an impetus to apply our techniques to selected petroleum basins.

The authors wish to thank Drs R. T. E. Schuttenhelm of the Dutch Geological Survey and P. P. E. Weaver of the Institute of Oceanographic Sciences Deacon Laboratory (IOSDL) for their help and resources provided during core sampling. Dr David Wray (University of Greenwich) and an anonymous referee provided helpful critiques of the original manuscript. This research was sponsored by the NERC (TJP; GT4/87/GS/39), Maurice Hill Research Fund of the Royal Society (IJ) and Kingston University.

References

AUFFRET, G. A., BUCKLEY, D. E., MULLER, C., ET AL. 1989. Geology, lithostratigraphy of STACOR cores from the Madeira (GME) and Southern Nares Abyssal Plains. In: SCHUTTENHELM, R. T. E., AUFFRET, G. A., BUCKLEY, D. E., CRANSTON, R. E., MURRAY, C. N., SHEPHARD, L. E. & SPIJKSTRA, A. E. (eds) Geoscience Investigations of Two North Atlantic Abyssal Plains – The Esope International Expedition, Commission of the European Community, Luxembourg, I, 117–171.

BHATIA, M. R. 1983. Plate tectonics and geochemical composition of sandstones. Journal of Geology, 91, 611–627.

——— 1985. Rare-earth element geochemistry of Australian Palaeozoic graywackes and mudrocks: provenance and tectonic control. Sedimentary Geology, 45, 97–113.

——— & CROOK, K. A. W. 1986. Trace-element characteristics of graywackes and tectonic setting discrimination of sedimentary basins. Contributions to Mineralogy and Petrology, 92, 181–193.

——— & TAYLOR, S. R. 1981. Trace-element geochemistry and sedimentary provinces: a study from the Tasman Geosyncline, Australia. Chemical Geology, 33, 115–125

BOUMA, A. H. 1962. Sedimentology of some Flysch Deposits, a Graphic Approach to Facies Interpretation. Elsevier, Amsterdam.

CHAMLEY, H., DIESTER-HAAS, L. & LANGE, H. 1977. Terrigenous material in East Atlantic sediment cores as an indicator of NW African climates. "Meteor" Forschungsergeb. Reihe C, 26, 44–59.

COLLEY, S. & THOMSON, J. 1985. Recurrent uranium relocations in distal turbidites emplaced in pelagic conditions. Geochimica et Cosmochimica Acta, 49, 2339–2348.

———, ———, WILSON, T. R. S. & HIGGS, N. C. 1984. Post-depositional migration of elements during diagenesis in brown clay and turbidites sequences in the North East Atlantic. Geochimica et Cosmochimica Acta, 48, 1223–1235.

DAVIS, J. C. 1986. Statistics and Data Analysis in Geology. Wiley, New York, 502–515.

DE LANGE, G. J., JARVIS, I. & KUIJPERS, A. 1987. Geochemical characteristics and provenance of late Quaternary sediments from the Madeira Abyssal Plain, N Atlantic. In: WEAVER, P. P. E. & THOMSON, J. (eds) Geology and Geochemistry of Abyssal Plains. Geological Society, London, Special Publication, 31, 147–165.

———, MIDDELBURG, J. J., JARVIS, I. & KUIJPERS, A. 1989. Geochemical characteristics and provenance of late Quaternary sediments from the Madeira and Southern Nares Abyssal Plains (North Atlantic). In: SCHUTTENHELM, R. T. E., AUFFRET,

G. A., BUCKLEY, D. E., CRANSTON, R. E., MURRAY, C. N., SHEPHARD, L. E. & SPIJKSTRA, A. E. (eds) *Geoscience Investigations of Two North Atlantic Abyssal Plains - The Esope International Expedition.* Commission of the European Community, Luxembourg, **II**, 785–851.

FLOYD, P. A. & LEVERIDGE, B. E. 1987. Tectonic environment of the Devonian Gramscatho Basin, south Cornwall: framework mode and geochemical evidence from turbiditic sandstones. *Journal of the Geological Society, London,* **144**, 531–542.

——, SHAIL, R., LEVERIDGE, B. E. & FRANKE, W. 1991. Geochemistry and provenance of Rhenohercynian synorogenic sandstones: implications for tectonic environment discrimination. *In:* MORTON, A. C., TODD, S. P. & HAUGHTON, P. D. W. (eds), *Developments in Sedimentary Provenance Studies,* Geological Society, London, Special Publication, **57**, 173–188.

HARTMANN, M., MULLER, P. J., SUESS, E. & VAN DER WEIJDEN, D. H. 1976. Chemistry of late Quaternary sediments and their interstitial water from the NW African continental margin. *"Meteor" Forschungsergeb. Reihe C,* **24**, 1–67.

HUMPHREYS, B., MORTON, A. C., HALLSWORTH, C. R., GATLIFF, R. W. & RIDING, J. B. 1991. An integrated approach to provenance studies: a case example from the Upper Jurassic of the Central Graben, North Sea. *In:* MORTON, A. C., TODD, S. P & HAUGHTON P. D. W. (eds) *Developments in Sedimentary Provenance Studies,* Geological Society, London, Special Publication, **57**, 251–262.

JARVIS, I. 1985. Geochemistry and origin of Eocene–Oligocene metalliferous sediments from the central equatorial Pacific. Deep Sea Drilling Project Sites 573 and 574: *Initial Report of the DSDP,* **85**, 781–804.

—— 1992. Sample preparation for ICP–MS *In:* JARVIS, K. E., GRAY, A. L. & HOUK, R. S. (eds) *Handbook of Inductively Coupled Plasma Mass Spectrometry,* Blackie, Glasgow, 172–224.

—— & HIGGS, N. C. 1987. Trace-element mobility during early diagenesis in distal turbidites: late Quaternary of the Madeira Abyssal Plain, N. Atlantic. *In:* WEAVER, P. P. E. & THOMSON, J. (eds) *Geology and Geochemistry of Abyssal Plains.* Geological Society, London, Special Publication, **31**, 145–151.

—— & JARVIS, K. E. 1992a. Plasma spectrometry in the earth sciences: techniques, applications and future trends. *In:* JARVIS, I. & JARVIS, K. E. (eds) *Plasma Spectrometry in the Earth Sciences,* Chemical Geology, **95**, 1–33.

—— & —— 1992b. Inductively coupled plasma-atomic emission spectrometry in exploration geochemistry. *In:* HALL, G. E. M. (ed.) *Geoanalysis,* Journal of Geochemical Exploration, **44**, 139–200.

——, PEARCE, T. J. & HIGGS, N. C. 1988. Early diagenetic geochemical trends in Quaternary distal turbidites. *Chemical Geology,* **70**, 10.

JONES, K. P. N. 1988. *Studies of fine-grained, deep-sea sediments.* PhD Thesis, Cambridge University,

UK.

——, McCAVE, I. N. & WEAVER, P. P. E. 1992. Textural and dispersal patterns of thick mud turbidites from the Madeira Abyssal Plain. *Marine Geology,* **107**, 149–173.

JØRGENSEN, N. O. 1986. Chemostratigraphy of Upper Cretaceous Chalk in the Danish Sub-basin. *American Association of Petroleum Geologists Bulletin,* **70**, 309–317.

KIDD, R. B., HUNTER, P. M. & SIMM, R. W. 1987. Turbidity-current and debris-flow pathways to the Cape Verde Basin: status of long-range side-scan sonar (GLORIA) surveys. *In:* WEAVER, P. P. E. & THOMSON, J. (eds) *Geology and Geochemistry of Abyssal Plains.* Geological Society, London, Special Publication, **31**, 33–48.

KUIJPERS, A. & WEAVER, P. P. E. 1985. Deepsea turbidites from the northwest African continental margin. *Dtsch. Hydrogr. Z.,* **38**, 147–164.

McCANN, T. 1991. Petrological and geochemical determination of provenance in the southern Welsh Basin. *In:* MORTON, A. C., TODD, S. P. & HAUGHTON, P. D. W. (eds) *Developments in Sedimentary Provenance Studies,* Geological Society, London, Special Publication, **57**, 215–230.

McCAVE, I. N. & JONES, K. P. N. 1988. Deposition of ungraded muds from high-density non-turbulent turbidity currents. *Nature,* **333**, 250–252.

MIDDELBURG, J. J. & DE LANGE, G. J. 1988. Geochemical characteristics as indicators of the provenance of Madeira Abyssal Plain turbidites. A statistical approach. *Oceanologica Acta,* **11**, 159–165.

NUCLEAR ENERGY AGENCY 1984. *Seabed Disposal of High-Level Radioactive Waste. A Status Report on the NEA Co-ordinated Research Programme,* Organisation for Economic Cooperation and Development, Paris.

PEARCE, T. J. 1991. *The Geology, Geochemistry, Sedimentology and Provenance of Late Quaternary Turbidites, Madeira Abyssal Plain.* PhD Thesis, CNAA, Kingston Polytechnic, UK.

—— & JARVIS, I. 1992a. Applications of geochemical data to modelling sediment dispersal patterns in distal turbidites: Late Quaternary of the Madeira Abyssal Plain. *Journal of Sedimentary Petrology,* **62**, 1112–1129.

—— & —— 1992b. Composition and provenance of turbidite sands: Late Quaternary, Madeira Abyssal Plain. *In:* MIDDELBURG, J. J. & NAKASHIMA, S. (eds) *The Geochemistry of North Atlantic Abyssal Plains.* Marine Geology, **109**, 21–53.

POTTS, P. J., TINDLE, A. G. & WEBB, P. C. 1992. *Geochemical Reference Material Compositions,* Whittles Publishing/CRC Press, Caithness/Boca Raton.

ROTHWELL, R. G., PEARCE, T. J. & WEAVER, P. P. E. 1992. Late Quaternary infilling of the Canary Basin, NE Atlantic. *Basin Research,* **4**, 103–131.

SEARLE, R. C. 1987. Regional setting and geophysical characterisation of the Great Meteor East area on the Madeira Abyssal Plain. *In:* WEAVER, P. P. E. & THOMSON, J. (eds) *Geology and Geochemistry of Abyssal Plains.* Geological Society, London,

Special Publication, **31**, 49–70.

SPEARS, D. A. & AMIN, M. A. 1981. A mineralogical and geochemical study of turbidite sandstones and interbedded shales, Mam Tor, Derbyshire, UK. *Clay Minerals*, **16**, 333–345.

STOW, D. V. A. & SHANMUGAM, G. 1980. Sequence of structures in fine-grained turbidites: comparison of recent deep-sea and flysch sediments. *Sedimentary Geology*, **25**, 23–42.

THOMSON, J., COLLEY, S., HIGGS, N. C., HYDES, D. J., WILSON, T. R. S. & SØRENSEN, J. 1987. Geochemical oxidation fronts in NE Atlantic distal turbidites and their effects in the sedimentary record. *In:* WEAVER, P. P. E. & THOMSON, J. (eds) *Geology and Geochemistry of Abyssal Plains.* Geological Society, London, Special Publication, **31**, 167–177.

——, HIGGS, N. C., JARVIS, I., HYDES, D. J., COLLEY, S. & WILSON, T. R. S. 1986. The behaviour of manganese in post-oxic Atlantic carbonate sediments. *Geochimica et Cosmochimica Acta*, **50**, 1807–1818.

TOTLAND, M. M., JARVIS, I. & JARVIS, K. E. 1992. An assessment of dissolution techniques for the analysis of geological samples by plasma spectrometry. *Chemical Geology*, **95**, 35–62.

WEAVER, P. P. E. 1993. High-resolution stratigraphy of marine Quaternary sequences. *In:* HAILWOOD, E. A. & KIDD, R. B. (eds) *High Resolution Stratigraphy.* Geological Society, London, Special Publication, **70**, 137–153.

——, BUCKLEY, D. E. & KUIJPERS, A. 1989a. Geological investigations of ESOPE cores from the Madeira Abyssal Plain. *In:* SCHUTTENHELM, R. T. E., AUFFRET, G. A., BUCKLEY, D. E., CRANSTON, R. E., MURRAY, C. N., SHEPHARD, L. E. & SPIJKSTRA, A. E. (eds) *Geoscience Investigations of Two North Atlantic Abyssal Plains – The ESOPE International Expedition.* Commission of the European Community, Luxembourg, **I**, 535–555.

—— & KUIJPERS, A. 1983. Climatic control of turbidite deposition on the Madeira Abyssal Plain. *Nature*, **306**, 360–363.

——, MASSON, D. G. & KIDD, R. B. 1994. Slumps, slides and turbidity currents – sea-level change and sedimentation in the Canary Basin. *Geoscientist*, **4**, 14–16.

—— & ROTHWELL, R. G. 1987. Sedimentation on the Madeira Abyssal Plain over the last 300,000

years. *In:* WEAVER, P. P. E. & THOMSON, J. (eds) *Geology and Geochemistry of Abyssal Plains.* Geological Society, London, Special Publication, **31**, 71–86.

——, ——, EBBING, J., GUNN, D. & HUNTER, P. M. 1992. Correlation, frequency of emplacement and source directions of megaturbidites on the Madeira Abyssal Plain. *Marine Geology*, **109**, 1–20.

——, SEARLE, R. C. & KUIJPERS, A. 1986. Turbidite deposition and the origin of the Madeira Abyssal Plain. *In:* SUMMERHAYES, C. P. & SHACKLETON, N. J. (eds) *North Atlantic Palaeoceanography.* Geological Society, London, Special Publication, **21**, 131–143.

—— & THOMSON, J. 1993. Calculating erosion by deep-sea turbidity currents during initiation and flow. *Nature*, **364**, 136–138.

——, —— & JARVIS, I. 1989b. The Geology and Geochemistry of the Madeira Abyssal Plain Sediments: A Review. *In:* FREEMAN, T. J. (ed.) *Disposal of Radioactive Waste in Seabed Sediments.* Advances in Underwater Technology, Ocean Science and Offshore Engineering, **18**, Society for Underwater Technology (Graham & Trotman), 51–78.

WILSON, T. R. S., THOMSON, J., COLLEY, S., HYDES, D. J. & HIGGS, N. C. 1985. Early organic diagenesis: the significance of progressive subsurface oxidation fronts in pelagic sediments. *Geochimica et Cosmochimica Acta*, **49**, 811–822.

——, ——, HYDES, D. J., COLLEY, S., CULKEN, F. & SØRENSEN, J. 1986. Oxidation fronts in pelagic sediments: diagenetic formation of metal-rich layers. *Science*, **232**, 972–975.

WRAY, D. S. 1990. *The petrology of clay-rich beds in the Turonian (Upper Cretaceous) of the Anglo-Paris Basin.* PhD Thesis, CNAA, City of London Polytechnic, UK.

—— & GALE, A. S. 1993. Geochemical correlation of marl bands in Turonian chalks of the Anglo-Paris Basin. *In:* HAILWOOD, E. A. & KIDD, R. B. (eds) *High Resolution Stratigraphy.* Geological Society, London, Special Publication, **70**, 211–226.

WRONKIEWICZ, D. J. & CONDIE, K. C. 1987. Geochemistry of Archaean shales from the Witwatersrand Supergroup, South Africa: source-area weathering and provenance. *Geochimica et Cosmochimica Acta*, **51**, 2401–2416.

SHRIMP zircon age control of Gondwanan sequences in Late Carboniferous and Early Permian Australia

JOHN ROBERTS,[1] JONATHAN CLAOUE-LONG,[2] PETER J. JONES[2] &
CLINTON B. FOSTER[2]

[1]*Department of Applied Geology, University of New South Wales, Sydney, NSW 2052,
Australia*

[2]*Australian Geological Survey Organisation, GPO Box 378, Canberra, ACT 2601, Australia*

Abstract: Australia during the Late Carboniferous formed part of the Gondwana
supercontinent and was close to the South Pole. Resulting continental glacigene deposits
and cold water marine sequences in the Southern New England Orogen cannot be correlated
biostratigraphically with Late Carboniferous successions in the northern hemisphere
because they contain a low diversity biota endemic to Gondwana. Magnetostratigraphic
correlation via the Permian–Carboniferous reversed magnetic superchron is presently
uncertain. Sensitive high-resolution ion microprobe (SHRIMP) zircon dating of volcanics
horizons in the sedimentary sequence of the Southern New England Orogen is now
establishing relationships between the Gondwanan faunas and warm-climate European
equivalents. In Continental sediments, correlation of the major Gondwanan glaciation of
Late Carboniferous Australia is revised from Stephanian to early Namurian–Westphalian in
age, a timing that matches that of the Hoyada Verde glaciation in Argentina. A major hiatus
of > 15 Ma between Carboniferous rocks of the Southern New England Orogen and
Permian rocks of the Sydney Basin of eastern Australia probably reflects the first
deformational movements within the accretionary prism of the orogen. *Nothorhacopteris*,
traditionally a Late Carboniferous indicator in Australia, now ranges from late Viséan to
Westphalian, and an 'enriched' assemblage, previously correlated with the top of the
Nothorhacopteris flora, is confined to the late Viséan to early Namurian. Palynofloras
assigned to the *Spelaeotriletes ybertii/Diatomozonotriletes birkheadensis* Zones (Namurian–
early Westphalian), are now confined to the Namurian, and a mid to late Westphalian age is
indicated for a more diverse palynoflora identified as uppermost *D. birkheadensis* Zone
(? = *Asperispora reticulatispinosus* Zone). In the cold climate marine sequence, the
Levipustula levis brachiopod Zone appears to be confined to the early Namurian, and so
partly corresponds in age to the continental glaciation.

During the Late Carboniferous, Australia lay
close to the South Pole on the margin of
Gondwana. This general situation is illustrated
in Fig. 1 and is confidently inferred from polar
wandering curve information and the special
nature of glacigene sediments, but further detail
of the palaeogeography, and correlation of
Australian Late Carboniferous sediments, is
frustrated by the difficulty of correlating sequences that contain low diversity cold
faunas. The Late Carboniferous to Early Permian Gondwanan biota of eastern Australia and
Argentina is endemic to Gondwana with the
possible exceptions of some, as yet only briefly
described, foraminifera (Palmieri 1983), and
some palynofloras (Jones & Truswell 1992).
Correlation between continental glacigene deposits and corresponding cold climate marine
sediments is not established, and correlation
with the type of marine stratigraphy of Europe

and Russia is controversial. As a result, comparisons between the major continental masses of
Gondwana and Laurasia have lacked precision
in the timing of geological events and processes.

A solution to this problem is offered by recent
publications (Lippolt & Hess 1983; Hess &
Lippolt 1986) of accurately measured $^{40}Ar/^{39}Ar$
ages about the Carboniferous–Permian boundary and for Late Carboniferous stages in
Europe, and the recent development of zircon
dating with the sensitive high resolution ion
microprobe (SHRIMP), which makes reliable
age measurement possible for the first time in the
altered volcanic units of Late Carboniferous
Australia. The isotopic and biostratigraphic
accuracy of the European age measurements
makes them invaluable time markers to which
correlations can be made, if corresponding ages
can be measured in the otherwise unconstrained
Gondwanan sequences. This contribution re-

From Dunay, R. E. & Hailwood, E. A. (eds), 1995, *Non-biostratigraphical Methods of Dating and Correlation*
Geological Society Special Publication No. 89, pp. 145–174

145

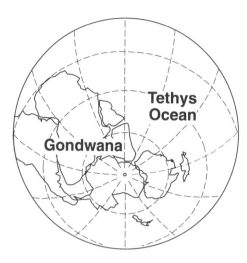

Fig. 1. Australia in the Late Carboniferous and Early Permian was close to the South Pole as part of the supercontinent of Gondwana. Much of southern Australia was covered by ice; glacigene and fluvioglacial sediments accumulated in terrestrial environments. Marine sediments in the subduction and island arc system off eastern Australia (now the New England Orogen) contain endemic cold-water faunas, and cannot be correlated with warm-climate European Carboniferous systems developing on the other side of Tethys. Adapted from Lackie & Schmidt (1993).

ports efforts to correlate the continental and marine Late Carboniferous successions of eastern Australia with each other, and with Europe, using the numerical ages of volcanic horizons within the sedimentary sequences. Zircon dating in the Early Carboniferous of eastern Australia is described elsewhere (Roberts *et al.* 1995). Zircon analytical results, and associated biostratigraphy, are reported in full in this contribution, which supersedes the interim account of the work given at the 1991 International Congress on the Carboniferous and Permian (Roberts *et al.* 1993*b*). Biostratigraphic zones and units used within this paper are based on eastern Australian palynofloras, macrofloras, brachiopods, conodonts, ammonoids and foraminifera. The biological affinity of zones or units within the first three categories is indicated in Fig. 13.

The Southern New England Orogen

During the greater part of the Carboniferous, an andesitic to dacitic volcanic arc, fore-arc basin and accretionary prism was generated at the eastern margin of Australia by a westerly dipping subduction zone (Aitchison *et al.* 1992) and is now preserved in eastern Australia as the Southern New England Orogen (SNEO). The

area studied in this contribution is located in the southern fore-arc basin of the SNEO within the Hunter–Myall region, NSW (Fig. 2).

Relationships between the Late Carboniferous and Permian sequences of eastern Australia, constructed from the mapping of Roberts & Engel (1987) and Roberts *et al.* (1991*b*) and scaled both numerically, from data in this paper, and biostratigraphically are illustrated in Fig. 3. The Late Carboniferous sediments comprise a continental glacigene facies in the west, extending from the early Namurian to the Westphalian, and a cold-water marine facies to the east that initially was transgressive but ended with regressive continental deposition in the Namurian. Sporadic ignimbritic volcanism contributed material to the western terrestrial sequence and periodically overlapped the marine sequence. The continental and marine areas are represented by completely different stratigraphic successions that crop out in well-separated areas (Roberts *et al.* 1991*b*). Hence, until the present study it was virtually impossible to correlate between the Late Carboniferous continental glacigene facies and the marine equivalents. Marked deterioration of climate is recorded towards the end of the Viséan. This was accompanied by uplift, waning of volcanism and a marked change in depositional pattern,

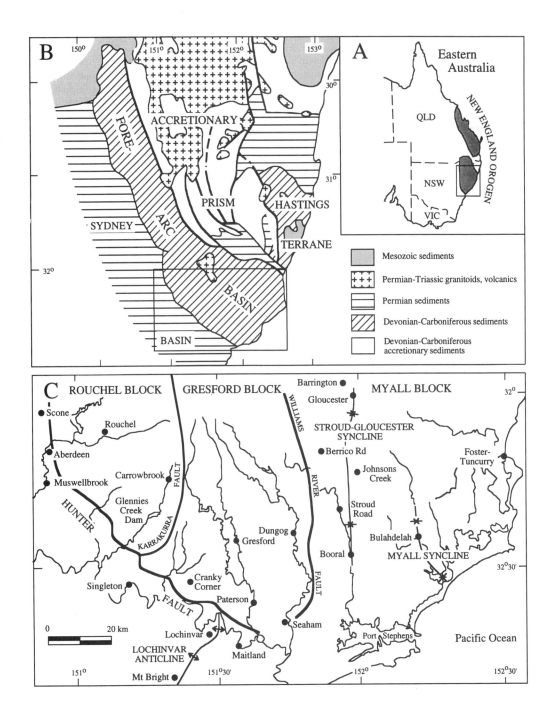

Fig. 2. Map showing locations referred to in this study. (**A**) Location of the Southern New England Orogen (SNEO) in eastern Australia; (**B**) major tectonic elements of the SNEO (adapted from Collins *et al.* 1993); (**C**) geography of the area outlined in (B) showing place names referred to in the text and the division of the area into major Blocks by fault systems. Modified from Roberts *et al.* (1995).

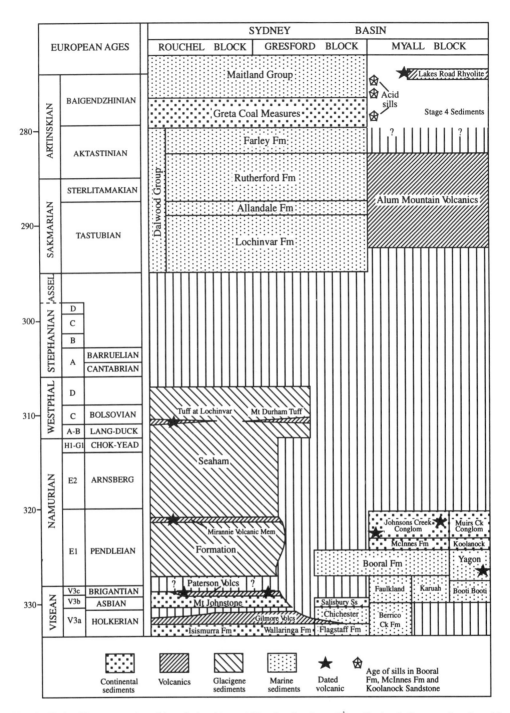

Fig. 3. Carboniferous stratigraphic relationships within the Southern New England Orogen. Stratigraphic relationships are based on those of Roberts *et al.* (1993*b*) adjusted for the biostratigraphic correlations revised by Roberts *et al.* (1993*c*) and numerical ages presented in this paper. Within the Viséan we use the Belgian chronostratigraphic notation rather than the formal stage nomenclature (see Conil *et al.* 1990; V2a + V2b = Livian; V3b + V3c = Warnantian). Within the Namurian: Chok-Yead, the Chokierian, Alportian, Kinder-scoutian, Marsdenian and Yeadonian Stages: Lang-Duck, Langsettian and Duckmantian Stages. Vertical lines indicate hiatus. Tournaisian and Viséan marine stratigraphy is controlled mainly by brachiopod zonation. Onset of glacigene sedimentation near the base of the Namurian (Seaham Formation) is matched in the marine sequence by low diversity cold water faunas (*Levipustula levis* Zone). Stars indicate volcanic horizon dated by SHRIMP zircon analysis.

that separated laterally coherent facies of the Late Carboniferous from Tournaisian–Middle Viséan conditions when deposition took place within three distinct blocks, each with its own major volcanic source and stratigraphic succession (cf. Roberts & Engel 1987; Roberts *et al.* 1993*b*).

Potential magnetostratigraphic correlations

Although the correlation of Palaeozoic sequences by magnetostratigraphic patterns is normally difficult, there is a well-established magnetically quiet period spanning the Late Carboniferous and Permian. This has great potential for correlation of Late Carboniferous Gondwanan sequences because it was originally defined in the SNEO of eastern Australia, and its length and consistency make it one of few magnetic zones likely to be recognized with confidence in the Palaeozoic. Irving & Parry (1963) and Irving (1966) originally named this period the Kiaman magnetic interval and defined its base in eastern Australia at the reversal between the normal-polarity Paterson Volcanics, then dated at 298 Ma on the basis of K–Ar dating (Evernden & Richards, 1962; subsequently revised to 308 Ma by Roberts *et al.* 1991*b*), and the reversed Seaham Formation glacigene sediments. As magnetic measurements became available from other parts of the world, frequent reversals were observed extending down into the Namurian. This resulted in confusion about the position of the Paterson Reversal and led Irving & Pullaiah (1976) to redefine the period as the Permian–Carboniferous Reversed (PCR) magnetic superchron with its base at a Namurian reversal in North America.

Correlation of the PCR between Australia and the northern hemisphere continents has been thrown into confusion by Claoué-Long *et al.* (1995). SHRIMP zircon dating has revised the age of the Paterson Volcanics to Viséan, 20 Ma older than previously supposed, and much older than the base of the PCR recognized in North America. The Paterson Volcanics must therefore belong to one of the several Early Carboniferous magnetically normal periods recognized in Europe and North America, and the PCR must begin somewhere within the overlying glacigene Seaham Formation. The present state of knowledge is summarized in Fig. 4. Even though this distinctive magnetically quiet period was originally recognized and defined in the SNEO, the location of its base in Australia is now unknown and awaits systematic magnetostratigraphic search.

The reference European timescale

A major development in the study of Carboniferous processes was the publication by Lippolt & Hess (1983) and Hess & Lippolt (1986) of $^{40}Ar/^{39}Ar$ ages for the Late Carboniferous and Early Permian of Europe. The isotopic and biostratigraphic precision of these new data effectively superseded the widely used reference Carboniferous timescales offered by Harland *et al.* (1982), De Souza (1982), and Forster & Warrington (1985). The later Harland *et al.* (1990) timescale does include the new $^{40}Ar/^{39}Ar$ ages in its database, but effectively dilutes their importance by giving equal weight to the large quantity of earlier, more ambiguous, constraints. It is therefore not useful to apply the Harland *et al.* (1990) timescale in the Late Carboniferous, and this study relates ages measured in Australia directly against the original Lippolt & Hess (1983) and Hess & Lippolt (1986) age measurements in Europe. The $^{40}Ar/^{39}Ar$ ages are plotted in Fig. 13, where they are used to calibrate the European stages as a reference scale against which to compare the Australian sequences.

In a separate contribution (Claoué-Long *et al.* 1995) the comparability of the $^{40}Ar/^{39}Ar$ sanidine ages are compared with our U–Pb zircon dates, and show that the two methods give indistinguishable age results, despite being based on different half-lives and different minerals measured in different laboratories against different standards. Our zircon ages for Australian sequences can therefore be referred directly to the European $^{40}Ar/^{39}Ar$ ages and this comparability is the foundation of the present study.

Age comparison of this type is achieved via inter-laboratory standards, and this introduces the uncertainty associated with the reference compositions of standard materials. In the case of SHRIMP zircon U–Pb ages, the inter-laboratory reproducibility of the reference $^{206}Pb/^{238}U$ ratio of the standard zircon (SL13) is within 0.06% (2σ), and this is a negligible contribution to the uncertainty of inter-laboratory comparisons (Claoué-Long *et al.* 1995). Relative to each other, $^{40}Ar/^{39}Ar$ ages are usually very precise because the age of the reference composition is not an issue with internal comparisons. However, rather large errors are attached to the reference compositions of $^{40}Ar/^{39}Ar$ standards and this affects inter-laboratory comparisons: Hess & Lippolt (1986) describe how this has the effect of expanding the ± 2 Ma (2σ) within-laboratory uncertainty in their ages to an external (inter-laboratory) uncertainty of ± 8 Ma (2σ), or more. For

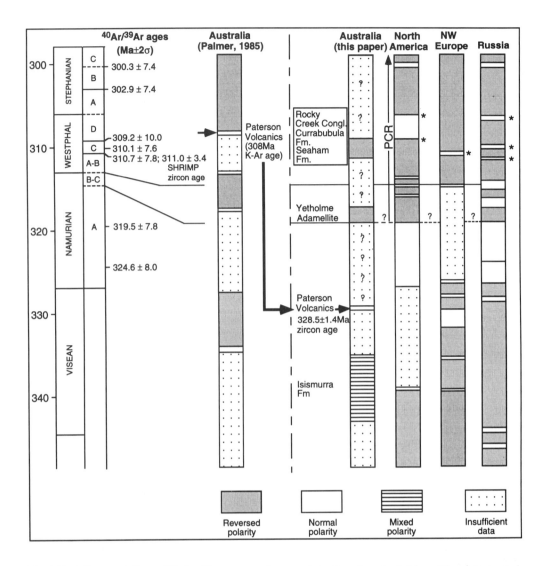

Fig. 4. Tentative correlations of Carboniferous magnetic reversal stratigraphy, adapted from Claoué-Long *et al.* (1995). Plotted against the European stratigraphic scale are the ^{40}Ar/^{39}Ar ages of Hess & Lippolt (1986). North American, European and Russian magnetostratigraphy follows Palmer *et al.* (1985), whose Australian scale was based on a Paterson Volcanics age of 308 Ma; stars indicate correlations of the Paterson Reversal that have been proposed on the basis of this age. Reassignment to the Viséan places the Paterson Volcanics significantly below the base of the PCR as recognized in North America. The location of the base of the PCR in Australia is now uncertain; a suggested correlation is indicated by the dashed line. The magnetostratigraphic scale for the late Namurian is expanded to accommodate the brevity of this interval. Magnetic data for Australia is from Irving (1966), Luck (1973), Facer (1976), Embleton (1976) and Palmer *et al.* (1985)

example, the sanidine ^{40}Ar/^{39}Ar age of the COTZ sample, used by Hess & Lippolt (1986) to date the Westphalian B, is 310.7 Ma with an internal reproducibility of ± 2.6 Ma (2σ), but the inter-laboratory uncertainty is ± 8 Ma (2σ). The SHRIMP U–Pb age for zircons in the same sample is 311.0 Ma with an inter-laboratory uncertainty of ± 3 A Ma (2σ) (Claoué-Long *et al.* 1995). The SHRIMP zircon Pb–U ages are therefore a factor of two more precise than the ^{40}Ar/^{39}Ar reference ages and the main limitation in the present comparison exercise is posed by

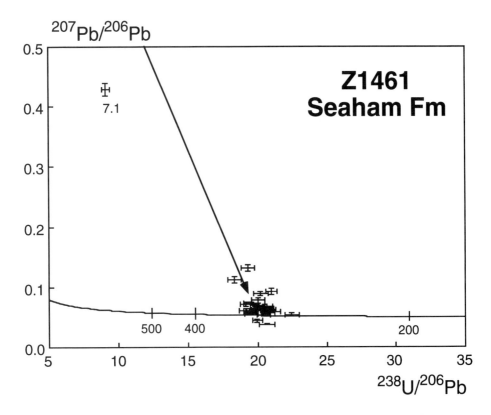

Fig. 5. Compositions of zircons in the Seaham Formation tuff at Lochinvar plotted without correction for common Pb on a $^{208}Pb/^{238}U$ v. $^{207}Pb/^{206}Pb$ Concordia diagram. The arrow indicates the vector of common Pb correction. Radiogenic $^{206}Pb/^{238}U$ compositions and corresponding ages are calculated by projection of the raw data on to Concordia along this vector. Dispersion of data to the right of this vector beyond measurement uncertainty indicates loss of radiogenic Pb in two grains. Grain 7 has an older apparent age and a very high content of common Pb. All the other zircons form a group of low common Pb compositions close to Concordia.

the uncertainty associated with $^{40}Ar/^{39}Ar$ reference standards.

Ages in the Australian sequences

SHRIMP analytical procedures

Attempts previously have been made to date volcanics in the SNEO using the K–Ar method of isotopic analysis, but this method of dating yielded inconclusive results. For example, Claoué-Long et al. (1995) show that the K–Ar apparent age of the Paterson Volcanics (which underlie the Seaham Formation glacigene sediments) is 20 Ma too young, a result attributable to an unresolved loss of Ar. It is possible that $^{40}Ar/^{39}Ar$ analysis would yield

more useful age information by allowing Ar loss components to be interpreted from Ar step-heating patterns, but rocks in the area have been subject to low grade metamorphism and deep surface weathering and few samples preserve intact the phases (such as sanidine feldspar) suited to Ar step-heating study. The mineral zircon, on the other hand, is a ubiquitous component of volcanic rocks in the SNEO and has survived alteration processes; frequently it is observed to have survived even the alteration of volcanic ashes to bentonite. The recent development of zircon U–Pb dating using the SHRIMP has opened zircon dating to the same interpretation methods offered by $^{40}Ar/^{39}Ar$ step-heating analysis, by the ability to probe and date separately the different altered, inherited and magmatic zones within single crystals. SHRIMP zircon analysis has therefore been applied to the problem of dating Carboniferous and Permian volcanic horizons in the SNEO.

Full discussion of the procedures for ion microprobe dating of Phanerozoic zircons is given by Claoué-Long *et al.* (1995). SHRIMP is used to target individual zones within zircon crystals at the 20 μm scale and a pattern of U–Pb ages for 20–40 grains is measured over a period of one or two days. In relatively young (i.e. Phanerozoic) zircons, ages are calculated from ^{206}Pb/^{238}U ratios that are calibrated by reference to a value of 0.0928 for the ^{206}Pb/^{238}U ratio of a homogeneous standard zircon; ages calculated in this way are insensitive to correction for common Pb. Individual zircons that may be inherited (older ages) or have lost radiogenic Pb (young apparent ages) are recognized from deviations in composition from the major population of ^{206}Pb/^{238}U data, an interpretation procedure very similar to the recognition of 'plateau ages' and Ar loss from ^{40}Ar/^{39}Ar step-heating patterns. In favourable circumstances, the primary crystallization age of the zircons can be interpreted reliably in complex and altered rocks of mixed age.

Data tabulation in this paper includes the raw total Pb compositions without correction for common Pb, and the radiogenic ^{206}Pb/^{238}U ratios and ages calculated by projection of the raw data on to Concordia, as shown in Fig. 5. This is mathematically equivalent to correcting for common Pb by mass balance between the ^{207}Pb/^{206}Pb ratio measured in each zircon and the theoretical radiogenic ratio appropriate to its age. This is the ^{207}Pb method of common Pb correction in which the fraction f of common ^{206}Pb$_c$ in the total measured ^{206}Pb$_m$ is given by:

$$f=\left[\left(\frac{^{207}\text{Pb}}{^{206}\text{Pb}}\right)_m - \left(\frac{^{207}\text{Pb}}{^{206}\text{Pb}}\right)^*\right] \Big/ \left[\left(\frac{^{207}\text{Pb}}{^{206}\text{Pb}}\right)_c - \left(\frac{^{207}\text{Pb}}{^{206}\text{Pb}}\right)^*\right],$$

where *m* denotes the measured ratio, *c* denotes the isotopic ratio of the common Pb and * denotes the ratio of radiogenic Pb appropriate to the age of the sample calculated as:

$$\frac{^{207}\text{Pb}^*}{^{206}\text{Pb}^*} = \frac{^{235}\text{U}}{^{238}\text{U}}\frac{e^{\lambda_{235}t}-1}{e^{\lambda_{238}t}-1}.$$

The radiogenic ^{206}Pb*/^{238}U ratio is then given by:

$$\frac{^{206}\text{Pb}*}{^{238}\text{U}} = (1-f)\left(\frac{^{206}\text{Pb}}{^{238}\text{U}}\right)_m$$

In clean zircons having low intrinsic common Pb, values very close to zero for *f* and ^{204}Pb will be obtained. Apparent negative values become possible as statistical fluctuation in very small counts close to zero, and some are indeed recorded in Tables 1–6; these are recordings within error of zero and they indicate that common Pb could not be detected with the available counting uncertainty.

Ages interpreted for each sample are calculated from the weighted mean of the group of ^{206}Pb*/^{238}U ratios identified as belonging to the primary magmatic population. Weightings are the inverse square of the errors on individual ratio measurements, and quoted errors are the weighted standard deviation of the mean. In samples where the errors on individual measurements are alike all the weightings are the same and this procedure becomes equivalent to quoting the mean and standard deviation of the mean. Samples Z1556, Z1590 and Z1046 were analysed by an abbreviated procedure in which measurement of element abundances was omitted.

Seaham Formation

The Seaham Formation is the best known continental glacigene sequence of Late Carboniferous Australia. It contains tillites and varved shales interlayered with tuffs, lithic sandstone and conglomerate; the varved shales contain dropstones and the conglomerates striated and faceted boulders. The stratigraphy of the formation is complex, and mappable units such as volcanics often have restricted local occurrence. Syn-depositional erosion is evident in lower parts of the formation in certain regions, and there is loss of section at the bounding upper unconformity with the Permian Lochinvar Formation of the Sydney Basin. The Paterson Volcanics, a laterally extensive unit underlying the Seaham Formation, bears glacial striations in some exposures; in other areas the volcanics have been completely eroded so that the glacigene deposits lie unconformably on lower parts of the Carboniferous succession.

Four different successions can be identified in the belt of outcrop between Muswellbrook and Seaham. At the eastern margin of the Cranky Corner Basin westwards towards Muswellbrook, the Paterson Volcanics are overlain, apparently conformably, by continental sediments containing the distinctive Mirannie Volcanic and Mount Durham Tuff Members (Roberts *et al.* 1991*b*). Using the Mirannie and Paterson Volcanics as datum planes, there is a measurable decrease in thickness of the glacigene sediments towards the eastern side of the Cranky Corner Basin.

Further to the east in a narrow central strip between the Cranky Corner Basin and 2 km west of Paterson the formation is thinner (c. 300 m), lacks volcanics and cannot be correlated with the sections to the west, east and south. In the Rosebrook Range north of Maitland the Seaham Formation lies on a major disconformity cut into the Gilmore Volcanic Group, removing the Mt Johnstone Formation and Paterson Volcanics. Elsewhere the Paterson Volcanics are eroded and glacially striated. Upper parts of the formation in this area have also been removed by erosion both before and during deposition of Permian rocks in the Mindaribba Syncline, south of Paterson (JR, unpublished data). The amount of erosion and the angular nature of the unconformity within the central outcrops suggest that the region was tectonically unstable during the Carboniferous glacial phase, perhaps in a similar fashion to the Lochinvar Anticline of the northern Sydney Basin that formed a horst-like high during the Late Permian (Glen 1993). The present axes of the central strip and the Lochinvar Anticline closely coincide, suggesting the existence of an ancient and long-lived structural high. Undifferentiated volcanics (ash falls and tuffs) reappear within the Seaham Formation at Paterson and are abundant in the type area at Seaham in the third succession. Unnamed felsic volcanics that cannot yet be related with those of other Seaham sequences are present in the upper part of a fourth succession at Winders Hill in the Carboniferous core of the Lochinvar Anticline. The youngest sediments of the Seaham Formation crop out at Mt Bright in the Mount View Range.

The age and correlation of the Seaham Formation are constrained by three dates for volcanic horizons in the western, central and southern sections of the formation, and an age for the Paterson Volcanics that underlie the glacigene sediments.

Paterson Volcanics

The Paterson Volcanics are a laterally continuous horizon underlying the Seaham Formation and so form the marker for the onset of Late Carboniferous Gondwanan glaciation. As described above, they have also been used as the Australian marker for the onset of the Kiaman reversed magnetic period, but this association is no longer tenable. The biostratigraphic associations and age of the Paterson Volcanics are described in detail by Claoué-Long et al. (1995): four independent zircon age measurements agree at a date of 328.5 ± 1.4 Ma (2σ).

Lochinvar tuff

Near Lochinvar (Fig. 2) the upper Seaham Formation contains a felsic volcanic unit c. 50 m below the Lochinvar Formation, the local base of the Permian. Sample Z1461 is from the uppermost volcanic unit in the Winders Hill section (southern region) of Browne (1927). This unit is local and cannot be linked with ignimbrites and tuffs in the western region, but a thick succession of more than 700 m intervening between the volcanic unit at Lochinvar and the underlying Winders Hill Granodiorite suggests that older parts of the formation in this section may be comparable with those in the west.

Zircon compositions measured in Z1461 are listed in Table 1 and plotted without correction for common Pb in a Tera & Wasserburg (1974) Concordia diagram in Fig. 5 (the geometry of Phanerozoic zircon compositions in this type of diagram is discussed by Claoué-Long et al. (1995)). With the exception of one grain, the zircon compositions group close to Concordia, indicating low contents of common Pb. The vector of common Pb correction is indicated by the arrow. Even without further processing, the raw data on Fig. 5 indicate that the main group of compositions control the age of the sample. The corresponding $^{206}Pb/^{238}U$ ages are calculated by projection of the raw data in this diagram on to the Concordia curve.

The homogeneity of the main population of analyses is tested in Fig. 6, where the $^{206}Pb/^{238}U$ ages of the zircons are plotted in a probability diagram. The x-axis of this diagram is scaled as probability, having the effect of normalizing data to the ideal normal distribution that appears on the diagram as a straight line of slope proportional to the standard deviation. The standard deviation plotted is that of the reproducibility of concurrent measurements of the standard zircon; the line therefore forms the reference for the statistical dispersion expected in replicate measurements of a homogenous composition. The age spectrum of the Lochinvar zircons marked in bold is indistinguishable from the Normal Distribution and these define a single population with a weighted mean radiogenic $^{206}Pb/^{238}U$ age of 310.6 ± 4.0 Ma (2σ), interpreted as the crystallization age of the volcanic horizon. Two zircon compositions have young apparent ages departing from the main population by >95% confidence; it is apparent that these grains have lost small amounts of radiogenic Pb. A single analysis has an apparent age significantly older than the main population: this is grain 7 whose anomalous common Pb content is highlighted in Fig. 5.

Table 1. *Shrimp U–Pb isotopic data for a tuff in the Seaham Formation at Lochinvar (Z1461)*

Grain area	U (p.p.m.)	Th (p.p.m.)	Th/U	$^{204}Pb^*$ (p.p.b.)	$f^{206}Pb^*$† (%)	$^{206}Pb/^{238}U$ ±1σ	$^{207}Pb/^{235}U$ ±1σ	$^{207}Pb/^{206}Pb$ ±1σ	$^{206}Pb/^{238}U$ ±1σ	Apparent age‡ (Ma) ±1σ
							Calibrated total Pb compositions		Radiogenic compositions*	
1.1	66	72	1.09	6	3.71	0.0495 ± 13	0.610 ± 30	0.0894 ± 35	0.0476 ± 13	300 ± 8
1.2	160	182	1.14	1	0.27	0.0508 ± 12	0.430 ± 18	0.0614 ± 20	0.0506 ± 12	318 ± 8
2.1	323	249	0.77	0	0.00	0.0473 ± 12	0.386 ± 14	0.0591 ± 12	0.0473 ± 12	298 ± 8
2.2	64	58	0.90	1	0.74	0.0495 ± 12	0.445 ± 27	0.0652 ± 27	0.0491 ± 12	309 ± 7
3.1	167	152	0.91	5	1.24	0.0500 ± 13	0.477 ± 19	0.0692 ± 19	0.0494 ± 13	311 ± 8
4.1	33	57	1.72	6	6.50	0.0545 ± 14	0.843 ± 49	0.1122 ± 55	0.0510 ± 14	320 ± 9
5.1	115	221	1.92	5	1.73	0.0512 ± 13	0.517 ± 22	0.0732 ± 22	0.0503 ± 13	316 ± 8
6.1	358	402	1.12	2	0.23	0.0499 ± 13	0.420 ± 14	0.0610 ± 12	0.0498 ± 13	313 ± 8
6.2	216	348	1.47	−2	−0.46	0.0512 ± 12	0.391 ± 16	0.0554 ± 17	0.0514 ± 12	323 ± 8
6.3	148	193	1.31	0	−0.10	0.0514 ± 12	0.414 ± 19	0.0584 ± 21	0.0515 ± 13	323 ± 8
6.4	173	182	1.05	0	0.12	0.0504 ± 12	0.417 ± 18	0.0601 ± 20	0.0503 ± 12	316 ± 7
7.1	36	62	1.72	87	45.25	0.1097 ± 30	6.481 ± 253	0.4284 ± 105	0.0601 ± 22	376 ± 13
8.1	68	58	0.85	1	0.48	0.0501 ± 13	0.435 ± 23	0.0630 ± 27	0.0499 ± 13	314 ± 8
9.1	134	214	1.60	−8	−2.56	0.0484 ± 13	0.255 ± 12	0.0382 ± 15	0.0496 ± 13	312 ± 8
9.2	167	282	1.69	2	0.41	0.0498 ± 12	0.429 ± 18	0.0625 ± 20	0.0496 ± 12	312 ± 7
9.3	159	273	1.72	−3	−0.68	0.0490 ± 12	0.362 ± 16	0.0536 ± 19	0.0493 ± 12	310 ± 7
9.4	131	211	1.61	−1	−0.19	0.0482 ± 12	0.383 ± 18	0.0576 ± 22	0.0483 ± 12	304 ± 7
9.5	52	62	1.18	3	2.27	0.0499 ± 12	0.534 ± 31	0.0776 ± 39	0.0488 ± 12	307 ± 7
11.1	226	330	1.46	4	0.70	0.0491 ± 9	0.439 ± 14	0.0649 ± 16	0.0488 ± 9	307 ± 5
12.1	51	62	1.23	5	4.10	0.0476 ± 9	0.608 ± 33	0.0926 ± 46	0.0457 ± 9	288 ± 5
13.1	148	279	1.89	2	0.58	0.0480 ± 8	0.423 ± 17	0.0639 ± 21	0.0478 ± 8	301 ± 5
14.1	43	35	0.81	3	2.46	0.0534 ± 9	0.583 ± 33	0.0792 ± 40	0.0521 ± 10	328 ± 6
18.2	112	149	1.32	1	0.45	0.0480 ± 12	0.416 ± 21	0.0628 ± 26	0.0478 ± 12	301 ± 7
18.3	90	90	1.00	3	1.49	0.0521 ± 13	0.512 ± 26	0.0713 ± 30	0.0513 ± 13	323 ± 8
26.1	76	82	1.09	17	8.92	0.0518 ± 13	0.942 ± 44	0.1320 ± 48	0.0472 ± 12	297 ± 7
27.1	107	132	1.24	1	0.21	0.0522 ± 13	0.438 ± 21	0.0608 ± 24	0.0521 ± 13	327 ± 8
28.1	138	128	0.93	−1	−0.21	0.0511 ± 12	0.404 ± 18	0.0574 ± 20	0.0512 ± 12	322 ± 8
29.1	150	219	1.46	−2	−0.61	0.0445 ± 11	0.332 ± 17	0.0541 ± 23	0.0448 ± 11	282 ± 7
30.1	80	120	1.49	2	1.02	0.0487 ± 12	0.453 ± 24	0.0675 ± 30	0.0482 ± 12	303 ± 7

* ^{207}Pb correction; † $f^{206}Pb$ indicates the percentage of common ^{206}Pb in the total measured ^{206}Pb; ‡ apparent age is the $^{206}Pb/^{238}U$ age.

Mirannie Volcanic Member

Sample Z845 is from the Mirannie Volcanic Member of the Seaham Formation, 110 m above the Paterson Volcanics in the Black Jack Range southeast of Glennies Creek Dam (Roberts *et al.* 1991*b*). Zircons in the Mirannie Volcanic have a very simple distribution of ages (Table 2). The spectrum of radiogenic $^{206}Pb/^{238}U$ ages is plotted in a probability diagram in Fig. 7 and overlain on the reference normal distribution defined from concurrent measurements of the standard zircons. None of the measured individual ages is detectably different from the statistical dispersion expected of replicate analyses of a homogeneous composition, and the group defines a weighted mean $^{206}Pb/^{238}U$ age of 321.3 ± 4.4 Ma that is interpreted as the crystallization age.

Matthews Gap Dacite

In the Mt Bright Inlier, sediments of the Seaham Formation disconformably overlie the Pokolbin Hills Volcanics (Brakel 1972). Zircons in the Matthews Gap Dacite Tuff Member of the Pokolbin Hills Volcanics have been dated using conventional isotope dilution analytical methods by Gulson *et al.* (1990) and give an age of 309 ± 3 Ma. This is indistinguishable from the age of 310.6 ± 4.0 Ma (2σ) measured at Lochinvar (described above). Seaham Formation units overlying the Matthews Gap Dacite must therefore be younger than other outcrops of the formation.

In biostratigraphic terms, both the Mirannie Volcanic Member and the volcanic unit near Lochinvar (Z1461) are younger than two biozones beneath the Paterson Volcanics: the late Viséan *R. fortimuscula* Zone from a tongue of the Chichester Formation within the Isismurra Formation; and palynofloras of the mainly late Viséan *Grandispora maculosa* Assemblage in the Mt Johnstone Formation (Playford & Helby 1968; Helby 1969). The Mirannie Volcanic Member is probably older than a palynoflora

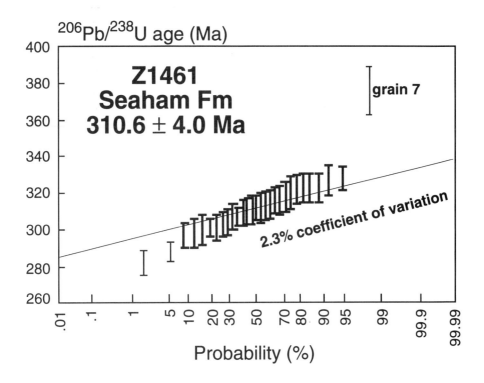

Fig. 6. Probability plot of the Seaham Formation tuff zircon ages. $^{206}Pb^{238}U$ ages have been calculated by projection of the raw measured data on to Concordia as described in Fig. 5. This plot refers the observed age population to the normal probability distribution indicated by the diagonal straight line, which is the locus of reproducibility of measurements of the standard zircon obtained during this analytical run (coefficient of variation 2.3%), and indicates the statistical dispersion expected of measurements of a homogeneous composition. Compositions in bold are indistinguishable from the normal distribution, indicating that they have the same age within error, and these group to give the mean age of the sample. Two zircons have lost small amounts of radiogenic lead and have young apparent ages departing from the main group by more than 95% confidence. The departure of grain 7 from the main group of compositions may be a function of its anomalous common lead content highlighted in Fig. 4.

within the Seaham Formation originally identified as belonging to the *Potonieisporites* Assemblage (Helby 1969). Sample 648 of Helby (1969), from *c.* 125 m above the Paterson Volcanics in quarries on the Woodville Road, 2.5 km SSE of Paterson, has been re-examined by Foster (1993). The palynoflora is relatively sparse, consisting predominantly of monosaccate, gymnospermous pollen of *Potonieisporites* spp., *Cannanoropollis* sp. and *Caheniasaccites* sp. Trilete spores from the assemblage include the following species, originally described by Jones & Truswell (1992): *Apiculatisporis pseudoheles, Brevitriletes leptoacaina, Dibolisporites disfacies, Reticulatisporites bifrons* and *Cyclogranisporites firmus.* The occurrence of the alga *Botryococcus* sp. suggests deposition in a freshwater environ-

ment. The key indices *Microbaculispora tentula* Tiwari and *Granulatisporites confluens* Archangelsky & Gamerro are not present within the assemblage. Using range zone data for the above species, a possible correlation exists with the *Asperispora reticulatispinosus* Oppel-zone (Zone D) of Jones & Truswell (1992), although the present assemblage lacks the eponymous species and other lycopod spores typical of this zone. Further work is needed to assess the significance of these differences; they may reflect either facies controls (most probably) or evolutionary (time) events. Jones & Truswell (1992) estimated that Zone D lay within the range Westphalian D to Late Autunian, but there is no independent evidence to support the younger age limit. In terms of the palynofloral succession from the

Table 2. *SHRIMP U–Pb Isotopic data for the Mirannie Ignimbrite Member*

Grain area	U (p.p.m.)	Th (p.p.m.)	Th/U	^{204}Pb* (p.p.b.)	f^{206}Pb*† (%)	Calibrated total Pb compositions			Radiogenic compositions*	
						^{206}Pb/^{238}U ±1σ	^{207}Pb/^{235}U ±1σ	^{207}Pb/^{206}Pb ±1σ	^{206}Pb/^{238}U ±1σ	Apparent age‡ (Ma)±1σ
1.1	76	65	0.85	12	6.53	0.0516 ± 10	0.745 ± 31	0.1047 ± 37	0.0483 ± 10	304 ± 6
1.2	67	54	0.80	2	1.14	0.0526 ± 19	0.445 ± 24	0.0615 ± 22	0.0520 ± 19	326 ± 12
2.1	44	42	0.95	7	6.46	0.0511 ± 10	0.733 ± 37	0.1041 ± 45	0.0478 ± 10	301 ± 6
2.2	56	65	1.17	1	1.07	0.0533 ± 19	0.448 ± 27	0.0609 ± 26	0.0527 ± 19	331 ± 12
3.1	96	77	0.80	9	3.73	0.0533 ± 10	0.605 ± 24	0.0822 ± 27	0.0513 ± 10	323 ± 6
4.1	100	111	1.11	7	3.03	0.0481 ± 9	0.508 ± 21	0.0766 ± 25	0.0466 ± 9	294 ± 6
4.2	64	88	1.38	3	1.86	0.0524 ± 19	0.485 ± 27	0.0672 ± 25	0.0514 ± 19	323 ± 12
5.1	61	53	0.87	6	4.03	0.0521 ± 10	0.608 ± 28	0.0846 ± 33	0.0500 ± 10	315 ± 6
5.2	90	117	1.30	3	1.43	0.0505 ± 19	0.444 ± 24	0.0638 ± 22	0.0498 ± 18	313 ± 11
6.1	149	173	1.16	6	1.68	0.0528 ± 10	0.479 ± 17	0.0658 ± 18	0.0519 ± 10	326 ± 6
7.1	74	66	0.89	7	3.79	0.0534 ± 11	0.609 ± 28	0.0827 ± 32	0.0514 ± 11	323 ± 6
8.1	124	169	1.36	5	1.80	0.0489 ± 17	0.450 ± 22	0.0668 ± 20	0.0481 ± 17	303 ± 11
9.1	88	51	0.58	4	2.00	0.0526 ± 19	0.495 ± 26	0.0684 ± 24	0.0515 ± 19	324 ± 11
9.2	63	58	0.93	2	1.36	0.0506 ± 19	0.442 ± 23	0.0633 ± 22	0.0499 ± 18	314 ± 11
10.1	78	59	0.76	5	2.52	0.0510 ± 18	0.510 ± 28	0.0725 ± 27	0.0497 ± 18	313 ± 11
11.1	117	70	0.60	5	1.74	0.0526 ± 19	0.481 ± 24	0.0663 ± 21	0.0517 ± 19	325 ± 11
12.1	144	199	1.38	4	1.12	0.0517 ± 18	0.437 ± 21	0.0613 ± 18	0.0511 ± 18	321 ± 11
12.2	78	106	1.36	2	1.25	0.0536 ± 20	0.461 ± 24	0.0624 ± 20	0.0529 ± 19	332 ± 12
13.1	134	113	0.84	5	1.66	0.0525 ± 19	0.475 ± 23	0.0656 ± 19	0.0516 ± 18	324 ± 11
14.1	66	192	2.92	4	2.32	0.0534 ± 19	0.522 ± 30	0.0709 ± 29	0.0521 ± 19	328 ± 11
14.2	157	171	1.09	1	0.24	0.0514 ± 19	0.385 ± 18	0.0543 ± 13	0.0513 ± 19	323 ± 12
15.1	41	61	1.47	3	2.67	0.0530 ± 19	0.539 ± 35	0.0737 ± 37	0.0516 ± 19	324 ± 11
16.1	625	935	1.50	142	8.60	0.0564 ± 20	0.943 ± 36	0.1213 ± 13	0.0515 ± 18	324 ± 11
16.2	248	271	1.09	221	26.98	0.0707 ± 25	2.619 ± 101	0.2686 ± 31	0.0516 ± 19	325 ± 11
16.3	465	628	1.35	293	20.53	0.0655 ± 23	1.959 ± 74	0.2170 ± 19	0.0521 ± 19	327 ± 11
16.4	295	354	1.20	2	0.28	0.0496 ± 18	0.373 ± 16	0.0546 ± 12	0.0494 ± 18	311 ± 11
17.1	94	103	1.10	4	1.85	0.0518 ± 19	0.479 ± 24	0.0671 ± 21	0.0508 ± 18	319 ± 11
18.1	74	84	1.14	3	1.41	0.0517 ± 19	0.453 ± 26	0.0637 ± 25	0.0509 ± 18	320 ± 11
18.2	106	141	1.33	2	0.73	0.0522 ± 19	0.419 ± 20	0.0582 ± 16	0.0518 ± 19	326 ± 12
19.1	161	186	1.16	4	1.06	0.0496 ± 18	0.416 ± 20	0.0608 ± 16	0.0491 ± 18	309 ± 11
20.1	157	167	1.06	7	1.84	0.0507 ± 18	0.469 ± 22	0.0671 ± 16	0.0498 ± 18	313 ± 11
21.1	75	44	0.59	4	2.21	0.0512 ± 18	0.494 ± 30	0.0700 ± 30	0.0501 ± 18	315 ± 11
22.1	112	108	0.96	5	1.94	0.0516 ± 19	0.483 ± 24	0.0679 ± 21	0.0506 ± 18	318 ± 11
23.1	54	88	1.63	3	2.39	0.0532 ± 19	0.524 ± 31	0.0715 ± 30	0.0519 ± 19	326 ± 12
24.1	123	180	1.47	3	1.13	0.0500 ± 18	0.424 ± 21	0.0614 ± 19	0.0495 ± 18	311 ± 11
25.1	143	149	1.04	2	0.52	0.0507 ± 18	0.395 ± 19	0.0565 ± 16	0.0505 ± 18	317 ± 11
26.1	59	73	1.25	2	1.58	0.0502 ± 18	0.450 ± 27	0.0650 ± 28	0.0494 ± 18	311 ± 11
27.2	43	55	1.27	2	2.03	0.0522 ± 20	0.522 ± 30	0.0686 ± 27	0.0541 ± 20	340 ± 12
29.1	78	74	0.95	5	2.80	0.0533 ± 19	0.549 ± 30	0.0747 ± 27	0.0518 ± 19	326 ± 11
29.2	73	103	1.41	2	1.07	0.0519 ± 19	0.436 ± 24	0.0609 ± 22	0.0514 ± 19	323 ± 12
30.1	225	330	1.47	3	0.66	0.0499 ± 18	0.397 ± 18	0.0577 ± 14	0.0496 ± 18	312 ± 11
32.1	252	709	2.81	3	0.55	0.0498 ± 18	0.390 ± 17	0.0567 ± 10	0.0495 ± 18	312 ± 11
33.1	57	70	1.24	4	3.17	0.0465 ± 17	0.498 ± 27	0.0777 ± 27	0.0450 ± 17	284 ± 10
34.1	119	124	1.04	4	1.30	0.0499 ± 18	0.432 ± 20	0.0628 ± 16	0.0492 ± 18	310 ± 11
35.1	134	105	0.78	3	0.96	0.0500 ± 18	0.413 ± 19	0.0600 ± 15	0.0495 ± 18	311 ± 11
36.1	145	110	0.75	2	0.71	0.0499 ± 18	0.399 ± 18	0.0580 ± 14	0.0496 ± 18	312 ± 11
37.1	63	83	1.33	2	1.44	0.0512 ± 19	0.451 ± 24	0.0638 ± 22	0.0505 ± 19	318 ± 11
38.1	83	123	1.48	2	0.75	0.0541 ± 20	0.436 ± 23	0.0584 ± 20	0.0537 ± 20	337 ± 12

* ^{207}Pb correction; † f^{206}Pb indicates the percentage of common ^{206}Pb in the total measured ^{206}Pb; ‡ apparent age is the ^{206}Pb/^{238}U age.

Bonaparte Basin, the present assemblage equates with the *Diatomozonotriletes birkheadensis* Zone, probably the uppermost part of that zone, in which case the age is likely to be late Westphalian.

Biostratigraphic control of successions overlying the Seaham Formation is available in the Cranky Corner Basin (Roberts *et al.* 1991*b*) and at Lochinvar. The Seaham Formation, as interpreted by Briggs & Archbold (1990), is confined

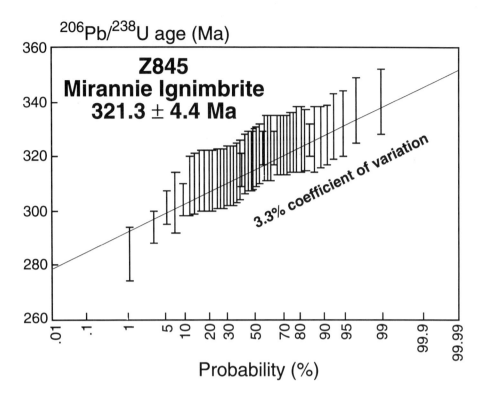

Fig. 7. Probability plot of zircon ages in the Mirannie Ignimbrite Member. Normal distribution is indicated by the diagonal line plotted from the concurrent reproducibility of the standard zircon. The data are not distinguishable beyond error from a normal distribution and the data group to give a simple homogeneous age.

to a level beneath the first influx of basaltic detritus and overlain, probably unconformably, by sediments of the Lochinvar Formation. The lower part of the Lochinvar Formation contains *Eurydesma playfordi* Dickins (D. Briggs, pers. comm.), *Trigonotreta* sp. nov. (Archbold & Dickins 1991) and palynofloras of the *Granulatisporites confluens* Zone (Foster 1993), identifying the unit clearly as Early Permian (Figs 3 & 13). Stage 3a palynofloras are present slightly higher in the Lochinvar Formation. *Eurydesma playfordi* is also present in the base of the Lochinvar Formation at Lochinvar, above sample Z1461.

Sills in Late Carboniferous formations

The Booral and lowermost McInnes Formations are part of the cold water marine facies containing the *Levipustula levis* Zone, the major Upper Carboniferous brachiopod zone of Gondwana. This zone is also recognized in Argentina, but lacks correlation to the northern hemisphere and to the Seaham Formation that defines the

Australian continental glacigene sequence. Zircon dates have been obtained for rhyolite units in three formations (the Booral and McInnes Formations and the Koolanock Sandstone) in an effort to constrain directly the age and correlation of the *L. levis* Zone, and these have indicated a surprisingly young (Permian) age for the upper limit of the Zone (Roberts *et al.* 1993*b*). Our subsequent field investigations have now shown that the dated units are thin sills rather than erupted units, so the young ages that have been reported do not constrain the biostratigraphy. Detailed information is given here to supersede the correlations proposed by Roberts *et al.* (1993*b*) on the basis that the rhyolites were volcanics. The misidentification of these rhyolitic sills as eruptions is a cautionary tale for dating studies of poorly exposed felsic igneous rocks.

Booral Formation

The Booral Formation is part of the eastern marine facies, and conformably overlies the Karuah and Faulkland Formations (both with

Marginirugus barringtonensis Zone brachiopod faunas). It contains faunas of the *L. levis* brachiopod Zone and passes upwards into the mainly continental McInnes Formation. The formation consists predominantly of grey marine siltstone and in places contains rhyolite horizons previously identified as flow units (Roberts *et al.* 1991*a, b*). Biostratigraphic constraint on the maximum age of the Booral Formation is provided by the *M. barringtonensis* brachiopod Zone in the underlying Karuah and Faulkland Formations. The conodont *Gnathodus bilineatus* (Roundy) is present within the *M. barringtonensis* Zone in Queensland (Jenkins *et al.* 1993; Roberts *et al.* 1993*c*) indicating a late Viséan age. An ammonoid from the *M. barringtonensis* Zone in the Yagon Siltstone that is referred to *Beyrichoceras bootibootiense* Campbell *et al.* (1983) may not belong to that genus (N. Riley, pers. comm.) and so cannot provide confirmation of the conodont age. The age of the uppermost Booral Formation is constrained by palynofloras and faunas in the overlying McInnes Formation described below. In the Berrico road section (Roberts *et al.* 1991*b*), the Booral Formation is underlain by unfossiliferous marginal-marine to non-marine sediments of the Faulkland Formation and overlain by the non-marine McInnes Formation. Two rhyolite units bracket the *L. levis* zone within this section, and were sampled as Z1489 from below the zone and Z784 from above. *L. levis* Zone faunas are present between 475 and 630 m above the base of the formation, the highest occurrence being *c.* 5 m below the Z784 rhyolite. Above this rhyolite there is 5 m of dark siltstone with sparse bivalves, another rhyolite (40 m thick) and *c.* 50 m of no outcrop beneath the overlying McInnes Formation. *Levipustula levis* Maxwell is present within the lower part of the overlying McInnes Formation at Booral, 34 km to the south. Rhyolite sample Z1489 from 180 m above the base of the Booral Formation, and below the first appearance of *L. levis*, has an age of 274.9 ± 3.8 Ma. Rhyolite sample Z784 from *c.* 80 m below the top of the formation was interpreted to have an age close to 286 Ma (Roberts *et al.* 1991*a*, 1993*b*), but re-analysis of the sample has indicated high incidences of both Pb loss and inherited zircons, and this frustrates confident calculation of an age.

McInnes Formation

The McInnes Formation conformably overlies the Booral Formation and is non-marine in all areas except near Booral township, where there is a gradational transition from the Booral into the McInnes Formation (Isaacs Formation of

Campbell 1961). Here, the lower part of the McInnes Formation is marine and contains a brachiopod fauna originally termed the *Syringothyris bifida* fauna (Campbell & Roberts 1969). Roberts *et al.* (1976) included this fauna within the *L. levis* Zone because it contains the nominate species. The assemblage is not known elsewhere and the restriction of nine invertebrate species or subspecies to this horizon may indicate that this is a younger part of the *L. levis* Zone. Reprocessing of palynofloral material from carbonaceous shales in the upper part of the McInnes Formation, near Stroud Road has clarified the previous assignment of *D. birkheadensis* Zone (Roberts *et al.* 1991*b*). The palynoflora contains the monosaccate pollen *Potonieisporites* sp. together with trilete spores *Reticulatisporites bifrons*, *D. birk-headensis*, *Verrucosisporites* spp. and lycopod spores. The assemblages are equated with those from the *S. ybertii–D. birkheadensis* Zones. In terms of ranges from the Galilee Basin they are assigned to the *D. birkheadensis* Zone [Zone C of Jones and Truswell (1992)]. In the Bonaparte Basin in western Australia the eponymous species first appears in the older (Viséan) *Grandispora maculosa* Assemblage but without associated monosaccate pollen. We interpret the palynoflora as indicating a Namurian–Westphalian age. The McInnes Formation is conformably overlain by the Johnsons Creek Conglomerate, which contains an 'enriched' *Nothorhacopteris* flora (Roberts *et al.* 1991*b*), followed disconformably by the Permian (Stage 3–4) Alum Mountain Volcanics (McMinn 1987; Roberts *et al.* 1991*b*). Zircons from a rhyolite (sample Z798) 325 m above the base of the McInnes Formation in the Monkerai South section (Roberts *et al.* 1991*b*, fig. 33) north of Stroud Road, yielded an age of 278.7 ± 3.4 Ma.

Koolanock Sandstone

The Koolanock Sandstone is usually conformable with the Yagon Siltstone and is overlain by the Muirs Creek Conglomerate. It consists of a coarsening-upwards succession of possibly estuarine to fluvial volcanogenic sandstone, with minor siltstone, 'rhyolite', conglomerate and carbonaceous shale. Biostratigraphically, the Koolanock Sandstone lies above the *L. levis* Zone of the underlying Yagon Siltstone, and below a poorly preserved palynoflora identified in this study as within the range *S. ybertii–D. birkheadensis* zones (Namurian–early Westphalian) in the overlying Muirs Creek Conglomerate. The second of three 'rhyolites' within the lower part of the Koolanock Sandstone (cf. Clarkes Road section in Roberts *et al.* 1991*b*, fig.

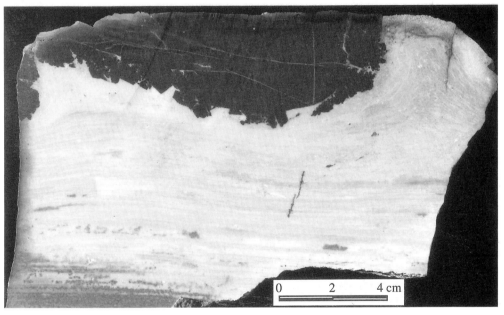

Fig. 8. Photographs of the upper contact of a 'rhyolite' in the Berrico Road section of the Booral Formation. Veining of the overlying siltstone, stoping of siltstone into the magma, and the warping of flow banding into cavities in the overlying contact all indicate that the unit is a sill, rather than an erupted unit as previously mapped.

43), was sampled as Z795. The sample was taken at the intersection of the Postmans Ridge and Koolanock Trails on the crest of Koolanock Range, 2 km north of the Clarkes Road section. The zircon age of sample Z795 is 276.3 ± 2.8 Ma, which is within error of the ages measured for rhyolites in the Booral and McInnes Formations.

Recognition as sills
The similarity of ages measured in the Booral and McInnes Formations and the Koolanock Sandstone poses a problem in interpreting the correlation of these units. The ages cluster in the range 279–275 Ma, close to the age of the Lakes Road Rhyolite Member of the Permian Alum Mountain Volcanics (described below). The

Table 3. *SHRIMP U–Pb isotopic data for a tuff in the Johnsons Creek Conglomerate at Stroud Road (Z1590)*

Grain area	$f^{206}Pb*†$ (%)	Calibrated total Pb compositions			Radiogenic compositions*	
		$^{206}Pb/^{238}U$ $\pm 1\sigma$	$^{207}Pb/^{235}U$ $\pm 1\sigma$	$^{207}Pb/^{206}Pb$ $\pm 1\sigma$	$^{206}Pb/^{238}U$ $\pm 1\sigma$	Apparent age‡ (Ma) $\pm 1\sigma$
1.1	0.44	0.0529 ± 12	0.413 ± 19	0.0566 ± 21	0.0527 ± 12	331 ± 7
2.1	1.54	0.0527 ± 12	0.475 ± 24	0.0654 ± 29	0.0519 ± 12	326 ± 7
3.1	1.04	0.0484 ± 11	0.410 ± 17	0.0614 ± 20	0.0479 ± 11	302 ± 6
4.1	0.51	0.0536 ± 12	0.422 ± 17	0.0571 ± 18	0.0534 ± 12	335 ± 7
5.1	0.64	0.0521 ± 11	0.418 ± 17	0.0582 ± 19	0.0518 ± 11	325 ± 7
6.1	1.50	0.0510 ± 11	0.458 ± 20	0.0651 ± 23	0.0503 ± 11	316 ± 7
7.1	0.52	0.0509 ± 11	0.402 ± 14	0.0572 ± 14	0.0507 ± 11	319 ± 7
8.1	0.47	0.0520 ± 11	0.407 ± 14	0.0568 ± 14	0.0517 ± 11	325 ± 7
9.1	0.33	0.0520 ± 11	0.399 ± 18	0.0557 ± 20	0.0518 ± 11	326 ± 7
10.1	0.71	0.0536 ± 12	0.434 ± 15	0.0588 ± 15	0.0532 ± 12	334 ± 7
11.1	1.18	0.0520 ± 12	0.449 ± 22	0.0625 ± 25	0.0514 ± 11	323 ± 7
12.1	0.75	0.0506 ± 11	0.412 ± 19	0.0590 ± 23	0.0502 ± 11	316 ± 7
13.1	0.50	0.0516 ± 11	0.405 ± 17	0.0570 ± 20	0.0513 ± 11	323 ± 7
14.1	1.08	0.0477 ± 10	0.406 ± 16	0.0617 ± 19	0.0472 ± 10	297 ± 6
15.1	0.61	0.0513 ± 11	0.410 ± 19	0.0579 ± 23	0.0510 ± 11	321 ± 7
16.1	0.46	0.0534 ± 12	0.417 ± 13	0.0567 ± 12	0.0531 ± 12	334 ± 7
17.1	0.63	0.0503 ± 11	0.403 ± 17	0.0581 ± 20	0.0500 ± 11	315 ± 7
18.1	1.03	0.0504 ± 11	0.426 ± 18	0.0613 ± 21	0.0499 ± 11	314 ± 7
19.1	5.98	0.0548 ± 12	0.764 ± 30	0.1011 ± 30	0.0516 ± 12	324 ± 7
20.1	0.18	0.0524 ± 12	0.394 ± 14	0.0545 ± 14	0.0524 ± 12	329 ± 7
21.1	0.36	0.0522 ± 11	0.402 ± 17	0.0559 ± 19	0.0520 ± 11	327 ± 7
22.1	0.18	0.0513 ± 11	0.365 ± 13	0.0516 ± 13	0.0514 ± 11	323 ± 7
23.1	1.50	0.0467 ± 10	0.419 ± 23	0.0650 ± 31	0.0460 ± 10	290 ± 6
24.1	0.41	0.0511 ± 11	0.397 ± 13	0.0564 ± 13	0.0508 ± 11	320 ± 7
25.1	0.34	0.0521 ± 11	0.401 ± 17	0.0558 ± 18	0.0520 ± 11	326 ± 7
26.1	0.16	00.522 ± 11	0.391 ± 18	0.0543 ± 20	0.0521 ± 12	327 ± 7
26.2	0.61	0.0507 ± 11	0.405 ± 20	0.0579 ± 23	0.0504 ± 11	317 ± 7
27.1	0.31	0.0502 ± 11	0.384 ± 18	0.0555 ± 22	0.0501 ± 11	315 ± 7
28.1	0.31	0.0513 ± 11	0.393 ± 21	0.0555 ± 26	0.0512 ± 11	322 ± 7
28.2	0.11	0.0521 ± 11	0.387 ± 14	0.0539 ± 14	0.0520 ± 11	327 ± 7
29.1	0.70	0.0503 ± 11	0.406 ± 15	0.0586 ± 16	0.0499 ± 11	314 ± 7
30.1	0.30	0.0520 ± 11	0.398 ± 15	0.0555 ± 16	0.0518 ± 11	326 ± 7
31.1	0.30	0.0510 ± 11	0.390 ± 17	0.0554 ± 19	0.0509 ± 11	320 ± 7
32.1	0.64	0.0519 ± 11	0.416 ± 22	0.0581 ± 26	0.0515 ± 12	324 ± 7
33.1	0.04	0.0522 ± 11	0.380 ± 12	0.0527 ± 11	0.0523 ± 11	328 ± 7
34.1	1.47	0.0561 ± 13	0.501 ± 34	0.0649 ± 40	0.0552 ± 13	347 ± 8
35.1	0.83	0.0524 ± 12	0.431 ± 19	0.0597 ± 22	0.0519 ± 12	326 ± 7
36.1	0.64	0.0482 ± 11	0.387 ± 18	0.0582 ± 22	0.0479 ± 11	302 ± 7

* ^{207}Pb correction; † $f^{206}Pb$ indicates the percentage of common ^{206}Pb in the total measured ^{206}Pb; ‡apparent age is the $^{206}Pb/^{238}U$ age.

rhyolites were dated on the basis of their identification as erupted flow units. However, the conflict between their ages and stratigraphic positions, and a previous report of sills north of this area (Campbell & McKelvey 1972) prompted us to re-examine the rhyolites to determine whether they are sills or flows. Outcrop in the area is restricted and the boundaries of the rhyolite units are not naturally exposed. However, we were able, by digging, to expose the upper boundary of one of the dated rhyolites in the Booral Formation on Berrico Trig Road, where it is in contact with marine siltstone. The contact is photographed in Fig. 8 and is clearly intrusive, with rhyolite penetrating the overlying siltstone along veins and fractures, igneous flow structures bending upwards into spaces within the lower surface of the siltstone, and angular blocks of siltstone stoped away from the contact to float within the magma. On this evidence we now interpret 'rhyolites' within Carboniferous formations in the northern Stroud–Gloucester and Myall Synclines as a suite of felsic sills intruded during a period of Permian volcanism. Their zircon ages, therefore, do not define the age of the associated sediments and faunas.

Table 4. *SHRIMP U–Pb isotopic data for a tuff in the Johnsons Creek Conglomerate at Johnson's Creek (Z1556)*

Grain area	$f^{206}Pb*$† (%)	Calibrated total Pb compositions			Radiogenic compositions*	
		$^{206}Pb/^{238}U$ $\pm 1\sigma$	$^{207}Pb/^{235}U$ $\pm 1\sigma$	$^{207}Pb/^{206}Pb$ $\pm 1\sigma$	$^{206}Pb/^{238}U$ $\pm 1\sigma$	Apparent age‡ (Ma) $\pm 1\sigma$
1.1	2.05	0.0504 ± 13	0.481 ± 31	0.0691 ± 39	0.0494 ± 13	311 ± 8
2.1	0.48	0.0517 ± 13	0.403 ± 15	0.0565 ± 13	0.0514 ± 13	323 ± 8
3.1	0.23	0.0542 ± 14	0.407 ± 15	0.0545 ± 13	0.0541 ± 14	339 ± 8
3.2	0.46	0.0532 ± 14	0.414 ± 14	0.0564 ± 11	0.0530 ± 13	333 ± 8
4.1	0.61	0.0528 ± 13	0.419 ± 17	0.0576 ± 17	0.0525 ± 13	330 ± 8
5.1	0.43	0.0513 ± 13	0.397 ± 14	0.0562 ± 13	0.0511 ± 13	321 ± 8
6.1	0.04	0.0505 ± 13	0.365 ± 16	0.0524 ± 18	0.0505 ± 13	318 ± 8
7.1	0.40	0.0499 ± 13	0.385 ± 14	0.0559 ± 12	0.0497 ± 13	313 ± 8
7.2	0.29	0.0498 ± 13	0.378 ± 14	0.0550 ± 12	0.0497 ± 13	313 ± 8
7.3	0.06	0.0509 ± 13	0.373 ± 12	0.0532 ± 10	0.0509 ± 13	320 ± 8
8.1	0.25	0.0575 ± 15	0.434 ± 15	0.0547 ± 12	0.0574 ± 15	360 ± 9
9.1	0.12	0.0535 ± 14	0.396 ± 12	0.0537 ± 8	0.0535 ± 14	336 ± 8
10.1	0.14	0.0487 ± 12	0.361 ± 16	0.0538 ± 17	0.0486 ± 12	306 ± 8
11.1	0.56	0.0539 ± 14	0.425 ± 14	0.0572 ± 11	0.0536 ± 14	336 ± 8
12.1	0.26	0.0506 ± 13	0.382 ± 16	0.0548 ± 16	0.0505 ± 13	317 ± 8
13.1	0.68	0.0504 ± 13	0.404 ± 18	0.0582 ± 20	0.0501 ± 13	315 ± 8
14.1	0.81	0.0497 ± 13	0.405 ± 24	0.0591 ± 30	0.0493 ± 13	310 ± 8
15.1	0.04	0.0539 ± 14	0.394 ± 14	0.0530 ± 12	0.0539 ± 14	339 ± 8
16.1	0.43	0.0552 ± 14	0.427 ± 17	0.0561 ± 16	0.0550 ± 14	345 ± 9
16.2	0.11	0.0531 ± 13	0.392 ± 15	0.0536 ± 13	0.0530 ± 13	333 ± 8
17.1	0.76	0.0529 ± 13	0.428 ± 20	0.0588 ± 21	0.0525 ± 13	330 ± 8
18.1	0.08	0.0491 ± 12	0.361 ± 12	0.0533 ± 9	0.0490 ± 12	309 ± 8
19.1	0.39	0.0531 ± 14	0.409 ± 17	0.0558 ± 17	0.0529 ± 14	333 ± 8
20.1	0.09	0.0529 ± 13	0.390 ± 14	0.0534 ± 12	0.0529 ± 13	332 ± 8
20.2	1.05	0.0491 ± 14	0.414 ± 34	0.0611 ± 45	0.0486 ± 14	306 ± 8
21.1	0.84	0.0508 ± 13	0.416 ± 17	0.0594 ± 16	0.0504 ± 13	317 ± 8
22.1	0.07	0.0516 ± 13	0.371 ± 15	0.0521 ± 14	0.0517 ± 13	325 ± 8
23.1	2.85	0.0495 ± 13	0.516 ± 35	0.0755 ± 45	0.0481 ± 13	303 ± 8
24.1	0.14	0.0539 ± 14	0.400 ± 16	0.0538 ± 14	0.0538 ± 14	338 ± 8
24.2	0.49	0.0544 ± 14	0.425 ± 18	0.0566 ± 18	0.0542 ± 14	340 ± 8
25.1	1.35	0.0471 ± 12	0.412 ± 12	0.0635 ± 8	0.0465 ± 12	293 ± 7
26.1	0.24	0.0497 ± 13	0.348 ± 14	0.0508 ± 14	0.0498 ± 13	313 ± 8
27.1	0.45	0.0520 ± 13	0.404 ± 15	0.0563 ± 14	0.0518 ± 13	325 ± 8
28.1	0.29	0.0528 ± 13	0.400 ± 15	0.0550 ± 14	0.0527 ± 13	331 ± 8
29.1	0.60	0.0521 ± 13	0.413 ± 16	0.0575 ± 16	0.0518 ± 13	325 ± 8
30.1	0.05	0.0495 ± 13	0.362 ± 11	0.0531 ± 9	0.0494 ± 13	311 ± 8
31.1	1.01	0.0647 ± 16	0.542 ± 22	0.0608 ± 18	0.0640 ± 16	400 ± 10
32.1	0.19	0.0535 ± 14	0.400 ± 16	0.0542 ± 15	0.0534 ± 14	336 ± 8
33.1	0.30	0.0513 ± 13	0.390 ± 13	0.0551 ± 10	0.0512 ± 13	322 ± 8
34.1	0.01	0.0495 ± 13	0.360 ± 13	0.0528 ± 12	0.0495 ± 13	311 ± 8
35.1	0.41	0.0501 ± 13	0.386 ± 13	0.0560 ± 10	0.0499 ± 13	314 ± 8
36.1	0.06	0.0519 ± 13	0.374 ± 12	0.0522 ± 9	0.0520 ± 13	327 ± 8

* ^{207}Pb correction; † $f^{206}Pb$ indicates the percentage of common ^{206}Pb in the total measured ^{206}Pb; ‡ apparent age is the $^{206}Pb/^{238}U$ age.

Johnsons Creek Conglomerate

An upper constraint to the age of the McInnes and Booral Formations, and therefore of the *L. levis* Zone, is provided by the Johnsons Creek Conglomerate, which is the highest Carboniferous unit in the Stroud–Gloucester Syncline. The conglomerate conformably overlies the McInnes Formation and is in turn disconformably overlain by the Permian Alum Mountain Volcanics.

Diamictite in some sections has been interpreted as glacial in origin (Roberts *et al.* 1991*b*), inferring correlation with the Seaham Formation, but the bulk of the formation consists of conglomerate and coarse lithic sandstone. Tuffs are present in some sections. The plant *Sphenopteridium* sp. from near the top of the Johnsons Creek Conglomerate is indicative of the 'enriched' *Nothorhacopteris* flora of Morris (1985), also present in the McInnes and upper Mt

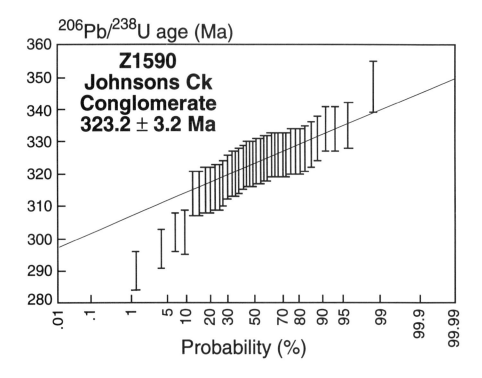

Fig. 9. Probability plot of zircon ages in the Z1590 sample of tuff from the Johnsons Creek Conglomerate. Most data are within error of the normal distribution defined from the concurrent reproducibility of the standard zircon, and group to give the mean age of the sample. Four grains have detectably young ages and are interpreted to have lost small amounts of radiogenic Pb. A single analysis has a slightly high apparent age and is interpreted as a xenocryst.

Johnstone Formations. The correlative Muirs Creek Conglomerate in the Myall Syncline contains a poorly preserved palynoflora equivalent to that in the McInnes Formation (i.e. within the range *S. ybertii–D. birkheadensis* Zones). The Johnsons Creek Conglomerate was previously considered to be latest Carboniferous in age on the basis of its plant fossils and stratigraphic position above the *L. levis* Zone (Roberts *et al.* 1991*b*).

Following the discovery that 'rhyolites' within the McInnes and Booral Formations are sills and not volcanics, two undoubtedly erupted pyroclastic tuffs in the Johnsons Creek Conglomerate were sampled for dating. Sample Z1556 is from 15 m from below the top of the formation at Johnsons Creek on the eastern limb of the Stroud–Gloucester Syncline; sample Z1590 is from *c.* 200 m below the top of the formation at the Stroud Road railway cutting on the western limb of the syncline (Roberts *et al.* 1991*b*, fig. 34). Zircon compositions for the two samples are listed in Tables 3 & 4, and radio-

genic $^{206}Pb/^{238}U$ ages of individual grains are plotted in probability diagrams in Figs 9 & 10. The samples contain simple groups of zircon compositions with indistinguishable ages. Sample Z1590 has a dominant population of zircons whose ages adhere to the Normal Distribution defined by concurrent analyses of the standard zircon. Four grains have distinctly young apparent ages that reflect minor Pb loss; one grain has an older age barely distinguishable from the main population and is identified as a xenocryst. The main population of analyses agree, within error, at an age of 323.2 ± 3.2 Ma (2σ), interpreted as the crystallization age of the tuff. Sample Z1556 also has a dominant population of zircon ages dispersed within error of the normal distribution; a single grain is detectably younger than the main group and is inferred to have lost a small amount of radiogenic Pb, and two zircons have distinctly old ages and are identified as xenocrysts. The remaining data agree within error at an age of 321.9 ± 4.0 Ma (2σ), which is interpreted as the crystallization

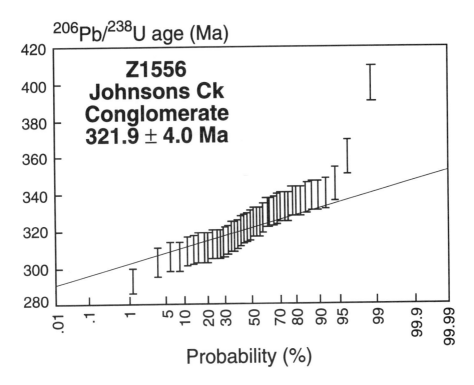

Fig. 10. Probability plot of zircon ages in the Z1556 tuff from the Johnsons Creek Conglomerate. As with the other Johnsons Creek sample, the majority of compositions adhere to the normal distribution defined from the concurrent reproducibility of the standard zircon, and these define the age of the sample. A single grain may have lost a small amount of radiogenic Pb and two xenocrysts have distinctly older ages.

age. (The apparent step in probability distribution within the main population of Z1556 zircons in Fig. 9 is interpreted as an artefact because it is not stepped beyond the error of the individual analyses.) By correlation to European $^{40}Ar/^{39}Ar$ ages, both of these volcanics are Namurian, which is consistent with the Namurian–Westphalian affinities of palynofloras from the underlying McInnes Formation and the correlative Muirs Creek Conglomerate.

Yagon Siltstone

The Yagon Siltstone in the Myall region at the eastern extremity of the Upper Carboniferous marine facies contains a thin unit of dacite, rhyodacite, tuffs and conglomerate named the Violet Hills Volcanic Member (Roberts *et al.* 1991*b*, fig. 41). The volcanics are restricted to the shallower western areas of outcrop. The Yagon Siltstone conformably overlies the Booti Booti Sandstone and is conformably overlain by the Koolanock Sandstone. Late Viséan *M. barring-*

tonensis Zone faunas within the Booti Booti Sandstone and lower Yagon Siltstone are followed by *L. levis* Zone faunas in the middle to upper part of the Yagon Siltstone. The last occurrence of *L. levis* is within the Violet Hill Volcanic Member at Violet Hill. An imprecise younger constraint on the *L. levis* Zone is provided by palynofloras assigned to the *S. ybertii–D. birkheadensis* Zones (Namurian–early Westphalian) in the Muirs Creek Conglomerate above the Koolanock Sandstone. An outcrop of the Violet Hills Volcanic Member on the Pacific Highway at Koolanock Range, *c.* 200 m below the base of the Koolanock Sandstone, was sampled as Z1046. Zircon compositions for this sample are listed in Table 5, and radiogenic $^{206}Pb/^{238}U$ ages of individual grains are plotted in a probability diagram in Fig. 11. No inherited grains are identified, nor is there any measurable loss of radiogenic Pb in any of the grains. The compositions adhere, within error, to the normal distribution calculated from concurrent analyses of the standard zircon, and define an age of

Table 5. *SHRIMP U–Pb isotopic data for the Violet Hill Volcanics*

Grain area	U (p.p.m.)	Th (p.p.m.)	Th/U	^{204}Pb* (p.p.b.)	f^{206}Pb*† (%)	Calibrated total Pb compositions			Radiogenic compositions*	
						^{206}Pb/^{238}U ±1σ	^{207}Pb/^{235}U ±1σ	^{207}Pb/^{206}Pb ±1σ	^{206}Pb/^{238}U ±1σ	Apparent age‡ (Ma) ±1σ
1.1	68	66	0.97	6	3.21	0.0567 ± 11	0.611 ± 28	0.0782 ± 31	0.0548 ± 11	344 ± 7
2.1	152	228	1.50	11	2.91	0.0556 ± 11	0.581 ± 22	0.0758 ± 23	0.0540 ± 10	339 ± 6
3.1	266	375	1.41	4	0.67	0.0505 ± 13	0.402 ± 16	0.0578 ± 16	0.0501 ± 13	315 ± 8
4.1	77	113	1.47	8	4.39	0.0519 ± 14	0.627 ± 33	0.0877 ± 36	0.0496 ± 14	312 ± 8
5.1	107	63	0.59	3	0.95	0.0568 ± 15	0.470 ± 24	0.0600 ± 24	0.0563 ± 15	353 ± 9
6.1	236	196	0.83	3	0.51	0.0551 ± 15	0.430 ± 17	0.0565 ± 16	0.0549 ± 15	344 ± 9
7.1	187	89	0.48	5	1.10	0.0551 ± 15	0.465 ± 20	0.0612 ± 18	0.0545 ± 15	342 ± 9
8.1	130	114	0.88	4	1.27	0.0519 ± 14	0.448 ± 21	0.0626 ± 22	0.0512 ± 14	322 ± 8
9.1	150	185	1.23	4	1.12	0.0533 ± 14	0.451 ± 20	0.0614 ± 20	0.0527 ± 14	331 ± 9
10.1	99	91	0.92	3	1.30	0.0524 ± 14	0.454 ± 24	0.0629 ± 26	0.0517 ± 14	325 ± 9
11.1	223	213	0.96	2	0.37	0.0558 ± 15	0.427 ± 17	0.0554 ± 15	0.0556 ± 15	349 ± 9
12.1	314	332	1.06	− 1	− 0.16	0.0526 ± 14	0.371 ± 14	0.0512 ± 13	0.0527 ± 14	331 ± 9
13.1	146	64	0.44	3	0.78	0.0535 ± 14	0.433 ± 20	0.0587 ± 21	0.0531 ± 14	333 ± 9
14.1	208	186	0.89	7	1.43	0.0521 ± 16	0.461 ± 21	0.0641 ± 18	0.0514 ± 16	323 ± 10
15.1	446	242	0.54	6	0.56	0.0523 ± 16	0.412 ± 16	0.0571 ± 12	0.0520 ± 16	327 ± 10
16.1	186	149	0.80	4	0.80	0.0530 ± 16	0.431 ± 20	0.0590 ± 18	0.0526 ± 16	331 ± 10
17.1	340	319	0.94	9	1.09	0.0506 ± 16	0.427 ± 17	0.0613 ± 14	0.0500 ± 15	315 ± 9
18.1	467	351	0.75	4	0.30	0.0530 ± 16	0.402 ± 16	0.0550 ± 11	0.0529 ± 16	332 ± 10
19.1	389	426	1.09	17	1.74	0.0545 ± 17	0.500 ± 20	0.0665 ± 14	0.0536 ± 17	336 ± 10
20.1	434	286	0.66	2	0.22	0.0533 ± 16	0.399 ± 16	0.0543 ± 11	0.0531 ± 16	334 ± 10
21.1	330	283	0.86	8	1.01	0.0519 ± 16	0.435 ± 18	0.0607 ± 14	0.0514 ± 16	323 ± 10
22.1	141	374	2.66	9	2.71	0.0520 ± 16	0.533 ± 25	0.0743 ± 24	0.0506 ± 16	318 ± 10
23.1	628	357	0.57	1	0.08	0.0539 ± 17	0.395 ± 15	0.0532 ± 9	0.0538 ± 17	338 ± 10
24.1	487	323	0.66	11	0.93	0.0529 ± 16	0.438 ± 17	0.0600 ± 12	0.0524 ± 16	329 ± 10
25.1	188	165	0.88	2	0.43	0.0525 ± 16	0.405 ± 20	0.0560 ± 20	0.0522 ± 16	328 ± 10
26.1	601	802	1.33	12	0.78	0.0532 ± 16	0.432 ± 17	0.0588 ± 13	0.0528 ± 16	332 ± 10
27.1	471	335	0.71	10	0.85	0.0512 ± 16	0.419 ± 17	0.0594 ± 13	0.0508 ± 16	319 ± 10
28.1	486	469	0.97	37	3.10	0.0518 ± 16	0.553 ± 23	0.0774 ± 19	0.0502 ± 16	316 ± 10

* ^{207}Pb correction; † f^{206}Pb indicates the percentage of common ^{206}Pb in the total measured ^{206}Pb; ‡ apparent age is the ^{206}Pb/^{238}U age.

330.6 ± 4.6 Ma (2σ). The relative imprecision of this mean is a function of the relatively small number of analyses that contribute to the age. Taking into account previously determined SHRIMP ages from the Early Carboniferous (Roberts *et al.* 1995), the age of the Violet Hills Volcanic Member is late Viséan (V3a). In biostratigraphic terms this age equates with the *R. fortimuscula* Zone, the zone characterized by the last of the Carboniferous warm water assemblages in NSW, and which is succeeded by the cold climate *M. barringtonensis* and *L. levis* assemblages. The latter two zones are present within a continuous succession beneath the Violet Hills Volcanic Member (Roberts *et al.* 1991*b*), with *L. levis* Zone brachiopods occurring in siltstone interbedded with volcanogenic units of the member. We are therefore forced to interpret the age of the Violet Hills Volcanic Member in terms of its upper (i.e. younger) limit of error.

Alum Mountain Volcanics

The Alum Mountain Volcanics, the earliest Permian rocks within the Hunter–Myall region, disconformably overlie Carboniferous units and are characterized by interbedded sediments containing the Gondwanan plants *Glossopteris* and *Gangamopteris*. Within the Myall Syncline the volcanics disconformably overlie the Muirs Creek Conglomerate. Previous mapping (Crane & Hunt 1980; Roberts *et al.* 1991*b*) suggested that the volcanics contained two members, the Burdekins Gap Basalt (base) and Lakes Road Rhyolite Members and were overlain by the Markwell Coal Measures. Recent drilling near Bulahdelah (Jenkins & Nethery 1992) exposed a lowermost rhyolite member (Sams Road Rhyolite Member) beneath the basalt, and coal seams below, within and above the volcanics. The Markwell Coal Measures are now regarded as a lateral equivalent of the Alum Mountain Volca-

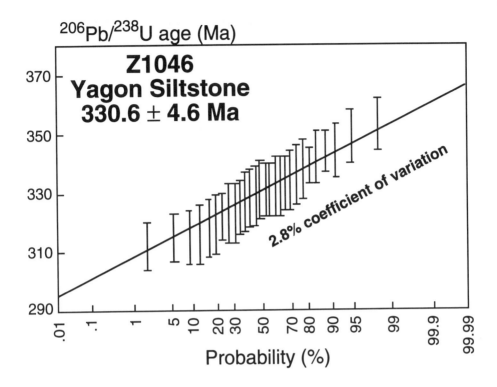

Fig. 11. Probability plot of zircon ages in the Violet Hills Volcanics. The data form a simple pattern adhering closely to the normal distribution line.

nics, a view supported by the occurrence of Stage 3a palynofloras within those measures (McMinn 1985). Jenkins & Nethery (1992) also reported coal seams interbedded with conglomerate below the Sams Road Rhyolite Member in two drill holes through basal parts of the Alum Mountain Volcanics on the southeastern limb of the Myall Syncline. Samples from these drill holes were barren of palynomorphs (Foster, unpublished data).

Stage 3a (*?Granulatisporites confluens* Zone) palynofloras are also present at the base of the Alum Mountain Volcanics in the Stroud–Gloucester Syncline near Stroud Road (Roberts *et al.* 1991*b*). A younger palynoflora, interpreted as Stage 4, was recorded from the Alum Mountain Volcanics by J. P. F. Hennelly in 1961 in an unpublished report to Australian Oil and Gas Corporation Limited. Replotting Hennelly's locality data indicates that the sample was taken from around the boundary between the Burdekins Gap Basalt and Lakes Road Rhyolite Members. According to Hennelly's data, *Granulatisporites trisinus* Balme & Hen-

nelly accounts for 35% of the total palynoflora; the occurrence of this species indicates that the sample cannot be older than Stage 3b. Also recorded, and accounting for 5% of the assemblage, is the distinctive trilete spore *Camptotriletes biornatus* which is characteristic species of the Greta Coal Measures (Balme & Hennelly 1956). Hennelly's locality needs to be relocated and resampled, but our present interpretation is that as a correlative of the Greta assemblages, this sample belongs to Stage 4. Hence, the Alum Mountain Volcanics are regarded as ranging in age from Stage 3 into Stage 4. A bleached weathering surface exposed by recent excavations on the upper surface of the Burdekins Gap Basalt Member at Burdekins Gap on the Pacific Highway suggests that extrusion of volcanics was not continuous throughout this interval.

Units beneath the Alum Mountain Volcanics contain both macrofloras and palynofloras. In the Myall Syncline the Muirs Creek Conglomerate contains a poorly preserved palynoflora that is equated with the *S. ybertii–D. birkheadensis* Zone (Namurian–early Westphalian) of the

Table 6. *SHRIMP U–Pb isotopic data for the Lakes Road Rhyolite*

Grain area	U (p.p.m.)	Th (p.p.m.)	Th/U	^{204}Pb* (p.p.b.)	$f^{206}Pb$*† (%)	Calibrated total Pb compositions			Radiogenic compositions*	
						$^{206}Pb/^{238}U$ ±1σ	$^{207}Pb/^{235}U$ ±1σ	$^{207}Pb/^{206}Pb$ ±1σ	$^{206}Pb/^{238}U$ ±1σ	Apparent age‡ (Ma)±1σ
1.1	85	40	0.47	5	3.26	0.0406 ± 8	0.436 ± 18	0.0778 ± 28	0.0393 ± 8	248 ± 5
2.1	86	40	0.47	4	2.53	0.0416 ± 8	0.413 ± 17	0.0720 ± 25	0.0406 ± 8	256 ± 5
3.1	138	87	0.63	3	1.23	0.0435 ± 8	0.369 ± 14	0.0616 ± 18	0.0429 ± 8	271 ± 5
4.1	78	38	0.49	4	2.18	0.0455 ± 9	0.434 ± 19	0.0692 ± 26	0.0445 ± 9	281 ± 5
5.1	83	41	0.50	3	1.97	0.0422 ± 8	0.393 ± 17	0.0675 ± 25	0.0414 ± 8	261 ± 5
6.1	82	46	0.56	2	1.01	0.0440 ± 9	0.363 ± 16	0.0598 ± 23	0.0436 ± 9	275 ± 5
7.1	59	22	0.37	3	2.44	0.0450 ± 9	0.442 ± 21	0.0713 ± 30	0.0439 ± 9	277 ± 5
8.1	98	57	0.58	1	0.72	0.0440 ± 8	0.349 ± 15	0.0575 ± 21	0.0437 ± 9	276 ± 5
9.1	86	43	0.50	2	1.11	0.0435 ± 8	0.364 ± 17	0.0606 ± 24	0.0430 ± 8	272 ± 5
10.1	87	42	0.49	2	1.37	0.0431 ± 8	0.372 ± 16	0.0627 ± 24	0.0425 ± 8	268 ± 5
11.1	89	48	0.54	3	1.51	0.0447 ± 9	0.394 ± 17	0.0639 ± 23	0.0441 ± 9	278 ± 5
12.1	59	23	0.40	2	1.86	0.0440 ± 9	0.404 ± 21	0.0667 ± 31	0.0432 ± 9	272 ± 5
13.1	81	41	0.50	2	1.43	0.0415 ± 8	0.361 ± 17	0.0632 ± 26	0.0409 ± 8	258 ± 5
14.1	69	33	0.47	2	1.78	0.0433 ± 8	0.394 ± 19	0.0661 ± 29	0.0425 ± 8	268 ± 5
15.1	76	38	0.50	1	0.93	0.0439 ± 9	0.358 ± 17	0.0592 ± 25	0.0435 ± 9	274 ± 5
16.1	99	64	0.65	3	1.40	0.0443 ± 9	0.384 ± 17	0.0630 ± 23	0.0437 ± 9	275 ± 5
17.1	86	42	0.48	3	1.50	0.0438 ± 8	0.385 ± 17	0.0638 ± 25	0.0432 ± 8	272 ± 5
18.1	96	56	0.58	13	6.28	0.0476 ± 9	0.669 ± 24	0.1020 ± 30	0.0446 ± 9	281 ± 5
19.1	88	51	0.58	2	1.21	0.0461 ± 9	0.391 ± 17	0.0615 ± 23	0.0456 ± 9	287 ± 5

* ^{207}Pb correction; † $f^{206}Pb$ indicates the percentage of common ^{206}Pb in the total measured ^{206}Pb; ‡ apparent age is the $^{206}Pb/^{238}U$ age.

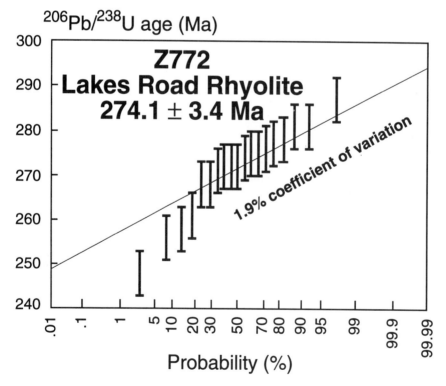

Fig. 12. Probability plot of zircon ages in the Lakes Road Rhyolite Member of the Alum Mountain Volcanics. The data form a simple pattern adhering closely to the normal distribution line. The few grains that drop below the line have detectably lost radiogenic Pb relative to the main population.

McInnes Formation [earlier preparations of this assemblage were identified as not older than *G. maculosa* Assemblage by Helby (in Roberts *et al.* 1991*b*)]; and in the Stroud–Gloucester Syncline the Johnsons Creek Conglomerate has an 'enriched' *Nothorhacopteris* flora which (from zircon ages in this paper) ranges in age from Late Viséan to early Namurian. Biostratigraphic control above the volcanics is limited to the Stroud–Gloucester Syncline where the Alum Mountain Volcanics are disconformably overlain by the marginal marine Duralie Road Formation (Dewrang Group) which contains palynofloras of upper Stage 4b–lower Stage 5a (Lennox 1991). The possibly equivalent Bulahdelah Formation in the Myall Syncline (Lennox 1991), which probably disconformably overlies the volcanics and coal-bearing sediments, contains marine faunas that are too poorly known to be correlated with those of other Permian units (Engel 1962).

The Lakes Road Rhyolite Member of the Alum Mountain Volcanics was sampled as Z772 at the top of the type section of the Alum Mountain Volcanics at Burdekins Gap on the Pacific Highway. The sample was collected from near the base of the rhyolite, 500 m above the Muirs Creek Conglomerate and 85 m below the base of overlying sediments previously identified as Markwell Coal Measures (Roberts *et al.* 1991*b*, fig. 44). Jenkins & Nethery (1992) have utilized complex flow folding textures and the thin but extensive distribution along the syncline to interpret the Lakes Road Rhyolite as an ignimbritic eruption; we agree with this assessment. Zircon compositions for this sample are listed in Table 6, and radiogenic $^{206}Pb/^{238}U$ ages of individual grains are plotted in a probability diagram in Fig. 12. Although a relatively small number of analyses was obtained, the relatively good precision (better than 2%) obtained for concurrent analyses of the standard zircon enables an age to be calculated with reasonable precision. Despite minor Pb loss in the sample, the major population of analyses agree at a mean age of 274.1 ± 3.4 Ma (2σ).

Correlations in the Late Carboniferous

A new correlation chart for the Late Carboniferous and Permian sediments of eastern Australia is constructed in Fig. 13. The Carboniferous biostratigraphic columns are based on correlations documented by Jones (1991) and Roberts *et al.* (1993*b*). European stages are calibrated to the $^{40}Ar/^{39}Ar$ ages of Lippolt & Hess (1983) and Hess & Lippolt (1986); Australian levels are calibrated to our zircon U–Pb ages. The Permian biostratigraphic columns follow Kemp *et al.* (1977) and Briggs (1993): relationships with the Russian stages are tenuous and based upon poorly documented correlations with Western Australia where sparse ammonoid faunas provide a measure of correlation with Russia (Archbold & Dickins 1991); radiometric age control is lacking. The lithological columns illustrate the palaeontological and isotopic evidence for both local and intercontinental correlation. Zircon ages in the Carboniferous part of the chart correlate for the first time the separately outcropping, and biostratigraphically distinct, continental and marine successions within eastern Australia, and provide correlation for both of these facies with the $^{40}Ar/^{39}Ar$-calibrated European stages.

Marine biozones

The *L. levis* brachiopod Zone of Australia and South America is the most significant marine biozone of the Carboniferous in Gondwanan continents. With one possible exception, it is absent from northern hemisphere continents and can be correlated only by the occurrence of the early Namurian ammonoid *Cravenoceras kullatinense* Campbell between two horizons with *L. levis* in the Hastings terrane, east of the accretionary prism of the SNEO. Tentative affinities between the *L. levis* Zone and a fauna in the Baikal region was recognized by Roberts *et al.* (1976), with at least five closely comparable species occurring in the Gondwanan and Baikal faunas. One of these, the productid originally identified as *Levipustula*, has since been assigned to a new genus *Lanipustula* (Klets 1983), but the difference between the two genera is minor. The first appearance of the *L. levis* Zone in the Booral Formation and Yagon Siltstone of the SNEO is within successions continuous with the underlying late Viséan *M. barringtonensis* Zone; conodonts described by Jenkins *et al.* (1993) correlate the *M. barringtonensis* Zone with most of the Asbian, the entire Brigantian and the lowermost Pendleian Stages of Britain.

Zircon ages reported here from the Johnsons Creek Conglomerate now effectively confine the *L. levis* Zone to the Namurian within the SNEO. In the Yarrol Syncline, Queensland, the *M. barringtonensis* Zone in the Baywulla Formation is accompanied by the first appearance of the conodont *Gnathodus bilineatus* (Jenkins *et al.* 1993) and is followed by the *L. levis* Zone in both the Branch Creek and Poperima Formations. The upper limit of the *L. levis* Zone is taken at the top of the Poperima Formation (Roberts *et al.* 1976), though a few remnant

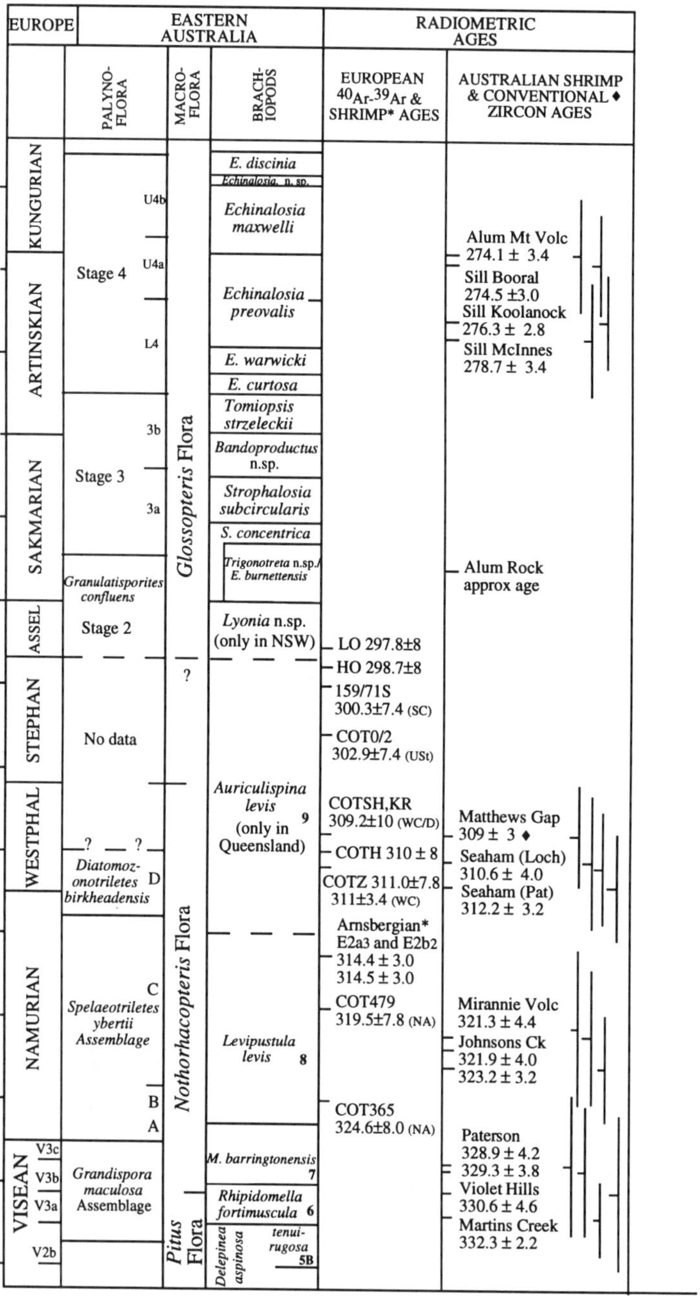
EUROPE
EASTERN AUSTRALIA
RADIOMETRIC AGES
PALYNO-FLORA
MACRO-FLORA
BRACH-IOPODS
EUROPEAN 40Ar-39Ar & SHRIMP* AGES
AUSTRALIAN SHRIMP & CONVENTIONAL ♦ ZIRCON AGES
270
275
280
285
290
295
300
305
310
315
320
325
330
335
KUNGURIAN
ARTINSKIAN
SAKMARIAN
ASSEL
STEPHAN
WESTPHAL
NAMURIAN
VISEAN
V3c
V3b
V3a
V2b
U4b
U4a
Stage 4
L4
3b
Stage 3
3a
Granulatisporites confluens
Stage 2
No data
? ?
Diatomozonotriletes birkheadensis
D
C
Spelaeotriletes ybertii Assemblage
B
A
Grandispora maculosa Assemblage
Glossopteris Flora
?
Nothorhacopteris Flora
Pitus Flora
E. discinia
Echinalosia. n. sp.
Echinalosia maxwelli
Echinalosia preovalis
E. warwicki
E. curtosa
Tomiopsis strzeleckii
Bandoproductus n.sp.
Strophalosia subcircularis
S. concentrica
Trigonotreta n.sp./ E. burnettensis
Lyonia n.sp. (only in NSW)
Auriculispina levis (only in Queensland)
9
Levipustula levis
8
M. barringtonensis
7
Rhipidomella fortimuscula
6
Delepinea aspinosa
tenuirugosa
5B
LO 297.8±8
HO 298.7±8
159/71S 300.3±7.4 (SC)
COT0/2 302.9±7.4 (USt)
COTSH,KR 309.2±10 (WC/D)
COTH 310 ± 8
COTZ 311.0±7.8
311±3.4 (WC)
Arnsbergian* E2a3 and E2b2
314.4 ± 3.0
314.5 ± 3.0
COT479 319.5±7.8 (NA)
COT365 324.6±8.0 (NA)
Alum Mt Volc 274.1 ± 3.4
Sill Booral 274.5 ±3.0
Sill Koolanock 276.3 ± 2.8
Sill McInnes 278.7 ± 3.4
Alum Rock approx age
Matthews Gap 309 ± 3 ♦
Seaham (Loch) 310.6 ± 4.0
Seaham (Pat) 312.2 ± 3.2
Mirannie Volc 321.3 ± 4.4
Johnsons Ck 321.9 ± 4.0
323.2 ± 3.2
Paterson 328.9 ± 4.2
329.3 ± 3.8
Violet Hills 330.6 ± 4.6
Martins Creek 332.3 ± 2.2

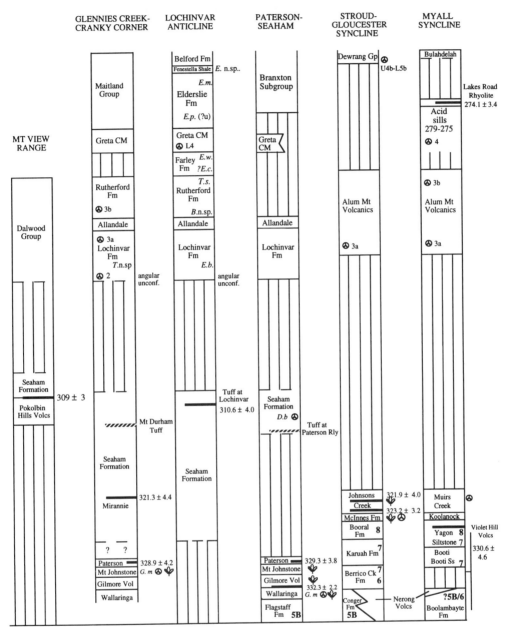

Fig. 13. Stratigraphic columns of Late Carboniferous and Permian sequences from the Southern New England Orogen and Sydney Basin plotted against biozones and SHRIMP ages from eastern Australia (this paper and Roberts *et al.* 1995). The corresponding European stages are scaled according to the $^{40}Ar/^{39}Ar$ ages of Lippolt & Hess (1983) and Hess & Lippolt (1986) from Western Europe (uncertainty values recalculated by Claoué-Long to give comparability with SHRIMP ages and our zircon age for the British Arnsbergian (Riley *et al.* 1995). Error bars are provided for the Australian SHRIMP ages. Dark lines indicate the positions of dated horizons in the Australian stratigraphy. Leaf symbols indicate significant plant localities; circles with tristar qualified by an abbreviation or number, palynological zones; and numbers (6–9) or abbreviations within columns brachiopod zones. Breaks in deposition are indicated by vertical lines. The Paterson Volcanic ages plotted are for the individual samples in the given sections, the mean measured age for this marker horizon is 328 ± 1.4 Ma. Within the Viséan the Belgian chronostratigraphic notion is used rather than the formal stage nomenclature (see Conil *et al.* 1990; V2a + V2b = Livian; V3b + V3c = Warnantian).

species persist through most of the sparsely fossiliferous and thick (> 1000 m) Rands Formation (*sensu* Dear *et al.* 1971), and is followed in the upper part of that unit by the *Auriculispina levis* Zone. The Rands Formation lacks volcanics and cannot be dated, but it is accompanied by Carboniferous rather than Permian bryozoans (Engel 1989).

The base of the L. levis Zone is constrained by evidence from Carrowbrook (Roberts *et al.* 1991*b*), where the *M. barringtonensis* Zone and at least part of the preceding *R. fortimuscula* Zone are underlain by the Martins Creek Ignimbrite Member, whose zircon U–Pb age is 332.3 ± 2.2 Ma (Roberts *et al.* 1995), and succeeded by the Paterson Volcanics (328.5 ± 1.4 Ma; Claoué-Long *et al.* 1995). The outcrops at Carrowbrook represent a marine tongue of Chichester Formation, with the *R. fortimuscula* Zone, intercalated within the non-marine Isismurra Formation. In marine regions of deposition north of Dungog and within the Stroud–Gloucester Syncline the *R. fortimuscula* Zone is followed by the *M. barringtonensis* and *L. levis* Zones (Roberts *et al.* 1991*b*). The *M. barringtonensis* Zone is no older than Asbian in age, as indicated by the first occurrence of the conodont *Gnathodus bilineatus* at the base of the *G. texanus–G. bilineatus* Zone in Queensland (Jenkins *et al.* 1993). Therefore, the *M. barringtonensis* Zone probably ranges in age between *c.* 330 and 326 Ma. The succeeding *L. levis* Zone must begin close to 326 Ma.

Close similarities exist with Argentina (Gonzalez 1990). Taboada's (1989) *Rugosochonetes–Bulahdelia* Zone contains species in common with, or closely comparable to, species in the *R. fortimuscula* and *M. barringtonensis* Zones of Australia, and is associated with the second Argentinian glacial episode of the El Paso Formation. The *L. levis* Zone is present in lower to middle parts of the overlying glacigene Hoyada Verde Formation. The Hoyada Verde glacial events can now be correlated as approximately synchronous with the onset of cold climate conditions in eastern Australia.

Palyno- and macrofloras of the continental succession

Zircon ages in non-marine parts of the SNEO succession have a significant effect on interpretation of the age and ranges of Carboniferous floras, particularly in the upper part of the succession.

Kemp *et al.* (1977) recognized two Carboniferous microfloras throughout Australia, the Early Carboniferous *Granulatisporites frustulentus* Microflora, and the *Secarisporites* Microflora, containing the *Grandispora maculosa, Spelaeotriletes ybertii* and *Potonieisporites* Assemblages, which ranged from the late Viséan to the latest Carboniferous. Powis (1984) subsequently proposed an additional *Diatomozonotriletes birkheadensis* Assemblage in the Galilee and Canning Basins which Foster & Waterhouse (1988) placed between the *S. ybertii* and *Granulatisporites confluens* Zones; the latter is in part equivalent to palynofloras of Stage 2 of Evans (1969). Stage 1 palynofloras were considered by Foster & Waterhouse (1988) to be facies controlled and of little time significance. Jones & Truswell (1992) recognized four Late Carboniferous palynofloral zones (A–D) in the Galilee Basin of Queensland that are approximately equivalent to the *S. ybertii* and *Potonieisporites* Assemblages of Kemp *et al.* (1977). As the first appearance of monosaccate pollen in the earliest Namurian (Clayton *et al.* 1990) also defines the base of the *S. ybertii* Assemblage, it follows that the base of the oldest biozone (Zone A) of Jones & Truswell (1992) is also earliest Namurian. The youngest zone (Zone D) of Jones & Truswell (1992) may be as young as Westphalian C, given the similarities between the palynofloras of Zone D and those in the Seaham Formation which from zircon data are unlikely to be younger than *c.* 309 or 310 Ma (Fig. 13). The palynoflora within the upper McInnes Formation is assigned to the *D. birkheadensis* Zone [Zone C of Jones & Truswell (1992)], interpreted as Namurian in age, and is constrained by zircon dates in the overlying Johnsons Creek Conglomerate to be older than 323 Ma. That within the Seaham Formation assigned to Zone D of Jones & Truswell (1992) or uppermost *D. birkheadensis* Zone, is likely to be older than 310 Ma, the age of zircons in the upper part of the Seaham Formation at Lochinvar.

Palynofloras within the Alum Mountain Volcanics range from Stage 3a (?*G. confluens* Zone) at the base to Stage 4. Similar palynofloral successions are present in the Bonaparte Basin, where Stage 3 palynofloras immediately follow the *G. confluens* Zone (Foster & Waterhouse 1988), and in the Early Permian sequence of the Cranky Corner Basin where the *G. confluens* Zone is associated with the *Trigonotreta* n. sp./*Eurydesma playfordi* Zone (D. J. C. Briggs, pers. comm.; Foster 1993). The Alum Mountain Volcanics had previously been considered to mark the base of the Permian System in eastern Australia on the basis of associated Stage 3 palynofloras (Roberts *et al.* 1993*b*).

However, our zircon date of 274 Ma for the Lakes Road Rhyolite Member, the uppermost unit within the volcanics, is too young to be basal Permian by correlation to the European ages. Our zircon ages (to be published elsewhere) date the *Trigonotreta* n. sp. and *Strophalosia subcircularis* Zones in southern Queensland and confine the base of Stage 3 to *c.* 293 Ma, which is above the base of the Permian as dated in Europe. The dated unit of the Alum Mountain Volcanics overlies a bleached contemporaneous weathering surface that may indicate an hiatus between eruptions within the volcanic sequence, with the Lakes Road Rhyolite being erupted during Stage 4 time, somewhat later than the underlying volcanics.

The *Nothorhacopteris* flora, once taken as indicative of the Late Carboniferous in eastern Australia, is present within the Mt Johnstone, Italia Road, Seaham, Booral and McInnes Formations, Johnsons Creek Conglomerate and Koolanock Sandstone of the SNEO, and the Majors Creek and Mingaletta Formations in the Hastings terrane. Zircon ages now constrain these formations to the interval late Viséan–Namurian, giving the *Nothorhacopteris* flora a range of *c.* 20 Ma, far less than suggested by Roberts *et al.* (1991*a*, 1993*b*). Morris's (1985) 'enriched' portion of the flora in the upper Mt Johnstone and McInnes Formations and the Johnsons Creek Conglomerate is confined within the lower half rather than the upper part of the range of *Nothorhacopteris* and has an age of late Viséan to Namurian or *c.* 332–320 Ma.

Regional geological correlations

A major conclusion of this study is that Carboniferous cold climate sequences in the SNEO are much older than previously recognized. Prior to this work the glacigene Seaham Formation was identified as Stephanian and taken to indicate that glaciation in eastern Australia commenced in the latest Carboniferous. On the basis of correlation to $^{40}Ar/^{39}Ar$ ages in Europe, the Australian glacial period is now dated as early Namurian at its base and extends into the Westphalian. This brings the Australian record more closely into line with events in Argentina where there are late Viséan and probably Namurian–Westphalian glacial episodes (Gonzalez 1990). A Late Carboniferous interglacial period in Argentina, before the major Permian glaciation, cannot be recognized in the SNEO because of a major hiatus between the Carboniferous and Permian successions.

Also, by correlation to the $^{40}Ar/^{39}Ar$ dates of European stages, the corresponding marine facies to the east, above the Viséan *Rhipidomella fortimuscula* Zone, is confined to the period late Viséan to early Namurian and is much older than indicated in previous palaeontological and stratigraphic studies (Roberts *et al.* 1985, 1991*b*, 1993*b*). In particular, marine sediments containing the low diversity *L. levis* marine Zone are shown to be equivalent to parts of the continental glacigene facies, a correspondence supported by the existence of glendonites, which are a cold climate indicator (Roberts *et al.* 1991*b*).

The new dates also indicate a major hiatus of at least 15 Ma duration between the marine and continental Late Carboniferous facies of the SNEO and Permian sediments of the Sydney Basin. This depositional break is represented by a slight angular unconformity near the Lochinvar Anticline in the Hunter region and probably reflects the first major deformation within the accretionary prism of the SNEO. According to Dirks *et al.* (1992) an eastward shift of the subduction zone forming the SNEO halted volcanism and caused the former accretionary prism to become located in a back-arc position, uplifted rapidly by compression and overthrusting and elevated in temperature. An episode of rifting in the earliest Permian formed the basins in the Manning and Hastings regions immediately before a more widespread extensional event that initiated Permian deposition in the Sydney–Bowen Basin. On our timescale these events took place between *c.* 300 and 295 Ma. The newly recognized hiatus coincides with widespread uplift throughout cratonic regions of Australia during the Late Carboniferous (Veevers & Powell 1990). Subsequent Early Permian extension caused the rejuvenation of older basins and the development of a new generation of Australian sedimentary basins (Roberts *et al.* 1985).

We thank A. T. Brakel for guidance in the field when sampling the Seaham Formation, D. Briggs for advice on Permian biostratigraphy and R. B. Jenkins for advice on the Alum Mountain Volcanics at Bulahdelah. This study was initiated by an Australian Research Council grant to JR and a Bureau of Mineral Resources Fellowship to JCL who conducted the work as a Fellow at the Research School of Earth Sciences, Australian National University. L. P. Black and G. C. Young are thanked for reviewing an early draft, and the expert assistance of S. Maxwell and D. Maidment is gratefully acknowledged. The contributions of JCL, PJJ and CBF are published with the permission of the Executive Director of the Australian Geological Survey Organisation.

References

AITCHISON, J. C., FLOOD, P. G. & SPILLER, F. C. P. 1992. Tectonic setting and paleoenvironment of terranes in the southern New England orogen, eastern Australia as constrained by radiolarian biostratigraphy. *Palaeogeography, Palaeoclimatology, Palaeoecology*, **941**, 31–54.

ARCHBOLD, N. W. & DICKINS, J. M. 1991. Australian Phanerozoic Timescales 6. Permian. A standard for the Permian System in Australia. *Bureau of Mineral Resources, Australia, Record*, 1989.

BALME, B. E. & HENNELLY, J. F. P. 1956. Trilete sporomorphs from Australian Permian sediments. *Australian Journal of Botany*, **4**, 240–260.

BRAKEL, A. T. 1972. The geology of the Mt View Range district, Pokolbin, NSW. *Journal and Proceedings of the Royal Society of New South Wales*, **105**, 61–70.

BRIGGS, D. J. C. 1993. Time control in the Permian of the Sydney–Bowen Basin and the New England Orogen. *In:* FINDLAY, R. H., UNRUG, R., BANKS, M. R. & VEEVERS, J. J. (eds) *Gondwana Eight. Assembly, Evolution and Dispersal.* A.A. Balkema, Rotterdam, 371–383.

—— & ARCHBOLD, N. 1990. Late Carboniferous Early Permian, Cranky Corner Outlier, northern Sydney Basin. *Newsletter on Carboniferous Stratigraphy*, **8**, 17–18.

BROWNE, W. R. 1927. The geology of the Gosforth district, NSW. *Journal and Proceedings of the Royal Society of New South Wales*, **60**, 213–277.

CAMPBELL, K. S. W. 1961. Carboniferous fossils from the Kuttung rocks of New South Wales. *Palaeontology*, **4**, 428–474

——, BROWN, D. A. & COLEMAN, A. R. 1983. Ammonoids and the correlation of the Lower Carboniferous rocks of eastern Australia. *Alcheringa*, **7**, 75–123.

—— & McKELVEY, B. C. 1972. The geology of the Barrington district, NSW. *Pacific Geology*, **5**, 7–48

—— & ROBERTS, J. 1969. Carboniferous. The faunal sequence and overseas correlations. *In:* PACKHAM, G. H. (ed.) *The Geology of New South Wales, Journal of the Geological Society of Australia*, **16**, 261–264.

CLAOUÉ-LONG, J. C., COMPSTON, W., ROBERTS, J. & FANNING, M. 1995. Two Carboniferous ages: a comparison of SHRIMP zircon dating with conventional zircon ages and $^{40}Ar/^{39}Ar$ analysis. *In: Geochronology, Time Scales and Stratigraphic Correlation.* SEPM Special Publication, **54**.

CLAYTON, G., LOBOZIAK, S., STREEL, M., TURNAU, E. & UTTING, J. 1990. Palynological events in the Mississippian (Lower Carboniferous) of Europe, North Africa and North America. *In:* BRENKLE, P. L. & MANGER, W. L. (eds) Intercontinental Correlation and Division of the Carboniferous System. *Courier Forschungsinstitut Senckenberg*, **130**, 79–84.

COLLINS, W. J., OFFLER, R., FARRELL, T. R. & LANDENBERGER, B. 1993. A revised Late Palaeozoic–Early Mesozoic tectonic history for the southern New England Fold Belt. *In:* FLOOD, P. G. & AITCHESON, J. C. (eds) *New England Orogen, eastern Australia.* Department of Geology and Geophysics, University of New England, Armidale, NSW, Australia, 69–84.

CONIL, R., GROESSENS, E., LALOUX, M., POTY, E. & TOURNEUR, F. 1990. Carboniferous guide foraminifera, corals and conodonts in the Franco–Belgian and Campine Basins: their potential for widespread correlation. *Courier Forschunginstitut Senckenberg*, **130**, 15–30.

CRANE, D. & HUNT, J. W. 1980. The Carboniferous sequence in the Gloucester–Myall Lake area, New South Wales. *Journal of the Geological Society of Australia*, **26**, 341–352.

DEAR, J. F., McKELLAR, R. G. & TUCKER, R. M. 1971. Geology of the Monto 1:250,000 sheet area. *Geological Survey of Queensland Report*, **46**, 1–124.

DE SOUZA, H. A. F. 1982. Age data from Scotland for the Carboniferous time scale. *In:* ODIN, G. S. (ed.) *Numerical Dating in Stratigraphy, Part I.* John Wiley, Chichester, 455–460.

DIRKS, P. H. G. M., HAND, M., COLLINS, W. J. & OFFLER, R. 1992. Structural-metamorphic evolution of the Tia Complex, New England fold belt; thermal overprint of an accretion–subduction complex in a compressional back-arc setting. *Journal of Structural Geology*, **14**, 669–688.

EMBLETON, B. J. J. 1977. Discussion: Palaeomagnetism, radiometric ages and geochemistry of an adamellite at Yetholme, NSW. *Journal of the Geological Society of Australia*, **24**, 121–123.

ENGEL, B. A. 1962. Geology of the Bulahdelah–Port Stephens district, NSW. *Journal and Proceedings of the Royal Society of New South Wales*, **95**, 197–215.

—— 1989. Carboniferous bryozoans as stratigraphic indicators in eastern Australia. *11ème Congrès International de Stratigraphie et de Géologie du Carbonifère Compte Rendu*, Beijing 1987, **3**, 33–40.

EVANS, P. R. 1969. Upper Carboniferous and Permian palynological stages and their distribution in Eastern Australia. *In: Gondwana Stratigraphy*, IUGS lst Gondwana Symposium, Buenos Aires, 1967, Unesco, 41–53.

EVERNDEN, J. F. & RICHARDS, J. R. 1962. Potassium–argon ages in eastern Australia. *Journal of the Geological Society of Australia*, **9**, 1–49.

FACER, R. A. 1976. Palaeomagnetism, radiometric ages and geochemistry of an adamellite at Yetholme, NSW. *Journal of the Geological Society of Australia*, **23**, 243–248.

FOSTER, C. B. 1993. Palynology. *In:* KORSCH, R. J. (ed.) *Eastern Australian basins mapping accord. AGSO 92, 20-21. Yearbook of the Australian Geological Survey Organisation.* Australian Government Publishing Service, Canberra.

—— & WATERHOUSE, J. B. 1988. The *Granulatisporites confluens* Oppel-zone and Early Permian marine faunas from the Grant Formation on the Barbwire Terrace, Canning Basin, Western Australia. *Australian Journal of Earth Sciences*, **35**, 135–157.

FORSTER, S. C. & WARRINGTON, G. 1985. Geochronology of the Carboniferous, Permian and Triassic. *In:* SNELLING, N. J. (ed.) *The Chronology of the Geological Periods.* Memoirs of the Geological Society of London, **10**, 99–113.

GLEN, R. A. 1993. The Lochinvar Anticline, the Hunter Thrust and regional tectonics. *In: Twenty-seventh Newcastle Symposium on Advances in the Study of the Sydney Basin,* April 1993, 33–38. Department of Geology, University of Newcastle, NSW.

GONZALEZ, C. R. 1990. Development of the Late Paleozoic glaciations of the South American Gondwana in western Argentina. *Palaeogeography, Palaeoclimatology Palaeoecology,* **79**, 275–287.

GULSON, B., DIESSEL, C. F. K., MASON, D. R. & KROGH, T. E. 1990. High precision radiometric ages from the northern Sydney Basin and their implication for the Permian time interval and sedimentation rates. *Australian Journal of Earth Sciences,* **37**, 459–469.

HARLAND, W. B., ARMSTRONG, R. L., COX, A. V., CRAIG, L. E., SMITH, A. G. & SMITH, D. G. 1990. *A Geologic Time Scale 1989.* Cambridge University Press, Cambridge.

——, COX, A. V., LLEWELLYN, P. G., PICKTON, C., SMITH, A. G. & WALTERS, R. 1982. *A Geologic Time Scale.* Cambridge University Press, Cambridge.

HELBY, R. J. 1969. Preliminary palynological study of Kuttung sediments in central eastern New South Wales. *Geological Survey of New South Wales Records,* **11**, 5–14.

HESS, J. C. & LIPPOLT, H. J. 1986. $^{40}Ar/^{39}Ar$ ages of tonstein and tuff sanidines: new calibration points for the improvement of the Upper Carboniferous time scale. *Chemical Geology (Isotope Geoscience Section),* **59**, 143–154.

IRVING, E. 1966. Paleomagnetism of some Carboniferous rocks from New South Wales and its relation to geological events. *Journal of Geophysical Research,* **71**, 6025–6051.

—— & PARRY, L. G. 1963. The magnetism of some Permian rocks in New South Wales. *Geophysical Journal of the Royal Astronomical Society,* **7**, 395–411.

—— & PULLAIAH, G. 1976. Reversals of the geomagnetic field, magnetostratigraphy, and relative magnituude of paleosecular variation in the Phanerozoic, *Earth Science Reviews,* **12**, 35–64.

JENKINS, R. B. & NETHERY, J. E. 1992. The development of Early Permian sequences and hydrothermal alteration in the Myall Syncline, central eastern New South Wales. *Australian Journal of Earth Sciences,* **39**, 223–237.

JENKINS, T. B. H., CRANE, D. T. & MORY, A. J. 1993. Conodont biostratigraphy of the Viséan Series in eastern Australia. *Alcheringa,* **317**, 211–283

JONES, M. J. & TRUSWELL, E. M. 1992. Late Carboniferous and Early Permian palynostratigraphy of the Joe Joe Group, southern Galilee Basin, Queensland, and implications for Gondwanan stratigraphy. *BMR Journal of Australian Geology & Geophysics,* **13**, 143–185.

JONES, P. J. 1991. *Australian Phanerozoic timescales 5. Carboniferous, biostratigraphic charts and explanatory notes.* Bureau of Mineral Resources, Australia Record 1989 /35.

KEMP, E. M., BALME, B. E., HELBY, R. J., KYLE, R. A., PLAYFORD, G. & PRICE, P. L. 1977. Carboniferous and Permian palynostratigraphy in Australia and Antarctica: a review. *BMR Journal of Australian Geology & Geophysics,* **2**, 177–208.

KLETS, A. G. 1983. A new Carboniferous productid genus. *Paleontological Journal,* **17**, 70–75.

LACKIE, M. A. & SCHMIDT, P. W. 1993. Remagnetisation of strata during the Hunter–Bowen Orogeny. *Exploration Geophysics,* **24**, 269–274.

LENNOX, M. 1991. Coal measures of the Stroud–Gloucester Syncline. *In:* ROBERTS, J., ENGEL, B. A. & CHAPMAN, J. (eds) 1991. *Geology of the Camberwell Dungog and Bulahdelah 1:100,000 sheets 9133, 9233, 9333.* Geological Survey of New South Wales, 167–190.

LIPPOLT, H. J. & HESS, J. C. 1983. Isotopic evidence for the stratigraphic position of the Saar–Nahe Rotliegend volcanism 1. $^{40}Ar/^{40}K$ and $^{40}Ar/^{39}Ar$ investigations. *Neues Jahrbuch für Geologie und Paläontologie Monatscheft,* **12**, 713–730.

LUCK, G. R. 1973. Palaeomagnetic results from Palaeozoic rocks of southeast Australia. *Geophysical Journal of the Royal Astronomical Society,* **32**, 35–52.

MCMINN, A. 1985. The age of the Markwell Coal Measures in the Markwell area. *Geological Survey of New South Wales Palynology Report,* 1985 /2 (GS1985/038) (unpublished).

—— 1987. Palynostratigraphy of the Stroud–Gloucester Trough, NSW. *Alcheringa,* **11**, 151–164.

MORRIS, L. N. 1985. The floral succession in eastern Australia. *In:* DIAZ, C. M. (ed.) *The Carboniferous of the World. II Australia, Indian subcontinent, South Africa, South America & North Africa.* IUGS Publication **20**, 118–123. Instuto Geologico y Minero de Espana and Empressa Nacional Adaro de Investigaciones Mineras SA, Spain.

PALMER, J. A., PERRY, S. P. G. & TARLING, D. H. 1985. Carboniferous magnetostratigraphy. *Journal of the Geological Society of London,* **142**, 945–955.

PALMIERI, V. 1983. Biostratigraphic appraisal of Permian foraminifera from the Denison Trough–Bowen Basin (central Queensland). *In: Permian Geology of Queensland.* Geological Society of Australia, Queensland Division, Brisbane, 138–154.

PLAYFORD, G. & HELBY, R. 1968. Spores from a Carboniferous section in the Hunter Valley, New South Wales. *Journal of the Geological Society of Australia,* **15**, 103–119.

POWIS, G. 1984. Palynostratigraphy of the Late Carboniferous sequence, Canning Basin, WA. *In:* PURCELL, P. G. (ed.) *The Canning Basin WA.* Proceedings of the Geological Society of Australia and Petroleum Exploration Society of Australia Symposium, Perth, 429–438.

RILEY, N. J., CLAOUÉ-LONG, J. C., HIGGINS, A. C., OWENS, B., SPEARS, A., TAYLOR, L. E. & VARKER, W. J. 1995. Geochronometry and geochemistry of the European Mid-Carboniferous boundary stratotype proposal, Stonehead Beck, North Yorkshire, UK. *Annales de la Société géologique de Belgique*, **116**, 275–289.

ROBERTS, J., CLAOUÉ-LONG, J. C. & JONES, P. J. 1991a. Calibration of the Carboniferous and Early Permian of the Southern New England Orogen by Shrimp ion microprobe zircons analyses. *Newsletter on Carboniferous Stratigraphy*, **9**, 15–17.

———, ——— & ——— 1993a. Revised correlation of Carboniferous and Early Permian units of the Southern New England Orogen, Australia. *Newsletter on Carboniferous Stratigraphy*, **11**, 23–26.

———, ———& ——— 1993b. SHRIMP zircon dating and Australian Carboniferous time. *12ème Congrès International de Stratigraphie et de Géologie du Carbonifère, Compte Rendu,* Buenos Aires, 1991, **2**, 319–338.

———, ——— & ——— 1995. Australian Early Carboniferous time. *In: Geochronology, Time Scales and Stratigraphic Correlation.* SEPM Special Publication, **54**.

——— & ENGEL, B. A. 1987. Depositional and tectonic history of the southern New England Orogen. *Australian Journal of Earth Sciences*, **34**, 1–20.

———, ——— & CHAPMAN, J. (eds) 1991b. *Geology of the Camberwell, Dungog and Bulahdelah 1:100,000 sheets 9133, 9233, 9333.* Geological Survey of New South Wales, 167–190.

——— *ET AL.* 1985. Australia. *In:* DIAZ, C. M. (ed.) *The Carboniferous of the World. II Australia, Indian subcontinent, South Africa, South America & North Africa.* IUGS Publication **20**, 9–145. Instuto Geologico y Minero de Espana and Empressa Nacional Adaro de Investigaciones Mineras SA, Spain.

———, HUNT, J. W. & THOMPSON, D. M. 1976. Late Carboniferous marine invertebrate zones of eastern Australia. *Alcheringa*, **1**, 197–225.

———, JONES, P. J. & JENKINS, T. B. H. 1993c. Revised ages for Early Carboniferous marine invertebrate zones of eastern Australia. *Alcheringa*, **17**, 353–376.

TABOADA, A. C. 1989. La fauna de la Formacion El Paso, Carbonifera Inferior de la Precordillera Sanjuanina. *Acta Geologica Lilloana*, **17**, 113–129.

TERA, F. & WASSERBURG, G. J. 1974. U–Th–Pb systematics on lunar rocks and inferences about lunar evolution and the age of the moon. *Proceedings of the Fifth Lunar Conference (supplement 5) Geochimica et Cosmochimica Acta*, **2**, 1571–1599.

VEEVERS, J. J. & POWELL, C. McA. 1990. Phanerozoic tectonic regimes of Australia reflect global events. *Journal of Structural Geology*, **12**, 545–551.

Direct Pb/Pb dating of Silurian macrofossils from Gotland, Sweden

JONATHAN RUSSELL

Department of Earth Sciences, University of Oxford, Oxford OX1 3PR, UK
Present address: Shell Research, KSEPL, PO Box 60, 2280 AB Rijswijk, The Netherlands

Abstract: Pb/Pb and U–Pb methods of direct isotopic dating, characteristically associated with igneous and metamorphic rocks, have been applied successfully to well-preserved stromatoporoidal carbonates from Gotland, Sweden. A Pb/Pb date of 432 ± 15 Ma for Wenlock–Ludlow boundary carbonates is in good agreement with independent chronometric estimates and is interpreted as the age of deposition/early diagenesis. A Pb/Pb date of 326 ± 11 Ma for older Middle Wenlock samples is *c.* 100 Ma too young for its stratigraphic position and is interpreted as a late diagenetic age associated with fluid movement along sub-surface faults. U–Pb systematics in both samples have been disturbed by U remobilization in response to recent weathering associated with post-glacial uplift in the Baltic.

For many years, the direct age determination of sediments has been an irritating problem to earth scientists. Direct Pb/Pb dating of carbonates should be considered a credible source of chronometric information. Application of the method to poorly constrained Precambrian stromatolitic carbonates is potentially powerful since host sediment ages may be only poorly constrained through indirect reference to basement and intrusive rocks. A broad geo-chemical review is presented to explain the suitability of certain carbonates for this method of isotopic investigation.

The lack of reliable radiometric methods for determining directly the age of sediments has been a long-standing problem in the earth sciences. Although Rb–Sr and K–Ar studies on authigenic components of fine-grained clastic sediments have met with limited success (e.g. Clauer 1981), both systems tend to yield dates that are younger than the inferred depositional age owing to the loss of radiogenic Ar and mobility of Rb during burial metamorphism. As a consequence, sediment ages have been constrained characteristically by indirect reference to underlying basement rocks, intrusives and/or interstratal volcanics of known age.

The application of U–Pb systematics to carbonate materials began as early as 1962, when Wampler & Kulp (1962) determined Pb model ages for marbles from southern Rhodesia. The authors concluded that the model ages represented minimum ages for sedimentation (although more realistically related to the last phase of metamorphism) and stated that '. . . under suitable conditions the isotopic composition of Pb in carbonate rocks may give valuable information on the geological history of an area.' Despite this early promise, Doe (1970) reviewed the potential of U–Th–Pb systematics for the direct dating of sediments and concluded that Pb/Pb isochrons more closely reflected average source age than the age of sedimenta-

tion. Consequently, interest in the subject diminished, not least because of the extremely low concentrations of U and Pb in sedimentary carbonates (characteristically < 5 p.p.m.), and more than a decade passed before the first direct Pb/Pb carbonate date was published. Moorbath *et al.* (1987) gave details of a 2839 ± 39 Ma Pb/Pb errorchron date for the Mushandike stromatolitic limestone of Zimbabwe (fully described by Orpen & Wilson 1981) and reported an extremely large range of Pb isotopic compositions ($^{206}Pb/^{204}Pb$ ratios from 31.47 to 140.50) in excess of that found in the majority of igneous and metamorphic rocks.

Since 1987 there has been renewed interest in the potential of direct dating of carbonate sediments and a number of publications have resulted. Jahn & Cuvellier (1989), Cuvellier (1989), Jahn *et al.* (1990), Russell (1991), Cuvellier (1992) and Russell (1992) have all published Pb/Pb dates for sedimentary carbonates, while Smith & Farquhar (1989), Smith *et al.* (1991) and DeWolf & Halliday (1991) have published U–Pb dates. This study seeks to apply Pb/Pb and U–Pb direct dating techniques to well-constrained Phanerozoic carbonates and to critically assess the results with a view to commenting on the interpretation of carbonate dates from Precambrian formations. The ability to date Precambrian carbonates directly is

From Dunay, R. E. & Hailwood, E. A. (eds), 1995, *Non-biostratigraphical Methods of Dating and Correlation*
Geological Society Special Publication No. 89, pp. 175–200

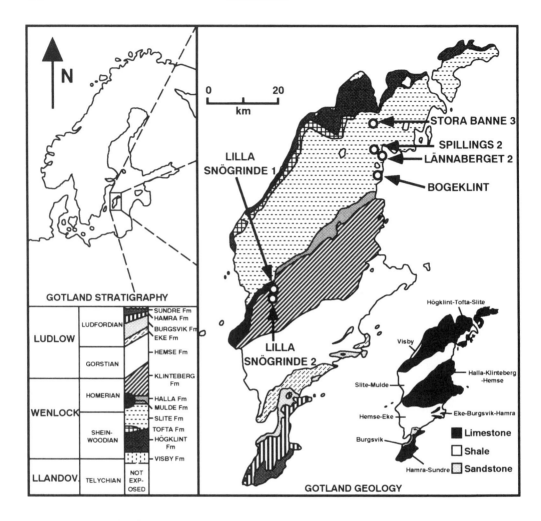

Fig. 1. Geological units of the Swedish island of Gotland, including localities for carbonte samples; after Kershaw 1993.

potentially powerful because detailed chrono-stratigraphies are seldom available and the ages of sediments tend to be imprecisely bracketed.

Methodology

Carbonate samples were prepared for analysis using ultra-pure reagents (calculated reagent Pb blanks: H_2O, $0.089\,ng\,ml^{-1}$; concentrated HCl; $0.053\,ng\,ml^{-1}$; concentrated HBr, $0.36\,ng\,ml^{-1}$; concentrated HNO_3, $0.98\,ng\,ml^{-1}$) and Teflon beakers under clean-room conditions at the Age Laboratory, Department of Earth Sciences, Oxford. Samples were broken into small fragments, between 0.8 and 1.5 g, and external surfaces/altered material was removed using a

diamond-studded grinding wheel lubricated by pure water. Dissolution of carbonate fractions was accomplished using ultra-pure 6 M HCl added in 1 ml volumes until the reaction had proceeded to completion. The chloride solutions were drawn off and centrifuged to remove insoluble suspended residues. The solutions were then aliquoted in the ratio 3:1 for isotopic composition (IC) and isotope dilution (ID) determinations. ID solutions were spiked for both Pb and U. IC and ID Pb separations were performed from bromide solutions using a modified three-column ion-exchange technique. Initially, a 2.25 ml column was used and then two smaller 0.3 ml columns. In each case AG 1-X8 100–200 mesh chloride exchange resin was

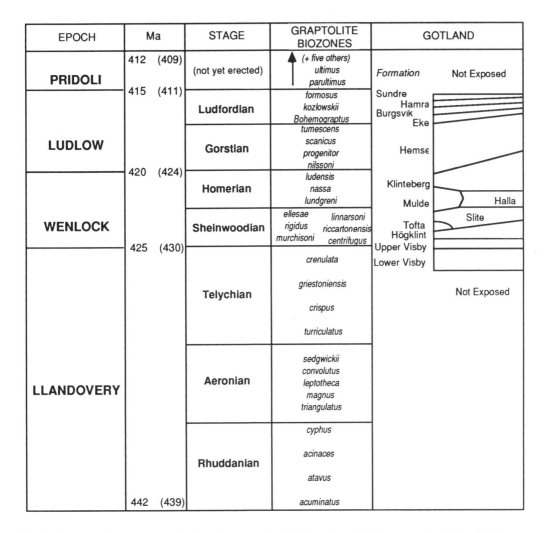

EPOCH	Ma	STAGE	GRAPHOLITE BIOZONES	GOTLAND	
PRIDOLI	412 (409) 415 (411)	(not yet erected)	▲ (+ five others) ultimus parultimus	*Formation*	Not Exposed
LUDLOW		**Ludfordian**	formosus kozlowskii *Bohemograptus*	Sundre Hamra Burgsvik Eke	
		Gorstian	tumescens scanicus progenitor nilssoni	Hemse	
WENLOCK	420 (424)	**Homerian**	ludensis nassa lundgreni	Klinteberg	
		Sheinwoodian	ellesae linnarsoni rigidus riccartonensis murchisoni centrifugus	Mulde Tofta Högklint	Slite Halla
	425 (430)		crenulata	Upper Visby Lower Visby	
LLANDOVERY		**Telychian**	griestoniensis crispus turriculatus		Not Exposed
		Aeronian	sedgwickii convolutus leptotheca magnus triangulatus		
		Rhuddanian	cyphus acinaces atavus		
	442 (439)		acuminatus		

Vertical scale is chronometric and relates to values for the time scale of McKerrow (pers. comm. 1992)

Values in brackets are from the time scale of Harland *et al.* (1989)

Diagrammatic representation of Gotland lithostratigraphy is interpretative

Fig. 2. Stratigraphy of the Silurian succession on the island of Gotland, Sweden.

first conditioned with 0.25 M HNO$_3$ and, after washing with 0.5 M HBr, the Pb was eluted in 6 M HCl. ID bromide wash solutions from the 2.25 ml columns were retained for U separation which was performed using the tri-*n*-butyl phosphate method of Arden & Gale (1974). Fully detailed analytical procedures are presented in Russell (1992).

Pb and U analyses were run on a fully automated VG Isomass 54E mass spectrometer following procedures published by Taylor *et al.*

(1984) and Kakazu *et al.* (1981). Pb isotopic ratios have been corrected for mass fractionation of $0.130 \pm 0.012\%$ ($2\sigma_{mean}$) amu^{-1}, based on replicate measurements of the NBS 981 standard at the Oxford Age Laboratory. U ratios have been corrected for fractionation of $0.132 \pm 0.012\%$ ($2\sigma_{mean}$) amu^{-1}, based on NBS standard U-500 measurements. The average procedural Pb blank for the period of investigation was 0.70 ± 0.30 ng (this figure may be considered negligible for Pb samples of $\geq 1 \mu$g) and a procedural U blank of 0.039 ± 0.01 ng has been adopted for the tri-*n*-butyl phosphate method at Oxford (Gale *et al.* 1980; negligible for U samples of ≥ 5 ng). Isochron regressions have been carried out using the Isoplot v.2.11 package of Ludwig (1990) and radiometric ages calculated using the decay constants listed in Steiger & Jäger (1977). Ages, model μ_1 values and isotope results are quoted at the 2σ (95% confidence limit) level.

Gotland stratigraphy

Gotland is Sweden's largest island and lies 110 km off the east coast of the mainland in the Baltic Sea (Fig. 1). The exposed Silurian sequence (late Llandovery–late Ludlow) is renowned for the rich fossil assemblages and carbonate reefs it contains and the sediments record the slow pulsatory regression of a pericratonic, tropical carbonate sea across the western margin of the Russian Precambrian shield (Bassett *et al.* 1989). The 12 lithostratigraphic divisions outlined in Fig. 1 (Hede 1921, 1925; Frykman 1989) are recognized to be both heterogeneous and diachronous in nature (Sundquist 1982), but they form the basis for stratigraphic studies and correlation on the island (see Fig. 2). Carbonate-dominated areas form topographic highs in the northwest, centre and southeast of the island (30–80 m elevation), while the interleaving marl-dominated beds form topographic lows. Lithologies are briefly described by Laufeld & Bassett (1981).

Biogenic carbonates (stromatoporoids, corals) were analysed from two separate horizons within the succession (Slite Formation unit g and lower Middle Klinteberg Formation), for which sample localities are indicated in Fig. 1. A listing of reference localities on the island is published in Laufeld (1974*b*) and an up-to-date listing is kept at 'Allekvia', the University of Lund's field station on the island. Biostratigraphic division of the succession is based upon abundant conodont fragments which are found in the shallow water sediments (Jeppsson 1983 and references therein) and rarer graptolite occurrences. Correlation with the British Silurian, which is zoned on the basis of graptolite taxa from deeper water sediments, can be made confidently to within one graptolite biozone (see Fig. 2).

In the northeast of the island, the Slite Formation (up to 100 m thick) consists of several generations of reefal carbonates and inter-reefal crinoidal/bryozoan grainstones and packstones, but to the southwest they pass into marl-rich facies. The formation is subdivided into seven lettered units plus the topmost, clastic-rich Slite Siltstone. Stromatoporoidal carbonates were analysed from widely separated unit g localities at Stora Banne 3 (1907895, unit g lower to basal part, *rigidus* biozone or older) and Spillings 2 (2307896, 2307897, unit g, *lundgreni* biozone) and specimens of the tabulate coral *Heliolites* sp. were analysed from unit g localities at Lännaberget 2 (1507892, unit g, *lundgreni* biozone) and Bogeklint 2 (1907899, unit g outlier). It is known that unit g of the Slite Formation (dominantly Sheinwoodian) contains limestone beds of varying age (Jeppsson, pers. comm.) and this results in uncertainty in precise correlation between localities.

The younger Klinteberg Formation, which spans the Wenlock–Ludlow boundary, is up to 70 m thick and in the southwest contains autochthonous stromatoporoid/tabulate coral framestones with coarse inter-reefal crinoidal grainstones and packstones. To the northeast, they pass into a complex of protected quieter water carbonate facies, containing abundant algal oncolites and bedded limestones. Hede (1958) divided the formation into six subunits in the northeast, but recognized only lower, middle and upper units in the southwest. Frykman (1989) interpreted the southwestern deposits as indicative of a shallow water, southerly-dipping ramp-like setting with numerous shoals and barriers. The Klinteberg Formation has been placed within the *ludensis, nilssoni* and *scanicus* graptolite biozones (topmost Wenlock–Ludlow, see Fig. 2) by Martinsson (1967) (from ostracode work) and Larsson (1979) (from tentaculid faunas). Stromatoporoidal carbonates were analysed from localities within the lower Middle Klinteberg Formation at Lilla Snögrinde 1 and Lilla Snögrinde 2, within the southwestern reefal tract of Klinteberg sedimentation. Both localities are from the upper part of the *ludensis* biozone (topmost Wenlock – Jeppsson, pers. comm.) and lie extremely close to the Wenlock–Ludlow boundary. Early marine cementation is common and evidence of meteoric diagenesis (Frykman 1984, 1985) suggests periods of Klinteberg emergence. 16078913 and LS were collected from Lilla Snögrinde 2 and are both large and domal forms of fresh appearance that show good preservation of internal structure and an amount of sparite cementation. 16078913 split into sequential growth zones along latilaminae (stromatoporoid growth discontinuities) allowing a more detailed isotopic study. 16078914 and 16078916 were collected some 56 m apart at Lilla Snögrinde 1 and both exhibit strongly recrystallized textures, with no preservation of internal structure.

Table 1. *Published dates for the Silurian period*

Epoch	Stage	Publication	Date (Ma)	Method
Early Devonian	Lochkovian	Thirlwall (1988)	413 ± 6	Rb–Sr on biotite
....				
SILURIAN				
Pridoli		Odin *et al.* (1982)	399 ± 9	K–Ar on biotite
....		McKerrow *et al.* (1985)	407 ± 6	Rb–Sr on biotite
Ludlow	Ludfordian	Odin *et al.* (1986*b*)	417 ± 10	K–Ar on biotite
....		Odin *et al.* (1986*b*)	423 ± 9	K–Ar on biotite
	Gorstian	McKerrow *et al.* (1985)	407 ± 14	Fission track on zircon
		Ross *et al.* (1982)	407 ± 18	Fission track on zircon
		Gale (1985)	415 ± 10	K–Ar on biotite/sanidine
		Ross *et al.* (1982)	419 ± 7	K–Ar on biotite
		Wyborn *et al.* (1982)	421 ± 2	K–Ar on biotite
		Gale (1985)	422 ± 6	Rb–Sr whole rock/biotite
		Kunk *et al.* (1985)	424 ± 4	Ar–Ar on biotite
		Odin *et al.* (1986*b*)	424 ± 8	K–Ar on biotite
	....	Armstrong (1978)	429 ± 8	Rb–Sr whole rock
Wenlock	Homerian	Harland *et al.* (1964)	413 ± 32	Rb–Sr whole rock
		Ross *et al.* (1982)	412 ± 24	Fission track on zircon
		Ross *et al.* (1982)	416 ± 18	Fission track on zircon
		Odin *et al.* (1986*a*)*	427 ± 6	K–Ar on biotite
	Sheinwoodian	Ross *et al.* (1982)	422 ± 14	Fission track on zircon
	....	Odin *et al.* (1986*a*)*	430 ± 6	K–Ar on biotite
Llandovery	Telychian	Armstrong (1978)	445 ± 30	Rb–Sr whole rock
	....			
	Aeronian	Odin *et al.* (1982)	436 ± 5	Ar–Ar on hornblende
	Rhuddanian	Compston *et al.* (1981)	429 ± 4	U–Pb on zircon
		Odin *et al.* (1982)	431 ± 6	Ar–Ar on hornblende
		Ross *et al.* (1982)*	437 ± 22	Fission track on zircon
		Tucker *et al.* (1990)	439 ± 2	U–Pb on zircon
		Armstrong (1978)	445 ± 26	K–Ar on biotite
	....	Armstrong (1978)	445 ± 40	Rb–Sr on biotite
ORDOVICIAN				
Late Ashgill		Tucker *et al.* (1990)	446 ± 3	U–Pb on zircon

* Data from Gotland, Sweden.

Silurian carbonate–marl sedimentation was interrupted frequently by the deposition of volcaniclastic bentonite beds, generally < 10 cm thick and composed principally of clay minerals (illite–smectite and kaolinite), biotite and feldspars (Odin *et al.* 1986*a*). Unaltered biotites separated from these bentonites have been dated successfully using K–Ar techniques.

Borehole and geophysical data (Flodén 1980; Winterhalter *et al.* 1982) indicate that beneath the 500 m of exposed latest Llandovery to Late Ludlow sediments, are 150 m of older Silurian rocks, 75–125 m of Ordovician sediments and 150–225 m of Cambrian–Precambrian sediments (Laufeld & Bassett 1981). Tectonic deformation of the carbonate-dominated Silurian succession has been minimal. Rocks strike northeast–southwest and have a maximum regional dip of 2–3° to the southeast. Only beneath reefal masses do dips exceed 5° and vary in orientation as the result of differential sediment compaction. Minor faulting is observed, particularly along the coastal cliff sections, but the downthrow of the predominantly normal faults is usually < 1 m. The only area in which the extent of structural disturbance increases significantly is in the region of the north–south Lärbo Valley in the north of the island. Organic matter within preserved conodonts shows no observable colour alteration and indicates that throughout their geological history, the sediments have never been subjected to elevated temperatures in response to deep burial (Epstein *et al.* 1977; Jeppsson 1983). Odin *et al.* (1984) suggested that a maximum burial depth of a few hundred metres was appropriate for the entire Silurian succession. Coupled with the occurrence of dated bentonite horizons within the succession (Odin *et al.* 1986*a*), this makes the Silurian rocks of Gotland ideal for testing the applicability of

Pb/Pb and U–Pb dating to Phanerozoic carbonates. Gotland's location on the stable western margin of the Russian Precambrian shield provides the explanation for the island's relatively simple geological history, which is so beautifully recorded in the exposed sequence of carbonates, marls and bentonites. Although much of the island was submerged at the end of the last glaciation, isostatic readjustment has subsequently produced uplift rates of between 0.7 and 2.0 m Ma^{-1} in the Gulf of Bothnia and the Baltic Sea region, resulting in a series of raised shorelines (Jeppsson 1982) and subjecting the sediments to the effects of Quaternary weathering.

Age constraints

Compared with the Archaean and Proterozoic, a relatively good consensus exists on Silurian timescale calibration. Indeed, the Silurian period was the first for which international agreement was reached on the stratigraphic definition and standardization of all four epochs and seven stages (Harland et al. 1989). The biostratigraphic division is based on the graptolite succession from Welsh Borderlands (Bassett et al. 1975) (see Fig. 2). McKerrow et al. (1980, 1985), Harland et al. (1982, 1989), Odin et al. (1982) and Jones et al. (1981) have all published timescales for this portion of geological time by selecting published dates from the available literature pool and applying a particular statistical treatment. The most recent timescales (Harland et al. 1989; McKerrow, pers. comm.) are inevitably more reliable than their predecessors because they incorporate the most recently published age data.

Epoch boundary ages from these scales are summarized in Fig. 2, while relevant radiometric data for the Silurian are detailed in Table 1. Published K–Ar dates for unaltered biotites separated from Silurian bentonites on Gotland provide an internal chronometric calibration for the lithostratigraphy. All errors are quoted at the 2 σ level.

Sample palaeontology

Detailed palaeontological studies for the island of Gotland have been published by Mori (1969, 1970) and Kershaw (1981, 1987, 1990) on stromatoporoids (now accepted as a class of phylum Porifera – after Galloway 1957; Stearn 1980); Martinsson (1967) on ostracodes; Laufeld (1974a) on chitinozoans; Larsson (1979) on tentaculids and Jeppsson (1974, 1979, 1983) on conodonts. Pb and U isotopic analyses per-

formed during this investigation have concentrated on stromatoporoidal carbonates.

Stromatoporoids were most abundant during the Ordovician, Silurian and Devonian when, along with tabulate corals, they were the principal reef builders. Related forms are also found in isolation during the Jurassic and Cretaceous. These colonial organisms built an open, highly porous, calacareous skeleton of laminar, domal, columnar, digitate or encrusting form. The skeleton had either a 3D framework or a reticulate geometry comprising structural elements both parallel and perpendicular to the growth surface (laminae and pillars) that linked to form porosity galleries. Growth discontinuities parallel to the main growth surface are marked by the development of latilaminae. Debate still exists over the nature of primary stromatoporoid biomineralogy, although aragonite, high magnesian calcite (HMC) or some combination thereof appear the most probable candidates (Stearn & Mah 1987; Rush & Chafetz 1991).

The classification of stromatoporoids is based on similarities in structural detail observed in thin section, rather than gross morphology, because it is known that the organisms demonstrated a significant phenotypic plasticity in response to genetic, functional and environmental controls (Stearn 1980). All forms show at least partial diagenetic recrystallization and for this reason, detailed microstructural classifications are not possible. In many examples, the faithful preservation of major structural elements and latilaminae, even after neomorphism, suggests that 'thin-film' diagenetic processes may have been responsible for the observed recrystallization (Pingitore 1976; Kershaw 1990). Structural elements are preserved as cloudy, inclusion-rich neomorphic low magnesian calcite (LMC), whereas later cement phases form a mosaic of clear, columnar to equant, primary LMC. Cathodoluminescence microscopy has proved an important tool in the petrographic study of such carbonate materials because it frequently illuminates relict primary structural detail, even after diagenetic recrystallization, and it allows diagenetic interpretation of cement stratigraphies (Frykman 1985; Russell 1992).

Isotopic results

Lower Middle Klinteberg Formation

Pb/Pb results
Four stromatoporoidal carbonates (16078913, 16078914, 16078916 and LS) were analysed from lower Middle Klinteberg Formation localities

Table 2. *Pb isotopic compositions of stromatoporoidal carbonates from the lower Middle Klinteberg Formation* (ludensis *biozone), Wenlock, Gotland*

Analysis	Sample	Corrected ratios*		
		$^{206}Pb/^{204}Pb$	$^{207}Pb/^{204}Pb$	$^{208}Pb/^{204}Pb$
K1	16078913A	68.976 ± 0.114	18.396 ± 0.032	38.247 ± 0.062
K2	16078913B	61.220 ± 0.068	17.988 ± 0.026	38.319 ± 0.058
K3	16078913AIII	70.648 ± 0.116	18.490 ± 0.032	38.351 ± 0.076
K4	16078913BI	66.529 ± 0.086	18.263 ± 0.026	38.375 ± 0.066
K5	16078913BII	60.470 ± 0.082	17.946 ± 0.026	38.356 ± 0.064
K6	16078913BIII	63.227 ± 0.110	18.076 ± 0.044	38.310 ± 0.104
K7	16078913a	71.660 ± 0.142	18.552 ± 0.042	38.305 ± 0.088
K8	16078913b	63.530 ± 0.126	18.143 ± 0.036	38.467 ± 0.080
K9	16078913c	71.726 ± 0.110	18.579 ± 0.032	38.482 ± 0.078
K10	16078913d	81.948 ± 0.122	19.139 ± 0.030	38.468 ± 0.070
K11	16078913e	76.536 ± 0.136	18.851 ± 0.036	38.283 ± 0.094
K12	16078913f	66.198 ± 0.084	18.264 ± 0.026	38.497 ± 0.062
K13	16078913g	64.158 ± 0.120	18.145 ± 0.046	38.420 ± 0.106
K14	16078913h	66.322 ± 0.096	18.255 ± 0.030	38.440 ± 0.072
K15	16078913i	60.036 ± 0.082	17.913 ± 0.028	38.311 ± 0.058
K16	16078913j	67.722 ± 0.076	18.348 ± 0.024	38.297 ± 0.050
K17	16078913k	68.244 ± 0.168	18.372 ± 0.046	38.359 ± 0.098
K18	160789131	69.178 ± 0.118	18.407 ± 0.034	38.285 ± 0.070
K19	16078913Z0	54.042 ± 0.056	17.563 ± 0.022	38.847 ± 0.048
K20	16078913Z1A	58.005 ± 0.018	17.805 ± 0.022	38.581 ± 0.048
K21	16078913Z1B	60.032 ± 0.062	17.922 ± 0.022	38.420 ± 0.048
K22	16078913Z2A	59.333 ± 0.062	17.878 ± 0.022	38.399 ± 0.048
K23	16078913Z2B	62.580 ± 0.066	18.058 ± 0.022	38.387 ± 0.048
K24	16078913Z2C	59.462 ± 0.062	17.874 ± 0.022	38.373 ± 0.048
K25	16078913Z3A	66.559 ± 0.070	18.244 ± 0.022	38.289 ± 0.048
K26	16078913Z3B	64.495 ± 0.068	18.161 ± 0.022	38.932 ± 0.048
K27	16078913Z3C	65.547 ± 0.068	18.206 ± 0.022	38.373 ± 0.048
K28	16078913Z4A	69.612 ± 0.072	18.435 ± 0.022	38.449 ± 0.048
K29	16078913Z4B	65.506 ± 0.072	18.230 ± 0.024	38.679 ± 0.050
K30	16078914	18.945 ± 0.020	15.688 ± 0.020	38.233 ± 0.050
K31	16078916	18.693 ± 0.020	15.660 ± 0.020	38.171 ± 0.048
K32	LS1L Top	28.397 ± 0.028	16.128 ± 0.020	38.143 ± 0.046
K33	LS1L Base	26.392 ± 0.028	16.016 ± 0.020	38.216 ± 0.048
K34	LS1S Top	28.808 ± 0.030	16.151 ± 0.020	38.182 ± 0.048
K35	LS1S Base	25.374 ± 0.026	15.979 ± 0.020	38.111 ± 0.046
K36	LS1S Base	25.839 ± 0.026	16.010 ± 0.020	38.153 ± 0.048

* Corrected for mass fractionation of $0.130 \pm 0.012\%$ ($2\sigma_{mean}$) amu^{-1} – errors are $2\sigma_{mean}$ for corrected ratios.

(*ludensis* biozone, topmost Wenlock) at Lilla Snögrinde 1 (16078914 and 16078916) and Lilla Snögrinde 2 (16078913 and LS), in the west of the island (see Fig. 1). Pb isotopic data are listed in Table 2 and plotted in Fig. 3. Twenty-nine analyses of specimen 16078913 (analyses K1– K29) exhibit radiogenic Pb isotopic compositions, a large colinear range in isotopic ratios ($^{206}Pb/^{204}Pb$ 54.0 to 81.9) and define a Pb/Pb isochron date of 441 ± 37 Ma (MSWD 1.42, model μ_1 value 8.14 ± 0.07). The isotopic range is remarkable in that it comes from analyses of a single fossil specimen. Five analyses of specimen LS (analyses K32–K36) are less radiogenic

and have a more limited Pb isotopic range ($^{206}Pb/^{204}Pb$ 25.4–28.8) which defines an imprecise date of 179 ± 270 Ma (MSWD 0.957, model μ_1 value 8.21 ± 0.09), only just within error of the 16078913 date. Specimens 16078914 and 16078916 are completely recrystallized and both have very similar, unradiogenic Pb isotopic compositions (analyses K30 and K31, respectively). The combination of all 36 data points produces a well-fitted Pb/Pb errorchron, corresponding to a date of 432 ± 15 Ma with an MSWD of 3.68 and a model μ_1 value 8.16 ± 0.03). These values are consistent with the regressed data for 16078913 alone and this more

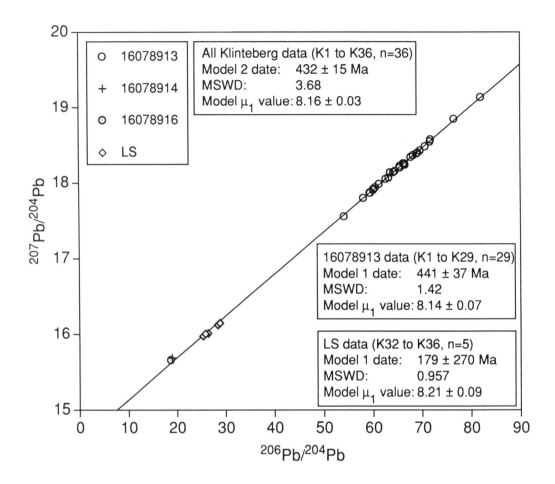

Fig. 3. Pb isotopic data for stromatoporoids from the lower Middle Klinteberg Formation (*lũdensis* biozone).

precise combined date is taken to be representative for the stratigraphic horizon.

U–Pb results

Potentially, U–Pb dating is capable of yielding more precise dates for Phanerozoic materials than Pb/Pb dating, but it is also more sensitive to disturbance of parent/daughter ratios. Results for U–Pb analyses of lower Middle Klinteberg stromatoporoids are detailed in Table 3 and illustrated in Fig. 4. While the Pb/Pb data plot along a well-defined mid-Silurian isochron, it is evident that U–Pb closed system conditions were not maintained. The isotopic disturbance must be a recent phenomenon, leaving insufficient time for the modified μ values to generate significant scatter in Pb isotopic ratios. The Pb/Pb method of radiometric dating is characteristically insensitive to recent changes in the

U/Pb ratio since the calculated date is based only on the measured radiogenic $^{207}Pb/^{206}Pb$ ratio (unlike the U–Pb method which requires determination of absolute U and Pb concentrations in addition to the Pb isotopic composition). Provided therefore that the lost Pb was of approximately the same isotopic composition as that remaining in the carbonate, there is no significant effect on the calculated date.

If one assumes that the least radiogenic analyses (K30 and K31) approximate to the Silurian initial Pb isotopic composition of a source with μ_1 value *c.* 8.2, it is possible to construct reference Silurian isochrons for 420 Ma (the date for the Wenlock–Ludlow boundary used by McKerrow, pers. comm.) and 430 Ma (the date obtained from Pb/Pb data). On Fig. 4, the majority of analyses lie to the right of these lines and so one may infer that U gain

Table 3. *U–Pb isotopic data for stromatoporoidal carbonates from the lower Middle Klinteberg Formation (ludensis biozone), Wenlock, Gotland*

Analysis	Sample	Concentrations (p.p.m.)		Corrected ratios*			
		U	Pb	$^{238}U/^{204}Pb$	$^{206}Pb/^{204}Pb$	$^{235}U/^{204}Pb$	$^{207}Pb/^{204}Pb$
K7	16078913a		0.059 ± 0.000		71.660 ± 0.142		18.552 ± 0.042
K9	16078913c		0.062 ± 0.000		71.726 ± 0.110		18.579 ± 0.032
K15	16078913i	0.870 ± 0.020	0.150 ± 0.001	585.764 ± 9.294	60.036 ± 0.082	4.246 ± 0.068	17.913 ± 0.028
K16	16078913j	0.915 ± 0.020	0.138 ± 0.001	715.246 ± 11.290	67.722 ± 0.076	5.185 ± 0.082	18.348 ± 0.024
K17	16078913K	0.923 ± 0.020	0.141 ± 0.001	712.513 ± 11.312	68.244 ± 0.168	5.165 ± 0.082	18.372 ± 0.046
K18	16078913l	0.898 ± 0.020	0.134 ± 0.000	732.629 ± 11.582	69.178 ± 0.118	5.311 ± 0.084	18.407 ± 0.034
K19	16078913O	0.900 ± 0.020	0.175 ± 0.001	495.280 ± 7.814	54.042 ± 0.056	3.590 ± 0.056	17.563 ± 0.022
K20	16078913Z1A	0.751 ± 0.020	0.129 ± 0.000	581.110 ± 9.162	58.005 ± 0.018	4.212 ± 0.066	17.805 ± 0.002
K21	16078913Z1B	0.717 ± 0.020	0.114 ± 0.000	637.280 ± 10.050	60.032 ± 0.062	4.619 ± 0.072	17.922 ± 0.022
K22	16078913Z2A	0.781 ± 0.020	0.122 ± 0.000	641.262 ± 10.098	59.333 ± 0.062	4.648 ± 0.074	17.878 ± 0.022
K23	16078913Z2B	0.729 ± 0.020	0.106 ± 0.000	711.640 ± 11.204	62.580 ± 0.066	5.158 ± 0.082	18.058 ± 0.022
K24	16078913Z2C	0.733 ± 0.020	0.107 ± 0.000	690.413 ± 10.870	59.462 ± 0.062	5.005 ± 0.078	17.874 ± 0.022
K25	16078913Z3A	0.824 ± 0.020	0.106 ± 0.000	832.927 ± 13.136	66.559 ± 0.070	6.038 ± 0.096	18.244 ± 0.022
K26	16078913Z3B	0.819 ± 0.020	0.104 ± 0.000	829.986 ± 13.078	64.495 ± 0.068	6.016 ± 0.094	18.161 ± 0.022
K27	16078913Z3C	0.835 ± 0.020	0.095 ± 0.000	929.493 ± 14.640	65.547 ± 0.068	6.738 ± 0.106	18.206 ± 0.022
K28	160789134A	0.806 ± 0.020	0.084 ± 0.000	1058.468 ± 16.660	69.612 ± 0.072	7.672 ± 0.120	18.435 ± 0.022
K29	16078913Z4B	0.741 ± 0.020	0.072 ± 0.000	1093.454 ± 17.198	65.506 ± 0.072	7.926 ± 0.124	18.230 ± 0.024
K30	16078914	0.031 ± 0.000	0.234 ± 0.001	8.378 ± 0.134	18.945 ± 0.020	0.061 ± 0.000	15.688 ± 0.020
K31	16078916	0.046 ± 0.000	0.597 ± 0.002	4.878 ± 0.084	18.693 ± 0.020	0.035 ± 0.000	15.660 ± 0.020

* Corrected for Pb isotopic mass fractionation of $0.130 \pm 0.012\%$ ($2\sigma_{mean}$) amu^{-1} and U isotopic mass fractionation of $0.132 \pm 0.012\%$ (2σ) amu^{-1} – errors are $2\sigma_{mean}$ for concentrations and corrected ratios.

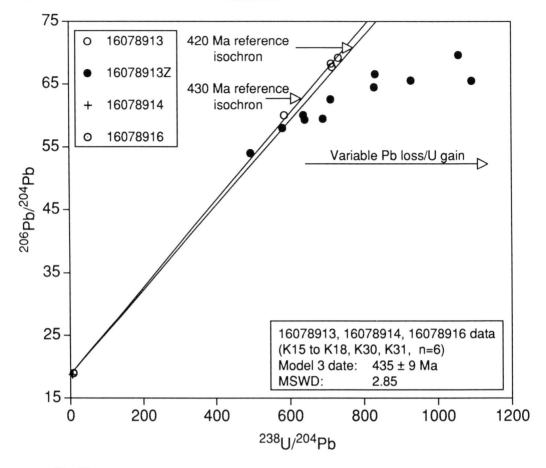

Fig. 4. ^{238}U–^{206}Pb isotopic data for stromatoporoids from the lower Middle Klinteberg Formation (*ludensis* biozone).

(or Pb loss) was responsible for the isotopic disturbance. Samples 16078913i–l (analyses K15–K18) were from a separate piece of stromatoporoidal carbonate to analyses K19–K29, while K30 and K31 were from completely separate organisms. If one merely considers analyses K15–K18, K30 and K31, they define a six-point ^{238}U–^{206}Pb errorchron (MSWD 2.85), corresponding to a model 3 date of 435 ± 9 Ma. [Model 3 Yorkfit dates in Isoplot version 2.11 (Ludwig 1990) are calculated assuming that data scatter is in part due to the assigned errors and in part due to some unknown, normally distributed variation in initial Pb isotopic ratio. This data treatment has been used where excess data scatter (MSWD > 2.5) suggests that initial ratio variation may have been a significant factor.] Within error, this is concordant with the Pb/Pb errorchron date (432 ± 15 Ma). The ^{235}U–^{207}Pb data for the same analyses exhibit a greater degree of scatter (MSWD 38.9) and define an older, less precise date of 490 ± 25 Ma (not plotted).

Slite Formation, unit g

Pb/Pb results

Three stromatoporoidal carbonates (1907895, 2307896 and 2307897) and two specimens of heliolitid coral (1507892 and 1907899) were analysed from Middle Wenlock Slite Formation unit g localities (*rigidus* and *lundgreni* biozones) at Lännaberget 2, Stora Banne 3, Bogeklint 2 and Spillings 2, in the northeast of the island (see Fig. 1). Pb isotopic data for the samples are listed in Table 4 and plotted in Fig. 5. Qualitatively, it is evident that each of the samples have distinct compositional ranges that plot along a well-defined linear trend, implying significant variation of *in situ* μ_2 values at

Table 4. *Pb isotopic compositions of carbonates from the Slite Formation, unit g* (rigidus *and* lundgreni *biozones*), *Wenlock, Gotland*

Analysis	Sample	Corrected ratios*		
		$^{206}Pb/^{204}Pb$	$^{207}Pb/^{204}Pb$	$^{208}Pb/^{204}Pb$
S1	1507892†	24.548 ± 0.048	15.963 ± 0.036	38.765 ± 0.080
S2	1507892A†	24.375 ± 0.028	15.956 ± 0.022	38.770 ± 0.054
S3	1907895a‡	21.434 ± 0.022	15.812 ± 0.020	38.417 ± 0.048
S4	1907895b‡	21.908 ± 0.022	15.810 ± 0.020	38.250 ± 0.048
S5	1907895Z1‡	21.736 ± 0.022	15.793 ± 0.020	38.292 ± 0.046
S6	1907895Z1a‡	21.105 ± 0.026	15.803 ± 0.022	38.464 ± 0.060
S7	1907895Z2‡	23.046 ± 0.024	15.864 ± 0.020	38.441 ± 0.048
S8	1907899†	18.536 ± 0.018	15.628 ± 0.020	38.156 ± 0.048
S9	1907899a†	18.546 ± 0.018	15.626 ± 0.020	38.150 ± 0.048
S10	2307897‡	28.401 ± 0.034	16.170 ± 0.022	38.592 ± 0.060
S11	2307897a‡	29.667 ± 0.030	16.222 ± 0.020	38.497 ± 0.048
S12	2307896a‡	80.107 ± 0.112	18.912 ± 0.030	38.733 ± 0.070
S13	2307896b‡	69.413 ± 0.094	18.327 ± 0.032	38.733 ± 0.068
S14	2307896c‡	65.688 ± 0.076	18.155 ± 0.034	38.660 ± 0.076
S15	2307896Q1A‡	75.169 ± 0.082	18.630 ± 0.024	38.820 ± 0.050
S16	2307896Q1B‡	76.235 ± 0.080	18.705 ± 0.024	38.881 ± 0.048
S17	2307896Q2A‡	69.545 ± 0.074	18.349 ± 0.022	38.786 ± 0.048
S18	2307896Q2B‡	67.885 ± 0.072	18.248 ± 0.022	38.773 ± 0.050
S19	2307896Q3A‡	63.725 ± 0.066	18.030 ± 0.022	38.652 ± 0.048
S20	2307896Q3B‡	55.141 ± 0.058	17.589 ± 0.022	38.659 ± 0.048
S21	230789Q4A‡	57.521 ± 0.060	17.711 ± 0.022	38.642 ± 0.048
S22	230789Q4B‡	73.264 ± 0.084	18.539 ± 0.024	38.760 ± 0.052
S23	2307896Z1‡	55.902 ± 0.058	17.585 ± 0.022	38.456 ± 0.048
S24	2307896Z1 RES‡	24.863 ± 0.046	15.969 ± 0.030	38.263 ± 0.076
S25	2307896Z2A‡	71.909 ± 0.074	18.453 ± 0.022	38.723 ± 0.048
S26	2307896Z2A RES‡	28.166 ± 0.150	16.199 ± 0.086	38.279 ± 0.206
S27	2307896Z2B‡	63.685 ± 0.066	18.044 ± 0.022	38.754 ± 0.048
S28	2307896Z3‡	74.606 ± 0.076	18.578 ± 0.022	38.823 ± 0.048
S29	2307896P4‡	49.084 ± 0.050	17.257 ± 0.022	38.688 ± 0.058
S30	230789P4 RES‡	32.955 ± 0.036	16.401 ± 0.020	38.411 ± 0.050

* Corrected for mass fractionation of 0.130 ± 0.012% ($2\sigma_{mean}$) amu^{-1} – errors are $2\sigma_{mean}$ for corrected ratios; † data for heliolitid corals; ‡ data for stromatoporoids.

isotopic closure. Carbonate analyses of stromatoporoid 2307896 (analyses S12–S23, S25 and S27–S29) have very radiogenic Pb isotopic compositions and exhibit a large isotopic range ($^{206}Pb/^{204}Pb$ 49.1–80.1). Analyses of HCl-insoluble residues (principally clays, <0.5% total sample weight) from the same stromatoporoid (analyses S24, S26 and S30) possess distinctly less radiogenic Pb isotopic compositions ($^{206}Pb/^{204}Pb$ 24.9–33.0), but plot along the same linear trend. Results for other stromatoporoids (analyses S3–S7, S10 and S11) and heliolitid corals (analyses S1, S2, S8 and S9) also plot along the same trend, but have moderate to unradiogenic Pb isotopic compositions ($^{206}Pb/^{204}Pb$ 18.5–30.0) and very limited isotopic ranges. Data for 2307896 carbonate samples (analyses S12–S23, S25 and S27–S29) define a 16

point Pb/Pb isochron, corresponding to a date of 324 ± 41 Ma (MSWD 2.34, model μ_1 value 8.21 ± 0.08), too young for the specimen's Middle Silurian stratigraphic position (which approximates to an age of 423 Ma; McKerrow, pers. comm.). All 2307896 data, including the residue analyses (S24, S26 and S30), define a more precise isochron date of 325 ± 23 Ma (MSWD 2.03, model μ_1 value 8.21 ± 0.04). The entire data set of 30 points defines an extremely well-fitted isochron date of 326 ± 11 Ma (MSWD 2.06, model μ_1 value 8.21 ± 0.01) and this value is taken as representative of the horizon.

U–Pb results

Results for U–Pb analyses of Slite Formation unit g carbonates are detailed in Table 5 and

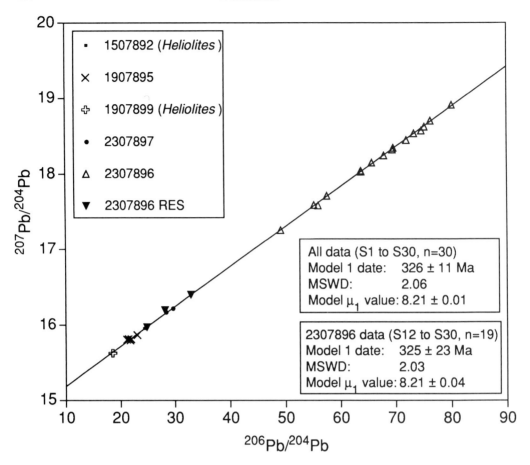

Fig. 5. Pb isotopic data for carbonates and residues from the Slite Formation unit g (*rigidus* and *lundgreni* biozones).

illustrated in Fig. 6. It is evident that closed system conditions have not been maintained. Although the U–Pb analyses do not give reliable information, it is possible to model the likely cause of isotopic scatter by considering the disposition of analyses relative to a 326 Ma reference isochron (the date obtained from the closed system Pb/Pb data). Assuming a model μ_1 value of 8.2, the initial $^{206}Pb/^{204}Pb$ isotopic ratio at 326 Ma would be c. 17.34 (see Fig. 6). U–Pb analyses plot, in general, to the left of this reference line, implying that the cause of isotopic scatter was U loss (or theoretically Pb gain) unlike Klinteberg Formation carbonates where Pb loss (or U gain) was responsible.

Interpretation

The reported Pb/Pb and U–Pb dates require meaningful geological interpretation before

their stratigraphic importance can be critically assessed. For this purpose it is necessary to consider other available geological information. Broadly speaking, the interpretation of carbonate dates is envisaged within the framework of the following major recrystallization events: primary deposition; early diagenetic recrystallization; late diagenetic recrystallization; or metamorphic/tectonic recrystallization.

Lower Middle Klinteberg Formation

Frykman (1984, 1985) documented a well-defined cement stratigraphy for the Klinteberg Formation and concluded that cementation had involved up to four generations of intergranular spar, in addition to fine-grained internal sediment produced by emergent erosion [crystal silt of Dunham (1969)]. Dull luminescent epitaxial overgrowths and bladed rims with inclusions

Table 5. *U–Pb isotopic data for carbonates from the Slite Formation, unit g (rigidus and lundgreni biozones), Wenlock, Gotland*

Analysis	Sample	Concentrations (p.p.m.)		Corrected ratios*			
		U	Pb	$^{238}U/^{204}Pb$	$^{206}Pb/^{204}Pb$	$^{235}U/^{204}Pb$	$^{207}Pb/^{204}Pb$
S3	1907895a‡		0.213 ± 0.001		21.434 ± 0.022		15.812 ± 0.020
S4	1907895b‡		0.145 ± 0.001		21.908 ± 0.022		15.810 ± 0.020
S5	1907895Z1‡		0.177 ± 0.001		21.736 ± 0.022		15.793 ± 0.020
S7	1907895Z2‡	0.294 ± 0.000	0.326 ± 0.001	61.036 ± 0.964	23.046 ± 0.024	0.442 ± 0.006	15.864 ± 0.020
S8	1907899†		0.242 ± 0.001		18.536 ± 0.018		15.628 ± 0.020
S10	2307897‡	0.261 ± 0.000	0.141 ± 0.001	134.835 ± 1.087	28.401 ± 0.034	0.977 ± 0.008	16.170 ± 0.022
S12	2307896a‡	0.584 ± 0.000	0.087 ± 0.000	710.983 ± 11.204	80.107 ± 0.112	5.154 ± 0.041	18.912 ± 0.030
S13	2307896b‡	0.471 ± 0.000	0.090 ± 0.000	733.516 ± 11.538	69.413 ± 0.094	5.317 ± 0.042	18.327 ± 0.032
S15	2307896Q1A‡	0.459 ± 0.000	0.074 ± 0.000	774.119 ± 12.170	75.169 ± 0.082	5.611 ± 0.044	18.630 ± 0.024
S16	2307896Q1B‡	0.478 ± 0.000	0.069 ± 0.000	707.756 ± 11.154	76.235 ± 0.080	5.130 ± 0.040	18.705 ± 0.024
S17	2307896Q2A‡	0.525 ± 0.000	0.074 ± 0.000	743.910 ± 11.718	69.545 ± 0.074	5.392 ± 0.042	18.349 ± 0.022
S18	2307896Q2B‡	0.514 ± 0.000	0.077 ± 0.000	667.525 ± 10.506	67.885 ± 0.072	4.839 ± 0.038	18.248 ± 0.022
S19	2307896Q3A‡	0.675 ± 0.020	0.081 ± 0.000	645.386 ± 10.176	63.725 ± 0.066	4.678 ± 0.037	18.030 ± 0.022
S20	2307896Q3B‡	0.614 ± 0.000	0.101 ± 0.000	719.985 ± 11.334	55.141 ± 0.058	5.219 ± 0.041	17.589 ± 0.022
S21	2307896Q4A‡	0.659 ± 0.020	0.084 ± 0.000	1094.591 ± 17.236	57.521 ± 0.060	7.934 ± 0.062	17.711 ± 0.022
S22	2307896Q4B‡	0.579 ± 0.000	0.068 ± 0.000	482.151 ± 7.608	73.264 ± 0.084	3.495 ± 0.028	18.539 ± 0.024
S23	2307896Z1‡	0.477 ± 0.000	0.117 ± 0.000	701.588 ± 11.034	55.902 ± 0.058	5.086 ± 0.040	17.585 ± 0.022
S25	2307896Z2A‡	0.662 ± 0.020	0.076 ± 0.000	662.104 ± 10.434	71.909 ± 0.074	4.799 ± 0.038	18.453 ± 0.022
S27	2307896Z2B‡	0.529 ± 0.000	0.105 ± 0.000	784.492 ± 12.344	63.685 ± 0.066	5.686 ± 0.045	18.044 ± 0.022
S28	2307896Z3‡	0.657 ± 0.020	0.077 ± 0.000	426.448 ± 6.728	74.606 ± 0.076	3.091 ± 0.024	18.578 ± 0.022
S29	2307896P4‡		0.141 ± 0.001		49.084 ± 0.050		17.257 ± 0.022

* Corrected for Pb isotopic mass fractionation of $0.130 \pm 0.012\%$ ($2\sigma_{mean}$) amu^{-1} and U isotopic mass fractionation of $0.132 \pm 0.012\%$ (2σ) amu^{-1} – errors are $2\sigma_{mean}$ for concentrations and corrected ratios; † data for heliolitid corals; ‡ data for stromatoporoids.

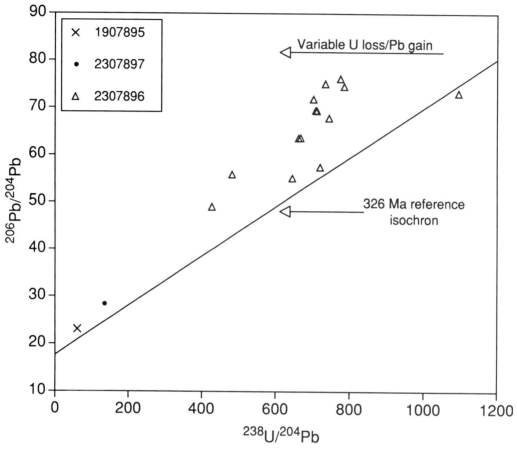

Fig. 6. $^{238}U-^{206}Pb$ isotopic data for carbonates from the Slite Formation unit g (*rigidus* and *lundgreni* biozones).

of microdolomite are interpreted as the closed system replacement of an early marine cement phase (stage 1) (Grover & Read 1983). They are succeeded by clear, non luminescent spar, deposited from meteoric oxidizing pore waters in the form of LMC (stage 2) and then clear, bright luminescent LMC spar deposited as progressively more reducing pore waters evolved upon reburial (stage 3) (Frank et al. 1982; Grover & Read 1983). Fine non-luminescent zonations indicate open system conditions with periodic reflux of oxidizing waters. Clear, subtly zoned, dull luminescent LMC spar occludes remaining porosity (stage 4). This stage is usually the most abundant and is interpreted as reflecting continued reducing conditions and burial (Frank et al. 1982). Frykman (1985) interpreted the low micrite content and the regularity and clarity of the growth zones observed in Klinteberg Formation cement stages 2–4 as evidence of crystal growth within primary porosity and minimal subsequent diagenetic

alteration of the primary LMC mineralogy.

Stromatoporoid 16078913, a large domal specimen collected in growth position from Lilla Snögrinde 2, preserves a vague laminar structure in thin section, with distinct latilaminae forming planes of weakness along which the carbonate could be easily split. Patchy preservation of structural detail indicates that imperforate, undulose laminar structures predominated, with pillars only weakly developed, suggesting classification as a *Clathrodictyon* sp. (Kershaw, pers. comm.) and implying that primary cementation must have been relatively early. Cathodoluminescence studies highlighted a cement stratigraphy broadly similar to that defined by Frykman (1985). A coarse, blocky, non-luminescent clear spar cement (stage 2) is formed on dull luminescent, cloudy structural components (stage 1) and succeeded firstly by a dull luminescent LMC phase (stage 4) and secondly by late, dull to bright luminescent dolomite. The stage 3 cement of Frykman (1984, 1985) is not

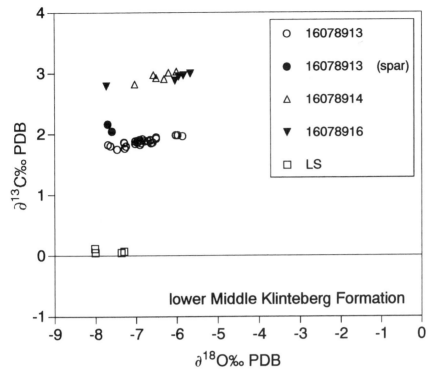

Fig. 7. Stable isotopic data for stromatoporoids from the lower Middle Klinteberg Formation (*ludensis* biozone).

developed. The late variably luminescent dolomite, observed by Frykman (1986) at Klinteberget 1, but not described by Frykman (1984, 1985) for Lilla Snögrinde localities, is interpreted as having formed under mixing zone conditions after slight marine transgression. While stromatoporoids 16078914 and 16078916 exhibited a domal structure in their outcrop at Lilla Snögrinde 1, in thin section they were found to be completely recrystallized and preserve no primary structural detail, making classification impossible.

Stable isotopic data for stromatoporoids from the lower Middle Klinteberg Formation are given in Fig. 7. Despite the fact that particular samples have distinct $\delta^{13}C$ compositional ranges, all values are consistent with a shallow marine origin followed by meteoric diagenesis. If one also considers the variation of Pb concentration, U concentration and isotopic composition between growth zones, as illustrated in Fig. 8, an interesting pattern begins to emerge. There is a steady increase in Pb concentration from the earliest formed carbonate (Z4, nearest the substrate) to the outermost growth zones, while the U concentration shows marked variability. The most uraniferous carbonate is associated with

the outermost growth zones. The $^{206}Pb/^{204}Pb$ ratio shows a general decrease from substrate to margin, a trend echoed by the $^{207}Pb/^{204}Pb$ ratio. The $^{208}Pb/^{204}Pb$ ratio, however, shows very little variation and hence it is concluded that Th was not a significant elemental contributant to stromatoporoid mineralogy. Th/Pb ratios for this sample were either close to zero, or constant (if the initial $^{208}Pb/^{204}Pb$ ratio was *c.* < 38). Unfortunately, the differentiation of relict depositional isotopic signatures from diagenetic signatures, and hence quantification of any primary biogenic isotopic/elemental fractionation, is not possible. However, it is clear that systematic variations in both concentration and isotopic composition do occur within the stromatoporoid and this phenomenon has to be explained in any geochemical model. From Fig. 4 it was concluded that either U gain or Pb loss was responsible for the isotopic scatter exhibited by the lower Middle Klinteberg carbonates. Considering the solubility of U in oxidizing waters (Klinkhammer & Palmer 1991) the well-documented evidence for emergence provided by raised Quaternary shorelines on Gotland (Jeppsson 1982) and the variability of U concentrations in Fig. 8, modification of the primary U

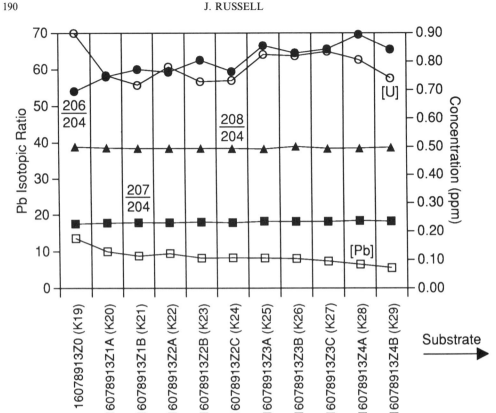

Fig. 8. Concentration and isotopic composition data for 16078913 zones from the lower Middle Klinteberg Formation (*ludensis* biozone).

signal by secondary enrichment appears to be the most plausible explanation.

It must be acknowledged that the scope of this study does not permit quantification of the potential effect that acid-leaching of HCl-insoluble residues may have had on the U and Pb concentration and composition of HCl-soluble 'carbonate' analyses. In theory, U and Pb leached from residues could have a significant effect where carbonate U and Pb concentrations are very low. It is possible that leaching could account for the observed scatter in stromatoporoid U–Pb data (assuming that the carbonate and residue were of approximately the same age and had similar model μ_1 values, since Pb/Pb data show very little scatter). As U–Pb radiometric dating of carbonates becomes more widely applied, this is an area that requires investigation and quantification.

The Pb/Pb date of 432 ± 15 Ma for lower Middle Klinteberg Formation stromatoporoids (*ludensis* biozone) (Fig. 3) is in good agreement with the proposed 424 Ma age of the Wenlock–Ludlow boundary (Harland *et al.* 1989). The six-

point $^{238}U/^{206}Pb$ date of 435 ± 9 Ma for analyses K15–K18, K30 and K31 is in excellent agreement with the Pb/Pb regression. The *c.* 430 Ma date is interpreted as the age of early carbonate diagenesis, most probably related to the recrystallization of stromatoporoids during meteoric diagenesis. Given the magnitude of the errors on both Pb/Pb and U–Pb dates, this value cannot be distinguished from the age of deposition. In terms of accuracy, the U–Pb date of 435 ± 9 Ma is 2 Ma outside error of the inferred 424 Ma age for the Wenlock–Ludlow boundary, while the Pb/Pb date of 432 ± 15 Ma is consistent.

Slite Formation, unit g

To date, no diagenetic studies have been published for the Slite Formation, although Sundquist (1982) and Laufeld *et al.* (1978) have commented on the petrography. Stromatoporoids 1907895, 2307896 and 2307897 show patch preservation of a broad, open, reticulate structure with micritic chevron-type laminae, suggesting classification as *Ecclimadictyon* sp.

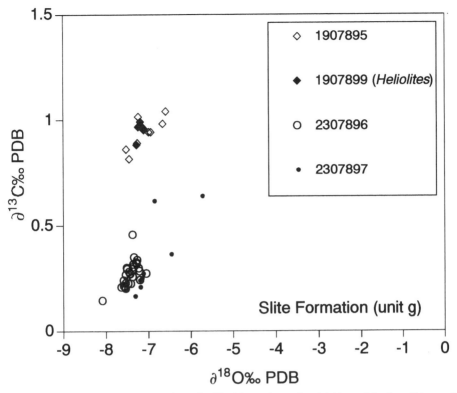

Fig. 9. Stable isotopic data for carbonates from the Slite Formation unit g (*rigidus* and *lundgreni* biozones).

(Kershaw, pers. comm.). Cathodoluminescence studies indicate that the preserved structural elements are essentially dull luminescent with bright luminescent micro-inclusions. A thin bright luminescent band at the outer margin of the skeleton is followed by blocky, non-luminescent fringing cement and thick dull luminescent calcite infill. A final bright luminescent cement is developed locally. The structural detail indicates that compression of the skeleton has not occurred and so early marine cementation is likely.

Stable isotopic data for Slite Formation unit g carbonates are grouped into two subsets; the first (1907895 and 1907899) has $\delta^{13}C$ clustered *c.* 1‰, while the second (2307896 and 2307897) had $\delta^{13}C$ values clustered *c.* 0.3‰ (Fig. 9). $\delta^{18}O$ values tend to lie between -6 and -8‰, suggestive of similar degrees of diagenetic alteration and implying that carbon isotopic values might reflect primary variations. Values of 1.4‰ for carbon and 0 to -4‰ for oxygen, quoted by James & Choquette (1983) for typical Silurian carbonates, reinforce the suggestion that carbon isotopic compositions may be primary and confirm the presence of meteoric

fluids during diagenetic recrystallization.

Systematic variations in Pb concentration, U concentration and isotopic composition between growth zones of 2307896 systematic variations are observed (Fig. 10). Although the Pb concentration shows little variation, the U concentration falls from *c.* 0.66 p.p.m. in the earliest formed zones to *c.* 0.54 p.p.m. in the outermost growth zones. The trend is opposite to that exhibited by stromatoporoid 16078913 from the Klinteberg Formation. Analysis S18 (2307896Q2B) appears anomalously high in respect of both Pb and U concentrations. $^{206}Pb/^{204}Pb$ ratios are highest at the core and tend to decrease toward the margin (with the exception of S22), a trend mirrored by $^{207}Pb/^{204}Pb$ values. More interestingly, the $^{206}Pb/^{204}Pb$ ratio shows an apparent inverse relationship to the U concentration, opposite to what one might expect, assuming no heterogeneity. $^{208}Pb/^{204}Pb$ values vary little from *c.* 38 (see earlier discussion).

From Fig. 6 it was implied that either Pb gain or U loss was responsible for the recent isotopic disturbance experienced by the Slite Formation carbonates. Recent Pb gain is most unlikely,

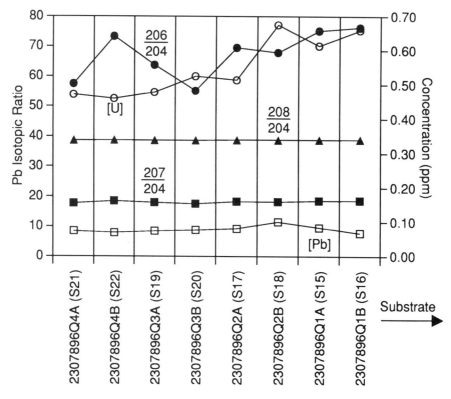

Fig. 10. Concentration and isotopic composition data for 2307896Q zones from the Slite Formation unit g (*rigidus* and *lundgreni* biozones).

since the Slite Formation Pb/Pb isochron is so well defined and any gain of Pb with a modern isotopic composition would generate scatter in excess of that observed in Fig. 5. Consequently, it is concluded that U loss was responsible for the recent isotopic disturbance. Considering the solubility and mobility of U in oxidizing waters (Klinkhammer & Palmer 1991) and the evidence for Quaternary emergence on Gotland (Jeppsson 1982), this is not unreasonable. It is not possible, however, to isolate unequivocal evidence for this explanation from the concentration and isotopic composition variations.

The Pb/Pb date of 326 ± 11 Ma for Slite Formation unit g stromatoporoids, residues and corals (Fig. 5) is significantly younger than the 423 Ma age for the Middle Wenlock proposed by McKerrow (pers. comm.). While the samples show a degree of stratigraphic variation (*rigidus* and *lundgreni* biozones), all analyses plot along the same Carboniferous Pb/Pb isochron (Middle Namurian). As no significant karstification of appropriate age is noted within the

Gotland succession, the isotopic resetting may have resulted from some sub-surface, deep-weathering event. However, Klinteberg Formation rocks from the west of the island, 70 m higher in the undisturbed stratigraphic section, yield chronostratigraphically accurate ages and so the isotopic resetting event appears to have been constrained to a particular stratigraphic interval.

It can be seen from Fig. 1 that the Slite Formation unit g sample localities (Stora Banne 3, Spillings 2, Lännaberget 2 and Bogeklint 2) all lie along a distinct north–south trend. This trend aligns with north–south sub-Cambrian faults identified by Flodén (1980) from seismic soundings north of Gotland. Jeppsson & Laufeld (1986) documented red weathering in association with Late Palaeozoic intrusions in Sweden (late Carboniferous to early Permian dykes in Skåne, sills in Västergötland) and found that this was confined to the vicinity of major fault and graben systems. The red weathering was shown to affect rocks down to the major sub-Cambrian peneplain, presumably by fluid percolation along fault systems. A similar mechanism is proposed in this study to explain late diagenetic recrystallization of the Slite Formation carbonates. This recrystallization event failed to affect the stratigraphically higher Klinteberg Formation because the two formations are separated by a zone of impermeable shales and mudstones (Upper Slite and Mulde Formations) which would have stratigraphically restricted fluid flow. Hence, the $c.330\,Ma$ date is interpreted in geological terms as the age of late carbonate diagenetic recrystallization during the early stages of the Late Carboniferous.

Geochemical review

Much work has been published on the geochemical behaviour of U (e.g. Langmuir 1978; Broecker & Peng 1982; Klinkhammer & Palmer 1991) and Pb (e.g. Schaule & Patterson 1981; Hamelin *et al.* 1990; Erel *et al.* 1991) but few have attempted to integrate the available data with respect to their behaviour in sediments. A broad geochemical review of the behaviour of U and Pb is presented to explain why carbonates can constitute suitable materials for direct dating by Pb/Pb and U–Pb methods. More specifically, why they were evidently capable of developing high, variable U/Pb ratios and maintaining closed system conditions for substantial periods of geological time.

The geochemical behaviour of U in solution is distinctive in that it converts to a more insoluble form on reduction. U in river waters is derived from the weathering of rocks at the Earth's surface and is transported into the marine realm as particulate matter, dissolved ionic species and/or adsorbed on to colloids. River waters have an average U concentration of 0.24 p.p.b. in solution and 3 p.p.m. in the form of particulate matter, whereas marine waters average 3.2 p.p.b. and marine clays 2 p.p.m. (Martin & Whitfield 1983). In the oceanic reservoir (pH > 6), U exists mainly as soluble U^{6+} in the form of the uranyl carbonate complexes $UO_2(CO_3)_2^{2-}$, $UO_2(CO_3)_3^{4-}$ and UO_2CO_3 and also as $UO_2(HPO_4)_2^{2-}$ (Langmuir 1978). These are probably the most important species for the uptake of U by carbonates since the charge and ionic radius discrepancy between U^{6+} and Ca^{2+} (0.08 and 0.099 nm, respectively) means that direct lattice site substitution of hexavalent U is insignificant (Möller *et al.* 1976). The oceanic residence time (ORT) (the average time an element resides in the marine reservoir prior to its removal by the sediment sink) of U is $c.$

4.9×10^5 years (Martin & Whitfield 1983), while the average mixing time of the open oceanic water mass is $c.$ 1600 years (Chester 1990). Therefore, U becomes homogenized rapidly within the oceans and the character of the lithological catchment area will not cause local perturbations.

Like U, weathered Pb is carried to the oceans in river waters as particulate matter, dissolved species and colloidal suspensions. Additional Pb is input from aeolian sources and as direct hydrothermal emissions within the ocean basins. River water has an average Pb concentration of 0.1 p.p.b. in solution and 100 p.p.m. in the form of particulate matter, while marine waters average 0.003 p.p.b. in solution and up to 200 p.p.m. as particulate marine clays (Martin & Whitfield 1983). In seawater, under oxidizing conditions, the main species of Pb [$PbCO_3$, $Pb(CO_3)^{2-}$ and $PbCl^+$] are concentrated in the surface microlayer and decrease in abundance with depth because Pb is rapidly scavenged from the oceans (Chester 1990). Scavenging has been defined by Goldberg (1954) as the adsorptive removal of trace element metals on to particles that sink through the water column. Unlike U, Pb is a reactive trace element in ocean waters, particularly at boundary layers (e.g. river–ocean, air–sea, sediment–ocean) (Schaule & Patterson 1981). Martin & Whitfield (1983) calculated the ORT of Pb to be 1.1×10^3 years, but because published estimates varied considerably, Wangersky (1986) proposed that ionic/complexed Pb and scavenged Pb had separate values of ORT. Estimates of the scavenged ORT generally fall in the range of 47–54 years (Balistieri *et al.* 1981; Whitfield & Turner 1987), while the ORT for ionic/complexed species is an order of magnitude greater (Brewer 1975). It is clear that both residence times are short with respect to the 1600 year mean mixing time for the oceanic water mass, and hence the concentration and composition of Pb is heterogeneous within the oceans and largely controlled by source (rivers, hydrothermal and aeolian input).

Concentration data for U and Pb indicate that river water has a dissolved U/Pb ratio of 2.4, while seawater has a much higher value of $c.$ 1100. These figures correspond to μ values of 170 and 80 000, respectively. Both fluvial and oceanic fine-grained clastic sediments have high Pb concentrations (> 100 p.p.m.). Therefore, sedimentary carbonates that are capable of variably sampling and concentrating dissolved U and Pb from the oceans, without the incorporation of significant amounts of clastic detrital material, have the potential to generate large ranges of *in situ* μ values at deposition, ideal for geochronological study using U–Pb systematics.

Carbonate sediments are deposited with preconcentrations of U and Pb, established authigenically within the water column through the formation of complexes with particulate or dissolved organic matter and particulate Fe–Mn oxyhydroxides (Disnar & Trichet 1983; Hsi & Langmuir 1985; Balistieri & Murray 1982, 1983, 1986; Trefry et al. 1985). U shows a greater affinity for non-labile organic complexing, while Pb is complexed preferentially by labile inorganic Fe–Mn oxyhydroxides. Small amounts of U and Pb may be incorporated into lattice related sites. Theoretically, both U^{3+} and Pb^{2+} can substitute for Ca^{2+} in the aragonite and, to a more limited extent, the calcite lattice, but U^{3+} is stable only under reducing conditions (Amiel et al. 1973; Möller et al. 1976). Lattice substitution is therefore unlikely to be a major factor as far as primary calcium carbonate precipitation is concerned because this occurs principally under oxidizing conditions.

Organic matter is capable of fixing UO_2^{2+} ions directly from seawater (Disnar 1981; Disnar & Trichet 1983). Moreover, the presence of organic matter within sedimentary carbonates controls redox conditions during diagenesis. Oxidation of organic matter during diagenesis can create suboxic conditions within the sediment and pore water that result in the reduction of uranyl carbonate complexes to more insoluble forms, causing a decrease in pore water U concentrations. Some fixation of Pb from suboxic boundary layer waters may also occur due to complexing with dissolved organic matter, thereby limiting the potential for the formation of inorganic complexes. Thus, a concentration gradient becomes established within the boundary layer and uranyl complexes are drawn down from the ocean (Klinkhammer & Palmer 1991). The flux is enhanced by U released from decomposition of organic materials within the boundary layer (Klinkhammer & Palmer 1991) and a certain amount of Pb may also be remobilized in this fashion. Initially, uranyl complexes are probably reduced to metastable U^{5+} intermediates (Kniewald & Branica 1988) at temperatures of 45–250°C (Meunier et al. 1990), but the ultimate sink for U from the oceans is most likely to be uranite (Klinkhammer & Palmer 1991), forming as temperatures extend into the 120–400°C range (Meunier et al. 1990). In low temperature diagenetic environments ($<120°C$), adsorption and complexing are more important processes than U mineral precipitation and so preconcentration is an important factor in determining total U concentration. The exact behaviour of Pb during early diagenesis remains unclear.

The following factors are important in the establishment of favourable geochemical conditions within sedimentary carbonates: chemical deposition from seawater resulting in potential for high μ values; intimate association with organic matter; suboxic conditions; and a low clastic content. Qualitatively, they all favour incorporation of U and discriminate against Pb. Biogenic carbonates (e.g. corals, stromatoporoids, stromatolites) possess each of the requisite characteristics and have been dated successfully using U–Pb and Pb/Pb techniques. Calculated in situ μ_2 values (a maximum of 1200 for stromatoporoids), however, are still almost two orders of magnitude lower than the μ value of seawater (80 000) and this reflects the relative efficiency of Pb removal from the oceans.

The short oceanic residence time of Pb and its highly reactive nature mean that the Pb isotopic composition of marine sediments is particularly sensitive to the character of the erosional hinterland. U, on the other hand, is conservative within the oceanic regime. Rapid fixation of Pb in the coastal environment may result in minor heterogeneities in initial Pb isotopic composition, contributing to data scatter on isochron diagrams. Indeed, Bruland & Franks (1983) and White et al. (1985) documented marked variations in the Pb isotopic composition of marine sediments, both between and within modern oceans, although such results may be complicated by mixing of source and anthropogenic Pb signals. Nevertheless, if large ranges in U/Pb ratios exist and the Pb isotopic heterogeneity is limited (likely because of rapid scavenging and fixation), useful age data may still be obtained. Model μ_1 values for sedimentary carbonates must also be controlled largely by the character of Pb input from the hinterland. The μ_1 values may be subject to alteration during early diagenesis, as U and Pb are fixed in suboxic environments, but they should still reflect the sedimentary source region. Under conditions of late diagenesis, metamorphism or igneous activity, significant alteration of μ_1 values is likely, but this will depend on the nature of the diagenetic system (whether open or closed), operative water/rock ratios and degree of contamination (see also discussion in Jahn & Cuvellier 1994).

Modification of primary elemental and isotopic signals usually occurs after deposition and this has implications for the interpretation of geochronological data. For isotopic ages to be reset to zero, rehomogenization of Pb isotopic ratios must occur without accompanying homogenization of U/Pb ratios (although these will inevitably be modified). Since U and Pb exhibit differing geochemical behaviour, events capable of crystallizing carbonates have the potential to effect such changes; for example, pressure solution and neomorphism, dolomitization, tectonic deformation, metamorphic recrystallization and igneous activity. Incomplete Pb isotopic homogenization will either bias or destroy age relationships, whereas U/Pb (and therefore μ value) homogenization will result in negligible data scatter (of no use for geochronology). One should also consider the possibility that U and Pb, or their intermediate decay products (such as ^{222}Rn), become decoupled during diagenesis and/or metamorphism, producing spurious age relationships. Models invoking such behaviour have been proposed to explain Pb/Pb ages for some Middle Proterozoic carbonates from China (Cuvellier 1992; Jahn & Cuvellier 1994). The nature of this study does not provide a sufficiently detailed data base to address this question directly, but most of the results obtained are consistent with independent age constraints and may be interpreted in simple geological terms, so any quantitative loss of intermediate daughter isotopes does not appear to have had a significant effect on carbonate dates.

U–Pb data for stromatoporoids (this study) and evidence from published fission track studies (Swart & Hubbard 1982) suggest that U mobility in oxidizing fluids is the dominant control on the modification of post depositional elemental and isotopic signatures (up to the onset of metamorphism). Each recrystallization will tend to lower *in situ* μ_2 values for carbonates. However, recent (i.e. post-Quaternary) loss of U and Pb, or gain of U will have had little appreciable effect on Pb/Pb systematics, because any Pb lost from the system has effectively the same isotopic composition as that remaining in the sample and insufficient time remains for modified U/Pb ratios to produce significantly differing Pb isotopic compositions. U–Pb geochronology, on the other hand, requires the quantitative retention of U and Pb and so any loss/gain of U or Pb will inevitably produce excess scatter of U–Pb data and will probably destroy age relationships.

Implications for Precambrian carbonates

A number of recent publications (Smith & Farquhar 1989; DeWolf & Halliday 1991; this study) have demonstrated that, under favourable conditions, U–Pb systematics can be applied meaningfully to the direct dating of Phanerozoic biogenic carbonates. Large ranges in U and Pb isotopic composition over sub-centimetre size domains are a characteristic feature of such materials (e.g. corals and stromatoporoids), and despite the mobility of U during diagenetic transformation, post-crystallization closed-system conditions have evidently been in operation for considerable periods of geological time.

Unfortunately, the slow accumulation of radiogenic ^{207}Pb during the Phanerozoic limits the precision and chronostratigraphic value of Pb/Pb carbonate dates for this portion of geological time. More precise constraints on sediment age can usually be obtained from K–Ar studies on potassic bentonites (Odin *et al.* 1984, 1986*a, b*) and U–Pb studies on separated volcanic zircons. Nevertheless, Pb/Pb studies are capable of identifying and temporally constraining late diagenetic and metamorphic recrystallization events in the Phanerozoic (Parnell & Swainbank 1990; Jahn 1988; Jahn *et al.* 1992; this study – see also discussion in Jahn & Cuvellier 1994). In the Precambrian, however, the apparent robustness of Pb/Pb system-atics in biogenic carbonates (in particular stromatolites), coupled with the extensive regional and temporal distribution of such materials has exciting potential for event dating, stratigraphic correlation and timescale calibration. Since Pb/Pb dates for Precambrian stromatolites may achieve a precision of $\pm 1\%$ (Moorbath *et al.* 1987; Cuvellier 1992; Russell 1992) this potential cannot be overemphasized in the light of the general paucity of reliable independent age constraints for Precambrian sedimentary sequences and basins around the World.

It has been demonstrated in this study that the interpretation of Pb/Pb data for Phanerozoic carbonates is by no means straightforward. Many processes capable of recrystallizing carbonates can disturb and even reset U–Pb and Pb/Pb systematics. Given that great reliance may come to be placed on Precambrian carbonate dates, it is vital that their interpretation is both realistic and consistent with other reliable geological constraints in the area (independent age data, geotectonic evolution, petrography, stable isotopes). In particular, one must be aware of the efficiency with which metamorphic recrystallization resets carbonate chronometers (see discussion in Jahn & Cuvellier 1994). Where metamorphism is obvious, as in the case of the Greenland Proterozoic marbles dated by Taylor & Kalsbeek (1990), interpretation is relatively simple, but where metamorphism is of lower grade (burial metamorphism) and the visible effects less discernible, erroneous interpretations could easily result. The careful selection of pristine sample materials and petrographic screening for signs of metamorphic alteration prior to analysis can be of considerable help in this respect.

Conclusions

The Pb and U isotope investigation of Silurian carbonates from the Swedish island of Gotland has demonstrated that Pb/Pb and, in certain cases, U–Pb systematics may be applied successfully to the dating of Phanerozoic sediments. A significant number of credible Pb/Pb and U–Pb dates for carbonates have now been published in the literature. These dates have been interpreted in terms of early diagenetic age (approximating to true stratigraphic age), late diagenetic age and age of metamorphic recrystallization. The potential of carbonate Pb/Pb and U–Pb geochronology for constraining the age of sedimentary and metamorphosed sequences is self-evident.

Analyses of four carbonate stromatoporoids from the lower Middle Klinteberg Formation (*ludensis* biozone, Wenlock–Ludlow boundary) give a Pb/Pb date of 432 ± 15 Ma and selected analyses give a ^{238}U–^{206}Pb date of 435 ± 9 Ma. Both dates are in good agreement with the inferred Wenlock–Ludlow age of 420 Ma and are interpreted as depositional/early diagenetic ages. In general, U–Pb data exhibit significant scatter and this is interpreted to be the result of recent isotopic disturbance (U loss) due to Quaternary weathering on the island. The scope of this study does not permit quantification of the effect that leaching of acid-insoluble residues may have on results. This is a topic that requires further investigation and quantification. A combination of results from the four separate stromatoporoids expands the overall Pb isotopic range and thereby enhances the precision of the regressed Pb/Pb date. This sampling strategy is recommended where single organism isotopic ranges are limited, or only small quantities of sample materials are available.

Analyses of three stromatoporoids and two heliolitid corals, collected from four localities within the diachronous Slite Formation unit g (*rigidus* and *lundgreni* biozones, Middle Wenlock), give a Pb/Pb date of 326 ± 11 Ma. The date is *c.* 100 Ma younger than the inferred stratigraphic age (423 Ma) and is interpreted as the age of late diagenetic recrystallization during the late Carboniferous. Stratigraphically higher carbonates in the Klinteberg Formation remained unaffected by this event because dia-

genetic waters, channelled along north–south faults at depth, were constrained stratigraphically by the largely impermeable, shale-dominated rocks of the Upper Slite and Mulde Formations. This represents the first reported chronometric control on the post-Silurian/pre-glacial geological history of the Gotland area.

A review of the geochemical behaviour of U and Pb has been presented to explain the suitability of certain carbonate materials for direct dating using U–Pb and Pb/Pb methods. Appropriate geochemical conditions are believed to be established where there is chemical deposition from seawater and concomitant potential for high μ values, intimate association with organic matter and suboxic conditions and a low clastic content. Carbonates that have been dated successfully thus far (corals, stromatoporoids and stromatolites) adhere to these criteria. Consequently, the criteria can be used as guidelines for the identification of other carbonates which have dating potential. Unpublished data from the author suggest that red algae and sponges merit further investigation.

The small number of reliable dates for Precambrian sedimentary sequences means that direct Pb/Pb dating of stromatolites offers a powerful tool for purposes of direct age determination and correlation. The increasing number of microfossil discoveries in such rocks suggests that an integrated Precambrian timescale may ultimately be possible. Pb/Pb dates for marbles, on the other hand, can help to constrain regional geotectonic episodes. Difficulties experienced in interpreting dates for Phanerozoic carbonates become amplified in the Precambrian and so a careful, realistic approach to interpretation is imperative. All available geological, chronometric, tectonic and geochemical information should be fully evaluated prior to interpretation.

The author wishes to acknowledge a generous grant from BP Exploration Operating Company Ltd, without which completion of this work would have been impossible. Analytical facilities in the Age Laboratory, Department of Earth Sciences at the University of Oxford were kindly made available by Professor Stephen Moorbath. In addition, my thanks go to Dr Lennart Jeppsson, who assisted in fieldwork arrangements; the University of Lund, which allowed access to its field station (Allekvia) during sample collection on Gotland; and Dr Martin Whitehouse, Professor Stephen Moorbath, Dr John Arden and Roy Goodwin for advice both during the analysis of samples and preparation of the manuscript. This work has benefited considerably from reviews by Bor-ming Jahn and Patrick Smith, for which the author is grateful.

Appendix

Brief descriptions of the reference localities cited on Gotland are included. For a full description, the reader is referred to Laufeld (1974b).

Bogeklint 2, CJ 6724 9417, c. 2175 m south–southeast of Boge church. Topographical map sheet 6 J Roma NO. Geological map sheet Aa 169 Slite.

Quarry in the SE-facing cliff c. 410 m south of the northeasternmost house at Klinte.

Slite Formation, unit g (outlier).

Klinteberget 1, CJ 3391 6340, c. 410 m north–northeast of Klinte church. Topographical map sheet 6 I Visby SO. Geological map sheet Aa 160 Klintehamn.

Inland cliff section facing northwest in the northernmost part of Klinteberget c. 30 m north of the triangulation point (52,04) at Klinteberget.

Klinteberg Formation, lower and middle parts.

Lännaberget 2, CJ 6928 9908, c. 840 m northeast of Slite church. Topographical map sheet 7 J Fårösund SO and NO. Geological map sheet Aa 169 Slite.

Section along the road to the oil storage tanks c. 60 m southwest of the main road and c. 35 m west of the northernmost industry building (electricity central) at Cementfabrik.

Slite Formation, unit g, lower part.

Lilla Snögrinde 1, CJ 3421 6315, c. 550 m east of Klinte church. Topographical map sheet 6 I Visby SO. Geological map sheet Aa 160 Klintehamn.

Section immediately east of the road, c. 290–415 m south-southeast of the crossroads just east of the northernmost end of the hill of Klinteberget.

Klinteberg Formation, lower middle part.

Lilla Snögrinde 2, CJ 3442 6199, c. 1340 m south-southeast of Klinte church. Topographical map sheet 6 I Visby SO. Geological map sheet Aa 160 Klintehamn.

Two sections immediately east of the road, c. 1485–1700 m south-southeast of the crossroads just east of the northernmost end of the hill of Klinteberget.

Klinteberg Formation, middle part.

Spillings 2, CJ 6672 9973, c. 2370 m northwest of Slite church. Topographical map sheet 7 J Fårösund SO and NO. Geological map sheet Aa 169 Slite.

Entrance wall of limestone quarry c. 435 m southeast of triangulation point at Klintsbackar.

Slite Formation, unit g.

Stora Banne 3, CK 6832 0987, c. 2580 m north-northwest of Lärbro church. Topographical map sheet 7 J Fårösund SO and NO. Geological map sheet Aa 171 Kappelshamn.

Quarry west of road, c. 1000 m north of the house west of the road at Simunds. Stora Banne 3 is the southern of the two quarries marked on the geological map sheet.

Slite Formation, unit g, lower part.

References

AMIEL, A. J., MILLER, D. S. & FRIEDMAN, G. M. 1973. Incorporation of uranium in modern coals. *Sedimentology*, **20**, 523–528.

ARDEN, J. W. & GALE, N. H. 1974. Separation of trace amounts of uranium and thorium and their determination by mass spectroscopic isotope dilution. *Analytical Chemistry*, **46**, 687–691.

ARMSTRONG, R. L. 1978. Pre-Cenozoic Phanerozoic time scale. *In:* COHEE *et al.* (eds) *Contribution to the Geological Time Scale*. American Association of Petroleum Geologists, Studies in Geology, **6**, 73–91.

BALISTIERI, L. S., BREWER, P. G. & MURRAY, J. W. 1981. Scavenging residence time of trace metals and surface chemistry of sinking particles in the oceans. *Deep Sea Research*, **28**, 101–121.

—— & MURRAY, J. W. 1982. The adsorption of Cu, Pb, Zn and Cd on goethite from major ion seawater. *Geochimica et Cosmochimica Acta*, **46**, 1253–1265.

—— & —— 1983. Metal–solid interactions in the marine environment: estimating apparent equilibrium binding constants. *Geochimica et Cosmochimica Acta*, **47**, 1091–1098.

—— & —— 1986. The surface chemistry of sediments from the Panama Basin: The influence of Mn oxides on metal adsorption. *Geochimica et Cosmochimica Acta*, **50**, 2235–2243.

BASSETT, M. G., COCKS, L. R. M., HOLLAND, C. H., RICKARDS, R. B. & WARREN, P. T. 1975. The type Wenlock series. *Institute of Geological Sciences Report*, **75/12**, 1–19.

——, KALJO, D. & TELLER, L. 1989. The Baltic Region. *In:* HOLLAND, C. H. & BASSETT, M. G. (eds) *A Global Standard for the Silurian System*. National Museum of Wales, Geological Series 9, Cardiff, 158–170.

BREWER, P. G. 1975. Minor elements in seawater. *In:* RILEY, J. P. & SKIRROW, G. (eds) *Chemical Oceanography*, Volume 1. Academic Press, London, 415–496.

BROECKER, W. S. & PENG, T. H. 1982. Tracers in the sea. Lamont-Doherty Geological Observatory Publication, New York.

BRULAND, K. W. & FRANKS, R. P. 1983. Mn, Ni, Zn and Cd in the western North Atlantic. *In:* WONG, C. S., BOYLE, E., BURLAND, K. W., BURTON, J. D. & GOLDBERG, G. D. (eds) *Trace Metals in Seawater*. New York, Plenum, 395–414.

CHESTER, R. 1990. Marine Geochemistry. Unwin

Hyman, London.

CLAUER, N. 1981. Rb–Sr and Kr–Ar dating of Precambrian clays and glauconies. *Precambrian Research*, **15**, 331–352.

COMPSTON, W., WILLIAMS, I. S., McCULLOCH, M. T., FOSTER, J. J., ARRIENS, P. A. & TRENDALL, A. F. 1981. A revised age for the Hamersley Group. *Geological Society of Australia, 5th Annual Convention, Abstracts* **3**, 40 (abstract).

CUVELLIER, H. 1989. La datation des roches carbonatées par la methode Pb/Pb. D.E.A. thesis, Rennes.

——— 1992. *Etude isotopique (U–Pb and Rb–Sr) des carbonates protérozoïques du Craton Sino–Coréen et du Briovérien de Bretagne centrale: Implications sur la datation directe des séquences sédimentaires.* Thèse de Doctorat de l'Université de Rennes.

DeWOLF, C. P. & HALLIDAY, A. N. 1991. U–Pb dating of a remagnetised Palaeozoic limestone. *Geophysical Research Letters*, **18**, 1445–1448.

DISNAR, J. R. 1981. Etude expérimentale de la fixation de métaux par un materiau sédimentaire actuel d'origine algaire II. Fixation 'in vitro' de UO_2^{2+}, Cu^{2+}, Ni^{2+}, Zn^{2+}, Pb^{2+}, Co^{2+}, Mn^{2+}, ainsi que de VO_3^-, MoO_4^{2-} et GeO_3^{2-}. *Geochimica et Cosmochimica Acta*, **45**, 363–379.

——— & TRICHETT, J. 1983. Pyrolyse de complexes organo-métalliques formés entre un matériau organique actuel d'origine algaire et divers cations métalliques divalents. (UO_2^{2+}, Cu^{2+}, Pb^{2+}, Ni^{2+}, Mn^{2+}, Zn^{2+} et Co^{2+}). *Chemical Geology*, **40**, 203–223.

DOE, B. R. 1970. Evaluation of U–Th–Pb whole rock dating on Phanerozoic sedimentary rocks. *Eclogae Geologica Helvetica*, **63**, 79–82.

DUNHAM, R. J. 1969. Early vadose silt in Townsend Mound (reef), New Mexico. *In:* FRIEDMAN, G. M. (ed.) *Depositional Environments in Carbonate Rocks*. Society of Economic Paleontologists and Mineralogists Special Publication, **14**, 139–181.

EPSTEIN, A. G., EPSTEIN, G. B. & HARRIS, L. D. 1977. Conodont colour alteration – an index to organic metamorphism. United States Geological Survey Profesional Paper, **995**, 1–27.

EREL, Y., MORGAN, J. J. & PATTERSON, C. C. 1991. Natural levels of lead and cadmium in a remote mountain stream. *Geochimica et Cosmochimica Acta*, **55**, 707–719.

FLODÉN, T. 1980. Seismic stratigraphy and bedrock geology of the central Baltic. *Stockholm Contributions to Geology*, **35**, 1–240.

FRANK, J. R., CARPENTER, A. B. & OGLESBY, T. W. 1982. Cathodoluminescence and composition of calcite cement in the Taum Saul Limestone (Upper Cambrian), southeast Missouri. *Journal of Sedimentary Petrology*, **52**, 631–638.

FRYKMAN, P. 1984. Cement stratigraphy in Silurian bioherm, Gotland, Sweden. International Association of Sedimentolgists, 5th European Meeting, Marseille 1984, 181–182.

——— 1985. Subaerial exposure and cement stratigraphy of a Silurian bioherm in the Klinteberg Beds, Gotland, Sweden. *Geologiska Föreningens i Stockholm Förhandlingar*, **107**, 77–88.

——— 1986. Diagenesis of Silurian bioherms in the Klinteberg Formation, Gotland, Sweden. *In:* SCHROEDER, J. H. & PURSER, B. H. (eds) *Reef Diagenesis*. Springer-Verlag, Berlin, Heidelberg, 399–423.

——— 1989. Carbonate ramp facies of the Klinteberg Formation, Wenlock–Ludlow transition on Gotland, Sweden. *Sveriges Geologiska Undersökning*, **C820**, 1–79.

GALE, N. H. 1985. Numerical calibration of the Palaeozoic time scale: Ordovician, Silurian and Devonian periods. *In:* SNELLING, N. J. (ed.) *The Chronology of the Geological Record*. Geological Society, Memoir, **10**, 81–88.

———, ARDEN, J. W. & ABRANCHES, M. C. B. 1980. Uranium–lead in the Bruderheim L6 chondrite and the 500 Ma shock event in the L-Group parent body. *Earth and Planetary Science Letters*, **48**, 311–324.

GALLOWAY, J. J. 1957. Structure and classification of the Stromatoporoidea. *Bulletin of American Paleontology*, **37**, 345–480.

GOLDBERG, E. D. 1954. Marine geochemistry I. Chemical scavengers. *Journal of Geology*, **62**, 249–265.

GROVER, G. & READ, J. F. 1983. Palaeoaquifer and deep burial related cements defined by regional cathodoluminescence patterns, Middle Ordovician carbonates, Virginia. *Bulletin of the American Association of Petroleum Geologists*, **67**, 1275–1303.

HAMELIN, B., GROUSSET, F. & SHOLKOVITZ, E. R. 1990. Pb isotopes in surficial pelagic sediments from the North Atlantic. *Geochimica et Cosmochimica Acta*, **54**, 37–47.

HARLAND, W. B., ARMSTRONG, R. L., COX, A. V., CRAIG, L. E., SMITH, A. G. & SMITH, D. 1989. A Geologic Time Scale, 1989. Cambridge University Press, Cambridge.

———, COX, A. V., LLEWELLYN, P. G., PICKTON, C. A. G., SMITH, A. G. & WALTERS, R. 1982. A Geologic Time Scale. Cambridge University Press, Cambridge.

———, SMITH, A. G. & WILCOCK, B. (eds) 1964. The Phanerozoic time scale. *Quarterly Journal of the Geological Society of London*, **120s**, 1–458.

HEDE, J. E. 1921. Gottlands Silurstratigrafi. *Sveriges Geologiska Undersökning*, **C331**, 1–100.

——— 1925. Beskrivning av Gotlands Silurlager. *In:* MUNTHE, H., HEDE, J. E. & VON POST, L. (eds) *Gotlands Geologi, en Översikt. Sveriges Geologiska Undersökning*, **C331**, 13–30.

——— 1958. Silurian entries. *In: Lexique Stratigraphique International 1*, Europe 2c, Suede–Sweden–Sverige, 498 pp.

HSI, C. K. D. & LANGMUIR, D. 1985. Adsorption of uranyl onto ferric oxyhydroxides: Application of the surface complexation site-binding model. *Geochimica et Cosmochimica Acta*, **49**, 1931–1941.

JAHN, B. M. 1988. Pb/Pb dating of young marbles from Taiwan. *Nature*, **332**, 429–432.

———, BERTRAND-SARFATI, J., MORIN, N. & MACÉ, J. 1990. Direct dating of stromatolitic carbonates from the Schmidtsdrif Formation (Transvaal Dolomite), South Africa, with implications on the age of the Ventersdorp Supergroup. *Geology*, **18**, 1211–1214.

——, CHI, W. R. & YUI, T. F. 1992. A Late Permian formation of Taiwan (marbles from Chia-Li well No. 1): Pb/Pb isochron and Sr isotopic evidence and its regional significance. *Journal of the Geological Society of China*, **35**, 193–218.

—— & CUVELLIER, H. 1989. Pb/Pb geochronology of carbonates and dating of sedimentary rocks. European Union of Geosciences, V, Geochronology Subsection OS06, (abstract).

—— & —— 1994. Pb/Pb and U–Pb geochronology of carbonate rocks: an assessment. *Chemical Geology (Isotope Geoscience Section)*, **115**, 125–151.

JAMES, N. P. & CHOQUETTE, P. W. 1983. Limestones – the seafloor diagenetic environment. *Geoscience Canada*, **10**, 159–179.

JEPPSSON, L. 1974. Aspects of late Silurian conodonts. *Fossils and Strata*, **6**, 1–54.

—— 1979. Conodont element function. *Lethaia*, **12**, 153–171.

—— 1982. *Third European conodont symposium (ECOS III), guide to excursion.* Publications from the Institutes of Mineralogy, Palaeontology and Quaternary Geology, University of Lund, **239**, 1–32.

—— 1983. Silurian conodont faunas of Gotland. *Fossils and Strata*, **15**, 121–144.

—— & LAUFELD, S. 1986. The Late Silurian Öved–Ramsåsa Group in Skåne, South Sweden. *Sveriges Geologiska Undersökning*, **58**, 1–45.

JONES, B. G., CARR, P. F. & WRIGHT, A. J. 1981. Silurian and early Devonian geochronology – a reappraisal with new evidence from the Bungonian Limestone. *Alcheringa*, **5**, 197–207.

KAKAZU, M. H., MORAES, N. M. P., IYER, S. S. & RODRIGUES, C. 1981. Reduction of oxide ions of uranium in single filament surface–ionisation mass spectrometry with application to rock samples. *Analytica Chimica Acta*, **132**, 209–213.

KERSHAW, S. 1981. Stromatoporoid growth form and taxonomy in a Silurian biostrome, Gotland. *Journal of Palaeontology*, **5**, 1284–1295.

—— 1987. Stromatoporoid–coral intergrowths in a Silurian biostrome. *Lethaia*, **20**, 371–380.

—— 1990. Stromatoporoid palaeobiology and taphonomy in a Silurian biostrome on Gotland, Sweden. *Palaeontology*, **33**, 681–705.

—— 1993. Sedimentation control on growth of stromatoporoid reefs in the Silurian of Gotland, Sweden. *Journal of the Geological Society, London*, **150**, 197–205.

KLINKHAMMER, G. P. & PALMER, M. R. 1991. Uranium in the oceans: Where it goes and why. *Geochimica et Cosmochimica Acta*, **55**, 1799–1806.

KNIEWALD, G. & BRANICA, M. 1988. Role of uranium (V) in marine sedimentary environments: A geochemical possibility. *Marine Chemistry*, **24**, 1–12.

KUNK, M. J., SUTTER, J., OBRADOVICH, J. D. & LANPHERE, M. A. 1985. Age of biostratigraphic horizons within the Ordovician and Silurian systems. *In:* SNELLING, N. J. (ed.) *The Chronology of the Geological Record.* Geological Society, Memoir, **10**, 89–92.

LANGMUIR, D. 1978. Uranium solution–mineral equilibria at low temperature with applications to sedimentary ore deposits. *Geochimica et Cosmochimica Acta*, **42**, 547–569.

LARSSON, K. 1979. Silurian tentaculids from Gotland and Scania. *Fossils and Strata*, **11**, 1–180.

LAUFELD, S. 1974*a*. Silurian Chitinozoa from Gotland. *Fossils and Strata*, **5**, 1–130.

—— 1974*b*. Reference localities for palaeontology and geology in the Silurian of Gotland. *Sveriges Geologiska Undersökning*, **C705**, 1–172.

—— & BASSETT, M. G. 1981. Gotland: The anatomy of a Silurian carbonate platform. *Episodes*, **2**, 23–27.

——, SUNDQUIST, B. & SJÖSTRÖM, H. 1978. Megapolygonal bedrock structures in the Silurian of Gotland. *Sveriges Geologiska Undersökning*, **C759**, 1–26.

LUDWIG, K. R. 1990. Isoplot – a plotting and regression program for radiogenic isotope data, for IBM-PC compatible computers. Version 2.11. United States Geological Survey, Open-File Report, 88–557.

McKERROW, W. S., LAMBERT, R. St J. & CHAMBERLAIN, V. E. 1980. The Ordovician, Silurian and Devonian time scales. *Earth and Planetary Science Letters*, **51**, 1–8.

——, & COCKS, L. R. M. 1985. The Ordovician, Silurian and Devonian periods. *In:* SNELLING, N. J. (ed.) *The Chronology of the Geological Record.* Geological Society Memoir, **10**, 73–80.

MARTIN, J.-M. & WHITFIELD, M. 1983. The significance of the river input of chemical elements to the ocean. *In:* WONG, C. S., BOYLE, E., BRULAND, K. W., BURTON, J. D. & GOLDBERG, E. D. (eds) *Trace Metals in Seawater.* New York, Plenum, 265–296.

MARTINSSON, A. 1967. The succession and correlation of ostracode faunas in the Silurian of Gotland. *Geologiska Foreningens i Stockholm Förhandlingar*, **89**, 350–386.

MEUNIER, J. D., LANDAIS, P. & PAGEL, M. 1990. Experimental evidence of uraninite formation from diagenesis of uranium-rich organic matter. *Geochimica et Cosmochimica Acta*, **54**, 809–817.

MÖLLER, M., RAJAGOPALIN, G. & GERMANN, K. 1976. A geochemical model for dolomitization. *Geologische Jahrbuch*, **20**, 57–76.

MOORBATH, S., TAYLOR, P. N., ORPEN, J. L., TRELOAR, P. & WILSON, J. F. 1987. First direct radiometric dating of Archaean stromatolite limestone. *Nature*, **326**, 865–867.

MORI, K. 1969. Stromatoporoids from the Silurian of Gotland, Part 1. *Stockholm Contributions to Geology*, **19**, 1–100.

—— 1970. Stromatoporoids from the Silurian of Gotland, Part 2. *Stockholm Contributions to Geology*, **22**, 1–152.

ODIN, G. S., CURRY, D., GALE, N. H. & KENNEDY, W. J. 1982. The Phanerozoic time scale in 1981. *In:* ODIN, G. S. (ed.) *Numerical Dating in*

Stratigraphy. Wiley, Chichester, 957–960.

———, HUNZIKER, J. C., JEPPSSON, L. & SPIELDANES, N. 1986*a*. Ages radiometriques K–Ar biotites pyroclastiques sedimentées dans le Wenlock de Gotland (Suède). *In:* ODIN, G. S. (ed.) *Calibration of the Phanerozoic Time Scale. Chemical Geology (Isotope Geoscience Section)*, **59**, 117–125.

———, HURFORD, J. A., MORGAN, D. J. & TOGHILL, P. 1986*b*. K–Ar biotite for Ludlovian bentonites from Great Britain. *In:* ODIN, G. S. (ed.) *Calibration of the Phanerozoic Time Scale. Chemical Geology (Isotope Geoscience Section)*, **59**, 127–131.

———, SPJELDNAES, N., JEPPSSON, L. & NIELSEN THORSHOJ, A. 1984. Field meeting in Scandinavia – time scale calibration. *Bulletin de Liaison et Informations*, I.G.C.P. Project 196, **3**, 6–23.

ORPEN, J. L. & WILSON, J. F. 1981. Stromatolites at *c.* 3500 Myr and a greenstone–granite unconformity in the Zimbabwean Archaean. *Nature*, **291**, 218–220.

PARNELL, J. & SWAINBANK, I. 1990. Pb/Pb dating of hydrocarbon migration into a bitumen-bearing ore deposit, North Wales. *Geology*, **18**, 1028–1030.

PINGITORE, N. E. 1976. Vadose and phreatic diagenesis, processes, products and their recognition in corals. *Journal of Sedimentary Petrology*, **42**, 985–1006.

ROSS, R. J., NAESER, C. W., IZETT, G. A., OBRADO-VICH, J. D., BASSETT, M. G., HUGHES, C. P., COCKS, L. R. M., DEAN, W. T., INGHAM, J. K., JENKINS, C. J., RICKARDS, R. B., SHELDON, P. R., TOGHILL, P., WHITTINGTON, H. B. & ZALZSIE-WICZ, J. 1982. Fision-track dating of British Ordovician and Silurian stratotypes. *Geological Magazine*, **119**, 135–153.

RUSH, P. F. & CHAFETZ, H. S. 1991. Skeletal Mineralogy of Devonian stromatoporoids. *Journal of Sedimentary Petrology*, **61**, 364–369.

RUSSELL, J. 1991. U–Pb systematics applied to dating of carbonates and carbonate macrofossils. *Terra Abstracts*, **3** (abstract).

——— 1992. Investigation of the potential of Pb/Pb radiometric dating for the direct age determination of carbonates. D.Phil thesis, University of Oxford.

SCHAULE, B. K. & PATTERSON, C. C. 1981. Lead concentrations in the northeast Pacific: Evidence for global anthropogenic perturbations. *Earth and Planetary Science Letters*, **54**, 97–116.

SCHOPF, J. W. (ed.) 1983. *Earth's Earliest Biosphere: Its Origin and Evolution*. Princeton University Press, Princeton.

SMITH, P. E. & FARQUHAR, R. M. 1989. Direct dating of Phanerozoic sediments by the $^{238}U-^{206}Pb$ method. *Nature*, **341**, 518–521.

———, & HANCOCK, R. G. 1991. Direct radiometric age determination of carbonate diagenesis using U–Pb in secondary calcite. *Earth and Planetary Science Letters*, **105**, 474–491.

STEARN, C. W. 1980. Classification of the Palaeozoic stromatoporoids. *Journal of Palaeontology*, **54**, 881–902.

——— & MAH, A. J. 1987. Skeletal microstructure of Paleozoic stromatoporoids and its mineralogical implications. *Palaios*, **2**, 76–84.

STEIGER, R. H. & JÄGER, E. 1977. Subcommission on geochronology: Convention on the use of decay constants in geo- and cosmochronology. *Earth and Planetary Science Letters*, **36**, 359–362.

SUNDQUIST, B. 1982. Carbonate petrography of the Wenlockian Slite Beds at Haganäs, Gotland. *Sveriges Geologiska Undersökning*, **C796**, 1–79.

SWART, P. K. & HUBBARD, J. A. E. B. 1982. Uranium in scleractinian corals. *Coral Reefs*, **1**, 13–19.

TAYLOR, P. N., CHADWICK, B., MOORBATH, S., RAMAKRISHNAN, M. & VISWANATHA, M. N. 1984. Petrography, chemistry and isotopic ages of Peninsular Gneiss, Dharwar acid volcanic rocks and the Chitradurga Granite with special reference to the Late Archaean evolution of the Karnataka Craton, southern India. *Precambrian Research*, **23**, 349–375.

——— & KALSBEEK, F. 1990. Dating the metamorphism of Precambrian marbles: Examples from Proterozoic mobile belts in Greenland. *Chemical Geology (Isotope Geoscience Section)*, **86**, 21–28.

THIRLWALL, M. F. 1988. Geochronology of Late Caledonian magmatism in northern Britain. *Journal of the Geological Society, London*, **145**, 951–967.

TREFRY, J. H., METZ, S., TROCINE, R. P. & NELSON, T. A. 1985. A decline in lead transport by the Mississippi River, *Science*, **230**, 439–441.

TUCKER, R. D., KROGH, T. E., ROSS, R. J. & WILLIAMS, S. H. 1990. Time scale calibration by high-precision U–Pb zircon dating of interstratified volcanic ashes in the Ordovician and Lower Silurian stratotypes of Britain. *Earth and Planetary Science Letters*, **100**, 51–58.

WAMPLER, J. M. & KULP, J. L. 1962. Isotopic composition and concentration of lead in some carbonate rocks. *Bulletin of the Geological Society of America*, Buddington Volume, 105–114.

WANGERSKY, P. J. 1986. Biological control of trace metal residence time and speciation: A review and synthesis. *Marine Chemistry*, **18**, 269–297.

WHITE, W. M., DUPRÉ, B. & VIDAL, P. 1985. Isotope and trace element geochemistry of sediments from the Barbados Ridge–Demerara Plain region, Atlantic Ocean. *Geochimica et Cosmochimica Acta*, **49**, 1875–1886.

WHITFIELD, M. & TURNER, D. R. 1987. The role of particles in regulating the composition of seawater. *In:* STUMM, W. (ed.) *Aquatic Surface Chemistry: Chemical Processes at the Particle–Water Interface*. Wiley, New York, 457–493.

WINTERHALTER, B., FLODÉN, T., IGNATIUS, H., AXBERG, S. & NIEMISTÖ, L. 1982. Geology of the Baltic Sea. *In:* VOIPIO, A. (ed.) *The Baltic Sea*. Elsevier Oceanography Series, Amsterdam, **30**, 1–121.

WYBORN, D., OWEN, N., COMPSTON, W. & MCDOU-GALL, I. 1982. The Laidlaw Volcanics: A late Silurian tie-point on the geological time scale. *Earth and Planetary Science Letters*, **59**, 99–100.

The application of samarium–neodymium (Sm–Nd) Provenance Ages to correlation of biostratigraphically barren strata: a case study of the Statfjord Formation in the Gullfaks Oilfield, Norwegian North Sea

A. DALLAND,[1] E. W. MEARNS[2] & J. J. McBRIDE[2]

[1]*Statoil, 5020 Bergen, N-5001 Bergen, Norway*
[2]*Isotopic Analytical Services Ltd, Campus 3, Aberdeen Science and Technology Park, Balgownie Drive, Bridge of Don, Aberdeen AB22 8GW, UK*

Abstract: Samarium–neodymium (Sm–Nd) isotope analyses indicate that the Statfjord Formation in the Gullfaks Oilfield was derived from two compositionally distinct clastic sediment source terrains. One provenance area produced sediment with 'low' Provenance Ages (< 1800 Ma), which are similar to those of numerous other North Sea formations. The sedimentary rocks of this group were deposited by palaeocurrents which flowed towards the NNE and it is likely that they were derived from Triassic, and possibly Devonian, strata to the southwest of the Gullfaks area (e.g. East Shetland Platform). The other provenance area produced sediment with anomalous 'high' Provenance Ages (> 1800 Ma). These sediments were transported predominantly towards the south and are compositionally less mature and have poorer reservoir properties than those with 'low' Provenance Ages. The source terrain for this provenance group is considered to lie to the north of the Gullfaks Oil Field area, where exposed Lewisian gneisses probably formed a major clastic sediment source terrain. Switching between these contrasting provenance areas gave rise to the interdigitating, low-to-high Provenance Age profile which characterizes the Statfjord Formation. The Provenance Age profiles are used in combination with conventional well logs and sedimentological core descriptions to construct a detailed reservoir zonation. The method has allowed confident correlation of channel sandstones with overbank mudstone deposits. The resulting correlation scheme has revealed the presence of intraformational erosional surfaces and normal faults that were previously undetected.

The samarium–neodymium (Sm–Nd) isotope stratigraphy technique

The Sm–Nd technique relies on the natural radioactive decay of ^{147}Sm to ^{143}Nd ($+^4He$) with a half-life of 106 Ga. The rare earth elements (REE), including Sm and Nd, are fractionated during the formation of continental crust. By determining the $^{143}Nd/^{144}Nd$ and $^{147}Sm/^{144}Nd$ isotope ratios of a mantle-derived crustal rock it is therefore possible to calculate a 'model age' for that sample which reflects the time elapsed since the rock first formed from the mantle (Faure 1986). Full discussions of Sm–Nd isotope systematics in sedimentary rocks have been presented by Mearns *et al.* (1989) and Mearns (1988, 1989, 1992).

The Sm–Nd isotope technique may be applied to study the average residence ages of exposed continental crust (e.g. McCulloch & Wasserburg 1978; Hamilton *et al.* 1983; Goldstein *et al.* 1984; Davies *et al.* 1985; Michard *et al.* 1985; Mearns 1988) and to provide information about provenance of sedimentary successions (e.g. O'Nions *et al.* 1983; Mearns *et al.* 1989; Mearns

1988, 1989, 1992). As a sedimentary rock is normally formed by reworking of pre-existing crustal rocks of varying ages, however, a Sm–Nd 'model age' determined from it will not represent a unique crust-forming event. It should, nevertheless, reflect the weighted average crustal residence times of the rocks in the sediment's source terrain, i.e. the Sm–Nd model age of a sedimentary rock reflects the **provenance** of the sedimentary material rather than its stratigraphic age (q.v. fig. 2 in Mearns 1992). For this reason, dates calculated by the Sm–Nd method for sedimentary rocks are termed 'Provenance Ages'.

The Gullfaks Oilfield

The Gullfaks Oilfield is located in block 34/10 of the Norwegian Sector of the North Sea (Fig. 1). The field contains estimated recoverable reserves of 1590 MMBO, with 265 MMBO (*c.* 17%) estimated to be within Statfjord Formation reservoir sandstones. Production from the Statfjord Formation comes from the eastern part of the field where the Brent Group (the principal

From Dunay, R. E. & Hailwood, E. A. (eds), 1995, *Non-biostratigraphical Methods of Dating and Correlation* Geological Society Special Publication No. 89, pp. 201–222

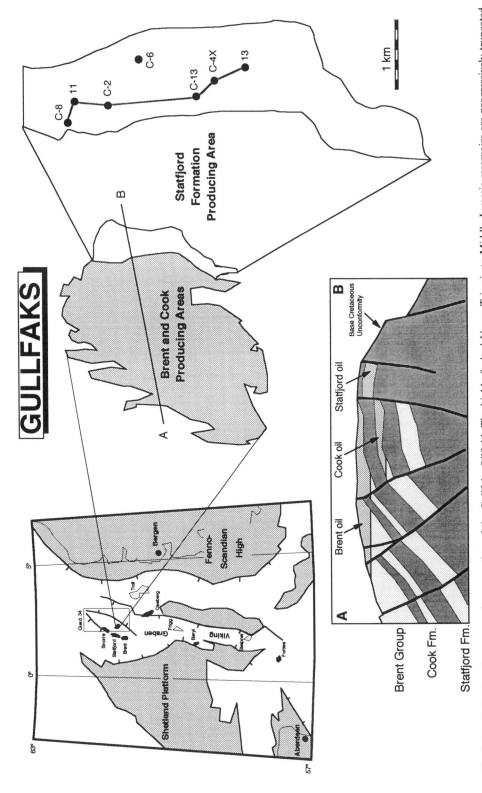

Fig. 1. North Sea location map and cross-section of the Gullfaks Oilfield. The highly faulted Upper Triassic to Middle Jurassic reservoirs are progressively truncated eastwards. Upper Cretaceous mudstones drape the Cimmerian Unconformity at the top of the structure. The Statfjord Formation produces mainly from a N–S elongated horst block in the eastern part of the field. Wells used in the Sm–Nd Provenance Age study are shown.

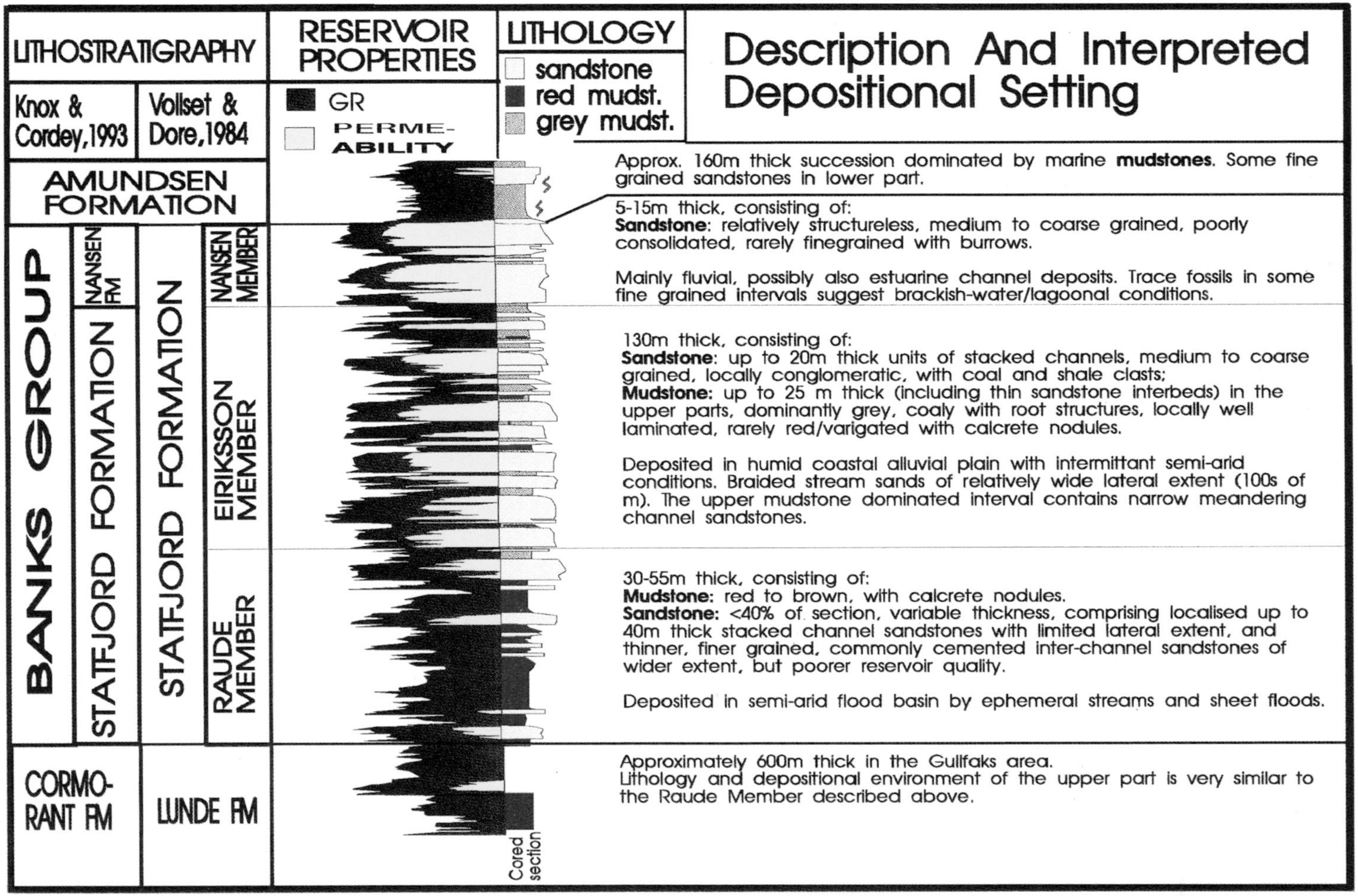

Description And Interpreted Depositional Setting

Approx. 160m thick succession dominated by marine **mudstones**. Some fine grained sandstones in lower part.

5-15m thick, consisting of:
Sandstone: relatively structureless, medium to coarse grained, poorly consolidated, rarely finegrained with burrows.

Mainly fluvial, possibly also estuarine channel deposits. Trace fossils in some fine grained intervals suggest brackish-water/lagoonal conditions.

130m thick, consisting of:
Sandstone: up to 20m thick units of stacked channels, medium to coarse grained, locally conglomeratic, with coal and shale clasts;
Mudstone: up to 25 m thick (including thin sandstone interbeds) in the upper parts, dominantly grey, coaly with root structures, locally well laminated, rarely red/variegated with calcrete nodules.

Deposited in humid coastal alluvial plain with intermittant semi-arid conditions. Braided stream sands of relatively wide lateral extent (100s of m). The upper mudstone dominated interval contains narrow meandering channel sandstones.

30-55m thick, consisting of:
Mudstone: red to brown, with calcrete nodules.
Sandstone: <40% of section, variable thickness, comprising localised up to 40m thick stacked channel sandstones with limited lateral extent, and thinner, finer grained, commonly cemented inter-channel sandstones of wider extent, but poorer reservoir quality.

Deposited in semi-arid flood basin by ephemeral streams and sheet floods.

Approximately 600m thick in the Gullfaks area.
Lithology and depositional environment of the upper part is very similar to the Raude Member described above.

LITHOLOGY
sandstone
red mudst.
grey mudst.

Cored section

RESERVOIR PROPERTIES
GR
PERME- ABILITY

LITHOSTRATIGRAPHY			
Knox & Cordey, 1993	Vollset & Dore, 1984		
AMUNDSEN FORMATION		NANSEN MEMBER	
STATFJORD FORMATION		EIRIKSSON MEMBER	
		RAUDE MEMBER	
			LUNDE FM
BANKS GROUP			
STATFJORD FORMATION	NANSEN FM		
			CORMO-RANT FM

Fig. 2. Lithostratigraphy and depositional environments of the Statfjord Formation succession of the Gullfaks Oilfield (modified from Olaussen et al. 1993).

reservoir sandstones) is severely truncated or totally absent as a result of erosion. The reservoir is capped by Upper Cretaceous mudstones which lie above the field-wide Base Cretaceous unconformity (Fig. 1).

The Statfjord Formation reservoir

The producing section of the Statfjord Formation in the Gullfaks Oilfield is *c.* 180 m thick. Its lithostratigraphy and interpreted depositional settings are summarized in Fig. 2. In this paper, the lithostratigraphic nomenclature proposed by Vollset & Doré (1984) for the Norwegian North Sea is used to subdivide the Statfjord Formation [although the twofold stratigraphy proposed for the UK sector by Knox & Cordey (1993) is also consistent with the provenance data]. In Gullfaks, the Eiriksson Member is the main reservoir unit.

Outline of the problem and previous work

As outlined in Fig. 2, the Statfjord Formation in the Gullfaks Oilfield was deposited in a fluvial to marginal marine environment (Roe & Steel 1985; Olaussen *et al.* 1993). The strata are virtually devoid of palynomorphs. Biostratigraphic data therefore provide very poor stratigraphic resolution. In addition, seismic resolution is extremely poor in this part of the field. Wireline log response and lithological criteria may, however, be applied to recognize stratigraphic boundaries within wells or to perform correlations between wells where well spacing is only a few hundred metres [e.g. applying similar criteria as suggested by Vollset & Doré (1984) the boundary between the basal Statfjord Raude Member and the Upper Triassic Lunde Formation may be picked at the base of a sand dominated interval; Fig. 2]. Experience from correlation attempts across the whole field, however, has shown that less confidence can be attached to log-based correlations when well spacings approach 1 km or more.

For these reasons, alternative stratigraphic techniques were considered in order to attempt reservoir correlation. A pilot Sm–Nd isotope study of the Brent Group and Statfjord Formation reservoirs in the Gullfaks Oilfield indicated that strata with low and high Provenance Ages were interdigitated in a highly systematic manner vertically through the Statfjord Formation of exploration well 34/10-13 (q.v. fig. 6 in Mearns 1989). This pattern of variation suggested that this technique had the potential to provide the resolution required for detailed reservoir zonation. This paper presents the results of the field-wide follow-up to this earlier study.

Results

The Sm–Nd isotope data

This paper presents the results of 217 Sm–Nd isotope analyses of samples collected from seven wells in the Gullfaks Oilfield (Fig. 1). This probably represents the highest density of Sm–Nd isotope analyses of sedimentary rocks ever to be reported. The results are tabulated in the Appendix and in Mearns (1989) for results from Well 34/10-13. The majority of the samples (204) were of conventional core with four side-wall cores and nine drill cuttings analysed over intervals where there was no core coverage (Appendix). The majority of samples (187) were from the Statfjord Formation, with 20 samples from the underlying Lunde Formation and ten from the overlying Amundsen Formation (Fig. 2) also analysed. The isotope analyses were performed by Isotopic Analytical Services Ltd. Analytical methods, data normalization and data reduction methods are summrized in the Appendix.

$^{147}Sm/^{144}Nd$ isotope ratios

The frequency distribution of $^{147}Sm/^{144}Nd$ isotope ratios (proportional to Sm/Nd) for the sandstone and mudstone samples is plotted in Fig. 3a. The majority of sandstone and mudstone samples fall in the range 0.09–0.12, and both lithologies have a mode at 0.10–0.11. This indicates that there is no evidence for fractionation of Sm/Nd ratios between the coarse- and fine-grained lithologies in this dataset. Therefore, Provenance Ages in the sandstones and mudstones should not be biased by mechanical or chemical fractionation of the Sm/Nd ratios. The Statfjord Formation samples fall within the range of published Sm/Nd ratios of a large number of sedimentary rocks and river sediments (e.g. Mearns, 1988), although the mode is biased towards a slightly lower value (Fig. 3a).

Nd concentration

The Nd concentration in the majority of samples is < 70 p.p.m. although concentrations do occasionally range up to 200 p.p.m. (Fig. 3b). The sandstones are clearly biased towards lower Nd concentrations compared with the mudstones which probably reflects the higher quartz content of the former.

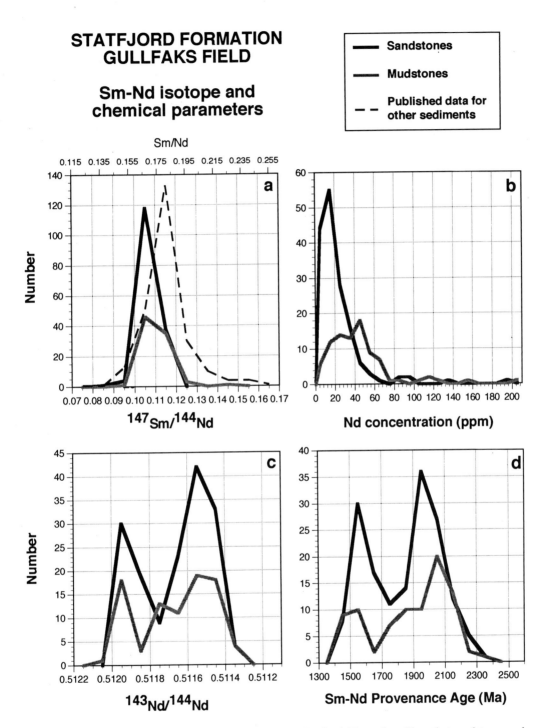

Fig. 3. Sm–Nd chemical and isotope parameters for the Statfjord Formation. Note that mudstones and sandstones exhibit a similar range and distribution of Sm/Nd ratios and Provenance Ages. The distinct bimodal distribution of provenance ages is recorded by both mud- and sandstones and suggests two distinct source areas for the Statfjord Formation in the Gullfaks Oilfield.

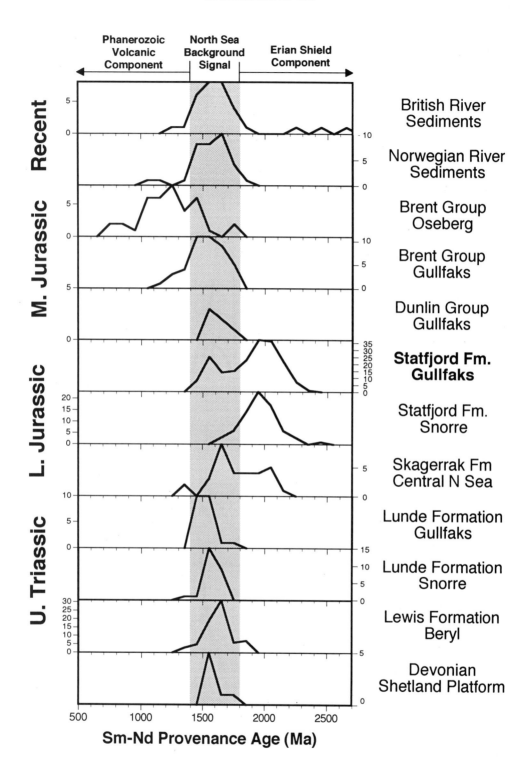

Sm-Nd Provenance Age (Ma)

Sm-Nd ISOCHRON DIAGRAM

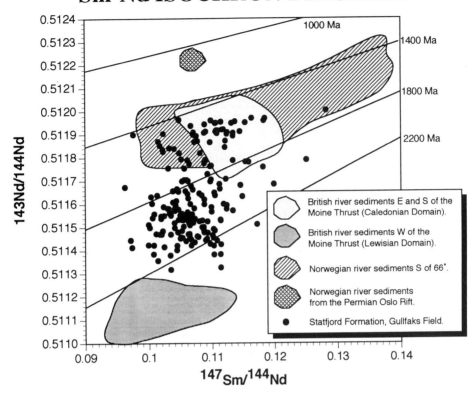

Fig. 5. Sm–Nd isochron plot of the Statfjord Formation samples compared with data for modern source areas represented by Norwegian and British river sediments (Fig. 4). The model isochrons converge on the depleted mantle point at $^{143}Nd/^{144}Nd = 0.51303$ and $^{147}Sm/^{144}Nd = 0.22$ (Appendix). The Statfjord Formation 'low' Provenance Age group falls within the field defined by British river sediments south and east of the Moine Thrust and the Norwegian river sediments to the left of this field. The Statfjord Formation 'high' Provenance Age group falls in a field between the British rivers east and south of the Moine Thrust (Caledonian provenance) and the rivers west of the Moine Thrust (Archaean Lewisian provenance). The 'high' Provenance Age group may therefore be produced by mixing these two provenance types. This may be achieved by mixing Lewisian type material derived from the area northwest of the Shetland Platform with Caledonian type material derived more proximally from the Shetland Platform itself.

Fig. 4. The frequency distribution of Provenance Ages in the Statfjord Formation compared with the Brent, Dunlin and Lunde reservoirs of Gullfaks and other North Sea formations. River sediments represent the compositions of modern source areas adjacent to the North Sea. The Lunde Formation in Gullfaks has broadly similar Sm–Nd provenance characteristics to the Lunde Formation in Snorre Field and the Lewis Formation in Beryl Field, and collectively these are similar to Devonian Old Red Sandstone deposits of the Shetland Platform. The latter therefore has an appropriate Sm–Nd composition to be the source material for the Upper Triassic strata. The Statfjord Formation of Snorre Field exhibits the high Provenance Ages seen in Gullfaks but lacks the low Provenance Age peak. The Skagerrak Formation of the central North Sea (several hundred kilometres south of Gullfaks), which is a broad chrono-stratigraphic equivalent, exhibits a similar bimodal distribution of Provenance Ages to the Statfjord Formation. Note that the high Provenance Age peak of the Statfjord Formation falls above the North Sea background provenance signal which is the basis for arguing for an Archaean, Erian Shield component in its provenance. See Fig. 1 for field locations. [Data sources: British rivers, Mearns (1988); Norwegian rivers, Mearns (unpublished data); Brent Oseberg, unpublished results reproduced with the kind permission of Statoil; Brent, Dunlin and Statfjord Gullfaks, Mearns (1989); Statfjord and Lunde Gullfaks, this paper; Statfjord and Lunde Snorre, Mearns *et al.* (1989); Skagerrak Formation of the central North Sea, unpublished results reproduced with the kind permission of Shell UK Expro and Esso Expro UK; Lewis Formation Beryl, unpublished results reproduced with the kind permission of Mobil North Sea Ltd; Devonian, Mearns & Trewin (unpublished data).]

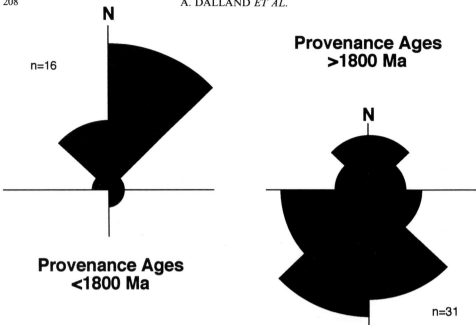

Fig. 6. Processed dipmeter data from two of the Gullfaks wells included in this Sm–Nd study (34/10-11 and 34/10-13) and one well from a study of Gullfaks South (34/10-30). Cross-bed directions in the sandstones are correlated with the average Provenance Ages of the same intervals. 'Low' Provenance Ages (< 1800 Ma) correlate with generally north-flowing palaeocurrents, while Provenance Ages > 1800 Ma correlate with generally southwards palaeocurrent directions. Dipmeter data are from channel facies sandstones. The Sm–Nd data are from the same sand intervals as the dipmeter data or, if no analyses were performed on the sandbody, from the finer grained abandonment or overbank facies associated with it. The dipmeter data were specially processed to show cross-bedded intervals and were calibrated against cores where available. About 70% of the main sandstone bodies contain dipmeter data interpreted as cross-bedding. The remaining 30% of the sandstones were too homogeneous, were dominated by current ripples, or did not give reliable data owing to poor hole conditions. The palaeocurrent measurements are grouped within 45° sectors. Measurements falling on sector boundaries are split between the two adjacent sectors.

$^{143}Nd/^{144}Nd$ isotope ratio and Provenance Age

The frequency distributions of the Sm–Nd Provenance Ages of the sandstone and mudstone samples are plotted in Fig. 3d. The overall range in sandstones is from 1400 to 2400 Ma and the results exhibit a clear bimodal distribution with peaks at 1500–1600 and 1900–2000 Ma. The range of Provenance Ages in mudstones is also 1400–2400 Ma, and these also exhibit a bimodal distribution with peaks at 1500–1600 and 2000–2100 Ma. There is the suggestion of a third peak in the mudstone data at 1800–1900 Ma which is more pronounced in the $^{143}Nd/^{144}Nd$ isotope ratio results (at 0.5117–0.5118; Fig. 3c). The distribution of Provenance Ages in the sandstones and mudstones is very similar and suggests that at least two distinct source areas provided sediment to the Statfjord Formation in the Gullfaks Oilfield.

The range in Provenance Ages in the Statfjord Formation is unusual for the North Sea (Fig. 4). The peak at 1500–1600 Ma falls within the

normal North Sea background range of 1400–1800 Ma. These rocks therefore have similar Provenance Age characteristics to numerous North Sea formations (Fig. 4). The peak at 1900–2100 Ma, however, falls outside the background range. These rocks are consequently interpreted to have 'anomalous' provenance, with only the Skagerrak Formation of the central North Sea having a similar range of 'high' Provenance Ages (Fig. 4).

Figure 5 shows the isotope data from this study together with those from Norwegian (Mearns, unpublished data) and British river sediments (Mearns 1988). The Statfjord Formation rocks with 'low' Provenance Ages (1400–1800 Ma) fall in the fields described by the British river sediments from south and east of the Moine Thrust and the Norwegian river sediments. This illustrates that the Statfjord Formation rocks with 'low' Provenance Ages (1400–1800 Ma) are compositionally similar to both the average Norwegian source area and the

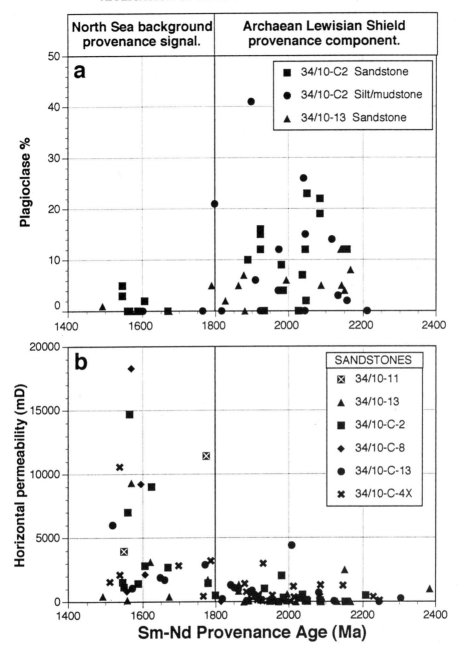

Fig. 7. (a) Cross-plot of plagioclase feldspar v. Provenance Age. The 'low' (< 1800 Ma) and 'high' (> 1800 Ma) Provenance Age drainage systems are characterized by different mineralogy, most clearly expressed by the plagioclase feldspar content. The difference is equally well expressed in sandstones and mudstone/siltstone samples. The XRD data have been tied to the nearest Sm–Nd sample (usually the vertical separation of the compared samples is < 2 m). Comparison across major erosional surfaces has been avoided. (b) Cross-plot of horizontal permeability v. Provenance Age. Core plug permeability values have been compared with the nearest Provenance Age values. Vertical difference between the data is generally < 20 cm. Permeability measurements on plugs that are not representative of the sand body (e.g. calcite nodules, thin shale horizons, etc.) have been excluded. Note that the correlation between low Provenance Age and high permeabilities would be even more pronounced than shown but for the failure of many friable, high permeability samples to yield reliable core analysis results.

British river source area south and east of the Moine Thrust.

In contrast, the samples with 'high' Provenance Ages (> 1800 Ma) plot between the Norwegian/British rivers south and east of the Moine Thrust, and the British rivers to the west of the Moine Thrust in Fig. 5. Archaen gneisses predominate west of the Moine Thrust which gives this area its distinct Sm–Nd composition (Figs 4 & 5). Although none of the Statfjord Formation samples fall within this Archaen gneiss compositional field, the observed isotope characteristics may be derived by mixing sediment from this terrain with sediment from a more typical North Sea source. The Archaen gneisses extend northwards from NW Scotland along the westerly margin of the Shetland Platform. The likely source terrain for this provenance group is therefore considered to lie to the NW of Gullfaks, off the northern tip of the Shetland Platform (Mearns *et al.* 1989).

Correlation of Provenance Ages with transport directions

Dipmeter data, processed to reveal cross-bed directions in the Statfjord Formation fluvial channels, have given consistent and probably reliable indications of transport directions in many of the main sandstone bodies of the formation. The transport directions are plotted in Fig. 6, showing that beds with 'low' Provenance Ages (< 1800 Ma) were transported predominantly towards the NNE while beds with 'high' Provenance Ages (> 1800 Ma) were transported predominantly towards the SSW.

This clear correspondence between transport directions and Provenance Ages provides strong support to the concept of two source areas for the Statfjord Formation and confirms that the low Provenance Age source probably lay to the southwest while the high Provenance Age source lay to the north of the Gullfaks Oilfield area.

Correlation of Provenance Ages with mineralogy

Provenance Ages are cross-plotted with plagioclase contents in Fig. 7a. Samples with 'low' Provenance Ages (< 1800 Ma) have low plagioclase contents, normally < 3%. Samples with 'high' Provenance Ages (> 1800 Ma) exhibit a much wider range in plagioclase contents that are commonly as high as 25%.

This significant difference in mineralogy between the two provenance groups is also consistent with markedly different source areas. The

low plagioclase contents of samples with Provenance Ages < 1800 Ma suggests that these sedimentary rocks are compositionally more mature and may represent recycled sediment. In contrast, the sedimentary rocks with Provenance Ages > 1800 Ma seem to be less compositionally mature and are perhaps, in part, first-cycle sediment.

Correlation of Provenance Ages with reservoir quality

Provenance Ages are cross-plotted with permeability (nitrogen measurements on core plugs) in Fig. 7b. 'High' Provenance Age (> 1800 Ma) sandstones have a maximum horizontal permeability of 4400 mD (average value 563 mD; $n = 62$). In contrast, 'low' (< 1800 Ma) Provenance Age sandstones exhibit horizontal permeabilities ranging up to 18 300 mD (average value 4168 mD; $n = 33$).

Interpretation

As discussed above, 'low' Provenance Age sandstones (< 1800 Ma) have variable and locally good horizontal permeabilities (up to 18 300 mD) and low plagioclase contents (generally < 3%). In contrast, 'high' Provenance Age sandstones have poorer permeabilities (up to 4400 mD) and generally higher plagioclase contents (up to 25%). We interpret this result to reflect differences in primary sediment composition between 'low' and 'high' Provenance Age sediments. We recognize, however, an alternative interpretation that these characteristics may reflect different secondary diagenetic processes in these rocks. If these properties were caused by diagenetic processes it would suggest the following: (1) all the sediments of the Statfjord Formation had similar primary feldspar composition and Sm–Nd isotope composition at the time of deposition; (2) at some stage in the burial history of the sequence, plagioclase feldspar was dissolved and removed from certain layers of the reservoir by a field-wide diagenetic process which also enhanced the permeability of the sandstones by creating secondary porosity and somehow altered the Sm–Nd isotopic composition of the sandstones. This model would therefore suggest that the 'low' Provenance Age layers represent zones of possible fluid flushing and diagenetic removal of feldspar, enhancement of permeability and alteration of Sm–Nd isotope composition. This model is not favoured, however, for the following reasons:

(1) The Sm–Nd isotope characteristics of the 'low' Provenance Age group are in no way considered 'anomalous' as would be expected if they were the result of some form of diagenetic alteration. In fact, it is the 'high' Provenance Age group of sediments that may be considered to have anomalously high Provenance Ages in the North Sea context.

(2) Sm and Nd are considered to be refractory in most sedimentary rocks and are unlikely to be exported to any great extent by circulating aqueous fluids. Furthermore, removal of plagioclase is not likely to alter the $^{143}Nd/^{144}Nd$ ratio of the bulk sediment and it is this isotope ratio which determines the Provenance Ages of the sediments.

(3) Finally, the bimodal distribution of Provenance Ages is observed in both mudstones and sandstones. It is considered to be highly unlikely that fluids flushing sandstones will have had a similar impact upon the Sm–Nd isotopes in the mudstones. The interpretation of the bimodal distribution of Provenance Ages in terms of dual provenance is, therefore, considered to be a much more plausible interpretation of the results.

Provenance of the Statfjord Formation

This study has confirmed the presence of two distinct, sedimentary provenance populations in the Statfjord Formation of the Gullfaks Oilfield distinguished by Provenance Age, palaeocurrent directions, feldspar content and reservoir permeabilities in the Gullfaks area. The Provenance Age and dipmeter data (Fig. 6) suggest that sediment was derived from at least two distinct source terrains and drainage systems: 'high' Provenance Age (> 1800 Ma) material from a terrain containing Archaean gneiss transported from the north of the Gullfaks area; and 'low' Provenance Age (< 1800 Ma) sediment most probably derived from a source terrain with similar Sm–Nd isotope characteristics as northern Britain (east and south of the Moine Thrust) and transported from the south of Gullfaks. It is likely that the intersection area of these two systems shifted with time, depending on tectonic movements in the area, thus giving rise to the interdigitating Provenance Age profile that characterizes this sequence (Fig. 8).

During deposition of the lower part of the Raude Member, the sediments were probably derived from the north of the Gullfaks Oilfield reflected by 'high' Provenance Ages and high plagioclase feldspar content. Both of these trends show a general decrease towards the top of the member. Dipmeter data indicate variable palaeocurrent directions, swinging mainly between southeast and southwest.

The lower to middle parts of the Eiriksson Member show an upward increase in Provenance Age and plagioclase content. Palaeocurrents throughout this section are mainly directed to the south. 'Low' Provenance Ages, northward-directed palaeocurrents and low plagioclase content dominate the upper part of the Eiriksson and Nansen members. This suggests that the main sediment-source area was to the south of the Gullfaks area at this time.

The generally low feldspar/high quartz content of the material brought in from the southern provenance area indicates a major contribution of recycled sediments. It is likely that these would have been derived from Triassic and possibly also from Devonian sediments to the southwest of Gullfaks (e.g. East Shetland Platform). In the northern provenance area, exposed Lewisian gneisses were probably a major source of the sedimentologically immature (high feldspar content) 'high' Provenance Age material.

This palaeogeographic model has implications for prediction of reservoir quality of the Statfjord Formation in the Tampen Spur area as it suggests that sandstone reservoirs with improved reservoir properties (low feldspar content and 'low' Provenance Age) have an increased likelihood of being deposited to the south of Gullfaks.

Reservoir zonation of the Statfjord Formation

A major advantage of the Provenance Age data in reservoir correlation is that the age values are totally independent of sample grain size (Fig. 3). This makes it possible to correlate a coarse-grained channel sandstone with its associated interfluvial fines and crevasse deposits. Figure 8 illustrates this point and details the interdigitating low-to-high Provenance Age profile that characterizes the Statfjord Formation and which is used to construct the reservoir zonation. Note how thick sandstone sequences in one well correlate with thin sandstones/mudstones in the other.

The Raude Member is characterized by variable but, generally high Provenance Ages (1800–2200 Ma). Several age trends and peaks may be correlatable within the Raude Member,

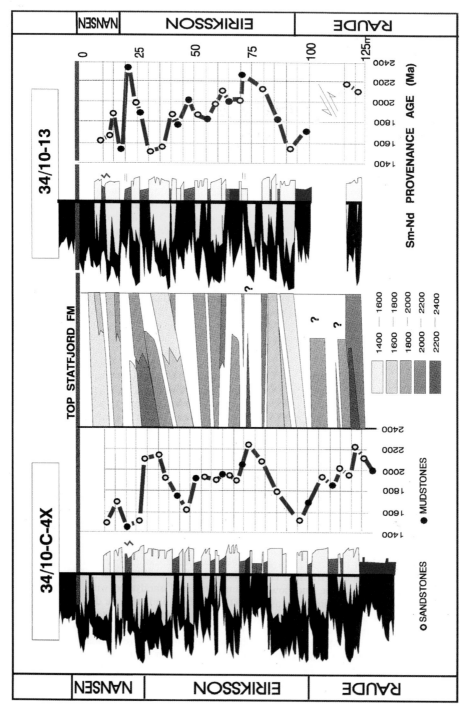

Fig. 8. Detail of the correlation between two wells to illustrate the concepts employed in Nd isotope stratigraphy. Because the Provenance Age values are independent of sediment grain size, it is possible to correlate channel sandstones with associated crevasse mudstones. Matching the low-to-high Provenance Age profiles between the wells allows a detailed reservoir zonation to be constructed.

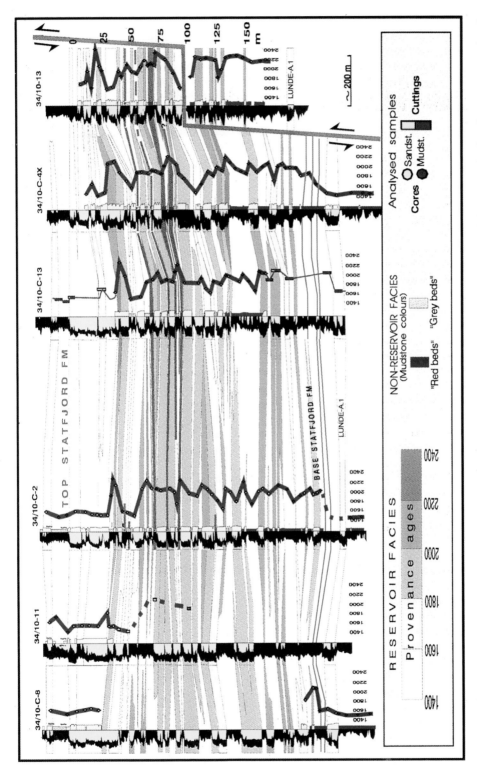

Fig. 9. Correlation of the Statfjord Formation reservoir facies sandstone bodies and red mudstone intervals within the Eiriksson Member based on Provenance Age data and well logs. The reservoir facies sandstones have been divided into five Provenance Age classes. In addition, the colours of the non-reservoir facies mudstones have been shown. Statoil's current Statfjord Formation reservoir zonation of the Gullfaks Oilfield is based on this correlation. See Fig. 1 for well locations.

but their recognition requires relatively dense sampling as age variations occur frequently within the lower part of the Statfjord Formation. A trend of decreasing Provenance Age occurs in the upper part of the Raude Member, and is particularly well established in Well 34/10-C-4X in the southern part of the field (Figs 1 & 8). This trend, however, becomes less distinct and then absent in northerly wells (Fig. 9). This has been interpreted as a major truncation. This interpretation is supported by the marked lithology transition from dominantly red beds and mudstone below to grey beds and sandstone above it. The reservoir properties of the sandstones are also significantly better above the boundary. In Statoil's reservoir zonation this unconformity defines the boundary between the Raude and Eiriksson members (Fig. 2). The Sm–Nd data have been important in correlating this boundary within the production area.

The Eiriksson Member starts with a major low-to-high Provenance Age trend, starting in the interpreted incised valley deposits (e.g. Fig. 8). Above the maximum peak of *c.* 2200 Ma is an interval of mudstones and sandstones with Provenance Ages of 1800–2100 Ma (Fig. 8). No field-wide Provenance Age markers are recognized within this interval and the reservoir properties of the sandstones in this section are generally poorer than the sand-dominated successions above and below. Above this interval is a sequence characterized by a change to considerably lower Provenance Ages (1500–1600 Ma), increased sand content and improved reservoir properties (Fig. 8). The upper part of the Eiriksson Member is characterized by a progressive increase in Provenance Age. The low Provenance Age boundary of this upper sequence may be another unconformity, but the Provenance Age data indicate less incision here than at the base of the Eiriksson Member.

The base of the Nansen Member corresponds within the inflexion point where the Provenance Age profile returns to lower values (e.g. Fig. 8). Above this level, Provenance Ages are characteristically between 1500 and 1600 Ma, with a possible anomaly towards higher ages in the lower part (Fig. 8). This profile characterizes the sequence of homogeneous, good reservoir-quality sandstones up to the marine ravinement surface that terminates the Statfjord succession (Fig. 2).

Figure 9 shows the resulting reservoir zonation when the Sm–Nd Provenance Age results are used in combination with conventional well logs and sedimentological core descriptions. In this correlation scheme, sandstone bodies have been classified into five Provenance Age groups.

The wireline log and Provenance Age data were generally in good agreement, and where the logs indicated two or more likely alternatives for correlation the Provenance Age data would frequently indicate which one to prefer. Correlation of red-bed mudstone intervals within the Eiriksson Member is shown in addition to the reservoir sandstone bodies.

Recognition of faults

The Triassic Lunde Formation, which underlies the Statfjord Formation in the Gullfaks Oilfield area, is characterized by a uniform low Provenance Age (*c.* 1500 Ma) which is capped by a transition zone which shows a progressive increase in Provenance Age into the lower part of the Statfjord Formation (e.g. Well 34/10-C-4X in Fig. 9). This characteristic profile gave additional support to the log correlation of the Lunde–Statfjord boundary throughout the Gullfaks Oilfield, but also provided the additional benefit of allowing the recognition of a major fault which had not been indicated by seismic interpretation or by available well data. Wireline logs indicate that the Statfjord Formation reservoir sequence in Well 34/10-C-6 is compatible with other wells (e.g. 34/10-C-2) from the top of the Statfjord Formation down to the upper Raude Member (Fig. 10). Below this level, however, the succession in 34/10-C-6 is considerably more sand-rich than in any of the other wells. Prior to the Sm–Nd study, this anomaly was attributed to lateral variation in sand content of the Raude Member, and the base of the Statfjord Formation was picked as illustrated in Fig. 10. The Provenance Age data from two cores cut in this lower section indicated that the sandstones had ages far younger than any Provenance Age previously obtained from the Raude Member. Furthermore, the data obtained were characteristic of the Lunde Formation (between 1400 and 1500 Ma). This result suggested that a major normal fault may intersect the well in the upper part of the Raude Member (at the arrow in Fig. 10), and as a result *c.* 80 m of the underlying Raude section is missing. Subsequent inspection of dipmeter data, reprocessed seismic lines and well logs supported this interpretation.

Conclusions

(1) The dataset presented in this paper probably represents the highest density of Sm–Nd isotope analyses of sedimentary rocks ever reported; (2) There is no evidence for fractionation of Sm/Nd ratios between sand- and mudstones. Prove-

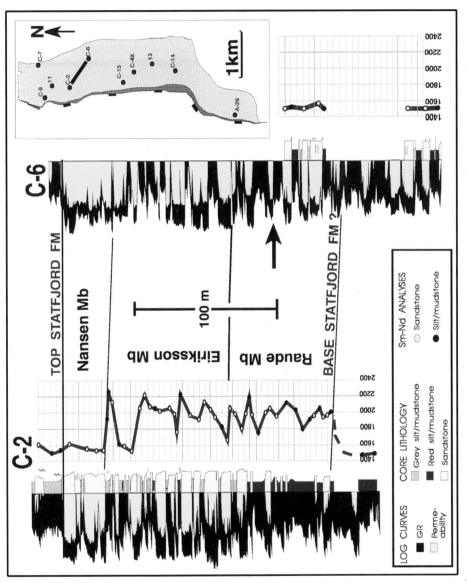

Fig. 10. The Statfjord Formation reservoir sequences encountered in Wells 34/10-C-2 and -C-6. Below the upper Raude Member, the succession in 34/10-C-6 is considerably more sand-rich than in -C-2. Prior to the Sm–Nd study, this anomaly was attributed to lateral variation in sand content of the Raude Member and the base of the Statfjord Formation was picked where shown. The Provenance Age data from two cores cut in the lower section of -C-6 were, however, characteristic of the Lunde Formation (between 1400 and 1500 Ma) suggesting that a major normal fault may intersect the well in the upper part of the Raude Member (arrowed).

nance Ages in the sand- and mudstones should not, therefore, be biased by mechanical or chemical fractionation of the Sm/Nd ratios; (3) A distinct bimodal distribution of Provenance Ages in the sand- and mudstones suggests that at least two distinct source areas provided sediment to the Statfjord Formation in the Gullfaks Oilfield; (4) Statfjord Formation sequences with 'low' Provenance Ages (< 1800 Ma) have similar Provenance Age characteristics as numerous North Sea formations. Sedimentary structures in these sequences indicate palaeocurrents flowed towards the NNE. These sedimentary rocks are compositionally mature and may represent recycled sediment. They exhibit permeabilites ranging up to 18 300 mD (average 4168 mD). It is likely that these were derived from Triassic, and possibly also from Devonian sediments to the south west of Gullfaks (e.g. East Shetland Platform); (5) Statfjord Formation sequences with 'high' Provenance Ages (> 1800 Ma) are interpreted to have 'anomalous' provenance. Dipmeter data suggest that these sequences were transported predominantly towards the SSW. Compared with the other parts of the Statfjord Formation, these strata seem to be less compositionally mature and have less good reservoir properties (average permeability 563 mD). The likely source terrain for this provenance group is considered to lie to the north of Gullfaks where exposed Lewisian gneisses may have formed a clastic sediment source terrain; (6) A major advantage of the Provenance Age data in reservoir correlation is that the age values are totally independent of sample grain size making it possible to correlate coarse-grained channel sandstones with their associated interfluvial fines and crevasse deposits; (7) An interdigitating low-to-high Provenance Age profile characterizes the Statfjord Formation and is used in combination with conventional well logs and sedimentological core descriptions to construct a reservoir zonation in which thick sandstone sequences in one well may be correlated with thin sandstones/mudstones in another; (8) A trend of decreasing Provenance Age in the upper part of the Raude Member becomes less distinct and then absent in northerly wells indicating the presence of a major truncation. This interpretation is supported by the marked lithology transition from dominantly red beds and mudstone below to grey beds and sandstone above it; (9) Provenance Age data in the lower section of Well 34/10-C-6 indicate that a major normal fault intersects this well in the upper part of the Raude Member and *c.* 80 m of the underlying Raude section is missing. Subsequent inspection of dipmeter data, reprocessed seismic lines and

well logs support this interpretation.

The authors wish to thank Statoil and their partners in the Gullfaks Licence PL050 (Norsk Hydro and Saga Petroleum) for permission to publish these results. We also wish to thank the following companies for their kind permission to publish Provenance Age data shown in Fig. 4: Statoil for the Brent of the Oseberg Field; Shell UK Expro and Esso Expro UK Ltd for the Skagerrak Formation of the central North Sea; and Mobil North Sea Ltd for the Lewis Formation of the Beryl Field. Gary Robertson and Hans Amundsen are acknowledged for their constructive reviews of the manuscript.

Appendix

Analytical methods

Core samples weighing 100–500 g were cleaned by scraping off all outer surfaces. Samples were then coarse crushed to pea-size in a jaw crusher followed by pulverization to rock powder in a tungsten carbide mill. A 10 g portion of the homogeneous powder was ignited at 500°C in atmosphere to remove all carbon. Following ignition samples were reground in an agate mortar.

Portions of the ignited powders (100 mg) were dissolved in 5 ml concentrated HF and 1 ml of concentrated HNO_3 inside Teflon digestion vessels at 150°C for 8 h. Following evaporation of the HF, samples were dissolved in dilute HNO_3 at 150°C. The HNO_3 sample solutions were split and weighed. One portion was used for determining the $^{143}Nd/^{144}Nd$ isotope ratio. A mixed $^{149}Sm–^{148}Nd$ spike was added to the second portion which was used for determining the Sm and Nd concentrations by isotope dilution.

After splitting and spiking, the HNO_3 was evaporated and samples redissolved in HCl. This was followed by a two-stage ion exchange procedure used to isolate Sm and Nd from the solutions. The first stage employs cation columns packed with Bio Rad AG50W *8, minus 400 mesh resin and separates the rare earth element group from the matrix elements. The second stage employs hydrogen phosphate-coated Teflon powder that gives clean separation of Nd (and the light REE) from Sm (and the middle REE). The purified Sm and Nd is then ready for mass spectrometer analysis.

Analyses were performed on a VG 354 mass spectrometer. Nd isotope ratios were measured employing the VG multi-dynamic data collection routine which yields $^{145}Nd/^{144}Nd$ in addition to $^{143}Nd/^{144}Nd$ ratios. The former ratio should be uniform in all terrestrial samples and therefore provides a good internal standard check on data quality (see tabulated isotope data). The mean $^{145}Nd/^{144}Nd$ ratio measured in samples is 0.348407 ± 0.000018 (2 SD, $n = 187$).

Five analyses of the USGS standard rock powder BCR-1 conducted during this work gave the following values; Sm = 6.612 ± 0.153 p.p.m.; Nd = 29.11 ± 0.70 p.p.m.; $^{147}Sm/^{144}Nd = 0.13830 \pm 0.00021$; $^{143}Nd/^{144}Nd = 0.512639 \pm 0.000012$; $^{145}Nd/^{144}Nd = 0.348411 \pm 0.000014$; (all uncertainties are two standard deviations).

Thirty-nine analyses of the Johnson and Matthey internal IAS standard gave $^{143}Nd/^{144}Nd = 0.511132 \pm 0.000012$ (2SD). The total Nd contamination of samples during analyses $< 250 \times 10^{-12}$ g.

Nd isotope ratios are normalized to $^{146}Nd/^{144}Nd = 0.7219$ (Faure 1986). Provenance Ages are calculated relative to a depleted mantle with present day $^{143}Nd/^{144}Nd = 0.51303$ and $^{147}Sm/^{144}Nd = 0.22$ (Mearns 1988). The decay constant for $^{147}Sm = 6.54 \times 10^{-12}$ y^{-1}.

Results

These are tabulated on the following pages.

References

DAVIES, G., GLEDHILL, A. & HAWKSWORTH, C. J. 1985. Upper crustal recycling in southern Britain: evidence from Nd and Sr isotopes. *Earth and Planetary Science Letters*, **75**, 1–22.

FAURE, G. 1986. *Principles of Isotope Geology.* 2nd ed. John Wiley & Sons, New York.

GOLDSTEIN, S. L., O'NIONS, R. K. & HAMILTON, P. J. 1984. A Sm–Nd isotopic study of atmospheric dusts and particulates from major river systems. *Earth and Planetary Science Letters*, **70**, 221–236.

HAMILTON, P. J., O'NIONS, R. K., BRIDGWATER, D. & NUTMAN, A. 1983. Sm–Nd studies of Archaean metasediments and metavolcanics from west Greenland and their implications for the Earth's early history. *Earth and Planetary Science Letters*, **62**, 263–272.

KNOX, R. W. O'B & CORDEY, W. G. 1993. *Lithostratigraphic nomenclature of the UK North Sea, Volume 3. Jurassic (Central and Northern North Sea).* British Geological Survey.

McCULLOCH, M. T. & WASSERBURG, G. J. 1978. Sm–Nd and Rb–Sr chronology of continental crust formation. *Science,* **200**, 1003–1011.

MEARNS, E. W. 1988. A Samarium–Neodymium isotopic survey of modern river sediments from northern Britain. *Chemical Geology (Isotope Geoscience Section)*, **73**, 1–13.

—— 1989. Neodymium isotope stratigraphy of Gullfaks Oil Field, In: COLLINSON, J. D. (ed.) *Correlation in Hydrocarbon Exploration.* Graham & Trotman, London, 201–215.

—— 1992. Samarium–neodymium isotopic constraints on the provenance of the Brent Group. In: MORTON, A. C., HASZELDINE, R. S., GILES, M. R. & BROWN, S. (eds) *1992, Geology of the Brent Group.* Geological Society, London, Special Publication, **61**, 213–225.

——, KNARUD, R., REASTAD, N., STANLEY, K. O. & STOCKBRIDGE, C. P. 1989. Samarium–Neodymium isotope stratigraphy of the Lunde and Statfjord Formations of Snorre Oil Field, northern North Sea. *Journal of the Geological Society, London,* **146**, 217–228.

MICHARD, A., GURIET, P., SOUDANT, M. & ALBAREDE, F. 1985. Nd isotopes in French Phanerozoic shales: external vs. internal aspects of crustal evolution. *Geochimica et Cosmochimica Acta,* **49**, 601–610.

OLAUSSEN, S., BECK, L., FÆLT, L. M., GRAUE, E., JACOBSEN, K. G., MALM, O. A. & SOUTH, D. 1993. Gullfaks Field – Norway; East Shetland Basin, Northern North Sea. In: *Atlas of Oil and Gas Fields of the World, Structural Traps VI.* American Association of Petroleum Geologists, Tulsa, 55–83.

O'NIONS, R. K., HAMILTON, P. J. & HOOKER, P. J. 1983. A Nd isotope investigation of sediments related to crustal development in the British Isles. *Earth and Planetary Science Letters,* **63**, 229–240.

ROE, S.-L. & STEEL, R. 1985. Sedimentation, sea-level rise and tectonics at the Triassic–Jurassic boundary (Statfjord Formation), Tampen Spur, northern North Sea. *Journal of Petroleum Geology,* **8**, 163–186.

VOLLSET, J. & DORÉ, A. G. 1984. A revised Triassic and Jurassic lithostratigraphic nomenclature for the Norwegian North Sea. *Norwegian Petroleum Directorate Bulletin,* **3**.

Appendix (continued) Results of Sm–Nd isotope analysis of samples

Well	Depth (m)	Lithology	Sm (p.p.m.)	Nd (p.p.m.)	$^{147}Sm/^{144}Nd$	$^{143}Nd/^{144}Nd$	±2SE	$^{145}Nd/^{144}Nd$	±2SE	Prov. Age (Ma)
34/10-C-8	2353.7	Massive, oil sat. F. sstn.	5.533	29.199	0.11537	0.511950	0.000010	0.348407	0.000005	1570
	2365.4	Massive pale grey siltstn. Caliche?	8.207	45.143	0.11068	0.511949	0.000006	0.348403	0.000005	1505
	2373.4	Unconsolidated, oil sat. sand.	3.773	20.574	0.11165	0.511920	0.000008	0.348405	0.000005	1558
	2380.0	Unconsolidated, oil sat. sand.	1.401	7.587	0.11241	0.511901	0.000010	0.348408	0.000007	1596
	2385.5	Unconsolidated, oil sat. sand	3.385	18.777	0.10976	0.511865	0.000010	0.348400	0.000005	1607
	2393.3	Oil saturated, F. sstn. Soft.	1.593	9.045	0.10722	0.511866	0.000008	0.348409	0.000007	1570
	2548.0	Massive, M. sstn, speckled, altered fspar, friable	6.346	36.914	0.10466	0.511653	0.000011	0.348411	0.000007	1815
	2557.4	Massive, M. sstn, pale grey, speckled, cemented.	2.607	14.910	0.10644	0.511458	0.000010	0.348407	0.000006	2102
	2559.6	Grey/red–brown laminated mudstn	3.985	21.400	0.11337	0.511570	0.000001	0.348397	0.000006	2079
	2562.9	Chocolate brown mudstn. Slickensides	9.138	50.445	0.11028	0.511878	0.000006	0.348401	0.000005	1597
	2568.0	Laminated, F. sstn/siltstn. Pale grey	6.587	36.751	0.10912	0.511838	0.000007	0.348401	0.000006	1635
	2574.9	Massive, M-C cemented sstn.	1.153	6.223	0.11284	0.511927	0.000008	0.348406	0.000008	1566
	2576.6	Laminated, impure sstn, siltstn, cemented	7.272	39.534	0.11199	0.511936	0.000008	0.348403	0.000006	1541
	2584.7	Massive, gr–br, silt/mud stn., slickensides	8.286	46.472	0.10855	0.511946	0.000007	0.348404	0.000006	1480
	2593.8	Massive, red–br, indurated siltstone	5.660	31.841	0.10823	0.511934	0.000007	0.348410	0.000004	1492
	2605.0	Massive, gr–br, indurated siltstone	12.823	69.534	0.11227	0.511954	0.000009	0.348409	0.000005	1520
34/10-11	1874.0	Laminated. F. sstn.	6.669	34.309	0.11835	0.511958	0.000007	0.348412	0.000006	1604
	1879.7	Massive, grey, indurated siltstn	5.061	24.022	0.12825	0.512001	0.000008	0.348409	0.000005	1705
	1889.0	Grey, indurated siltstone	4.435	31.579	0.10479	0.511962	0.000008	0.348419	0.000007	1411
	1892.8	M. impure sstn.	2.083	11.204	0.11318	0.511948	0.000009	0.348444	0.000008	1541
	1899.5	Brown F-sstn, w mica falkes	5.098	27.961	0.11099	0.511916	0.000008	0.348415	0.000005	1555
	1903.8	M. impure sstn.	3.633	19.989	0.11065	0.511911	0.000007	0.348422	0.000006	1557
	1913.2	M. sand, unconsolid	1.574	8.654	0.11074	0.511917	0.000014	0.348411	0.000006	1550
	1920.0	As above but with concretion	3.080	16.534	0.11340	0.511786	0.000008	0.348419	0.000006	1774
	1923.3	Grey siltstn, grading into F. sstn.	19.054	108.382	0.10703	0.511897	0.000013	0.348381	0.000006	1526
	1925.5	M–C impure sstn, rubble	1.780	9.525	0.11374	0.511954	0.000007	0.348428	0.000005	1541
	1927.5	SWC	1.238	7.365	0.10232	0.511902	0.000009	0.348371	0.000008	1459
	1934.0	SWC	2.278	14.210	0.09759	0.511898	0.000012	0.348408	0.000007	1408
	1954.0	SWC	1.159	6.417	0.10997	0.511566	0.000012	0.348405	0.000009	2021
	1980.5	SWC	1.150	6.379	0.10979	0.511696	0.000020	0.348409	0.000008	1840
34/10-C-2	2299.4	Sand	6.730	35.782	0.11451	0.511911	0.000010	0.348403	0.000007	1613
	2310.9	Massive grey siltstn	7.027	38.366	0.11150	0.511953	0.000017	0.348402	0.000008	1510
	2319.6	Massive impure siltstn	8.437	44.124	0.11640	0.511968	0.000009	0.348402	0.000005	1560
	2327.5	Sand	2.467	13.241	0.11341	0.511892	0.000010	0.348402	0.000006	1624
	2336.0	M., oil sat sand, rubble	5.417	29.071	0.11344	0.511905	0.000006	0.348407	0.000006	1606

Well	Depth (m)	Lithology	Sm (p.p.m.)	Nd (p.p.m.)	$^{147}Sm/^{144}Nd$	$^{143}Nd/^{144}Nd$	±2SE	$^{145}Nd/^{144}Nd$	±2SE	Prov. Age (Ma)
	2340.0	Sand	5.967	32.286	0.11252	0.511908	0.000010	0.348397	0.000008	1588
	2349.0	M., oil sat. sand, rubble	3.323	18.084	0.11188	0.511921	0.000008	0.348404	0.000006	1560
	2351.8	Massive M. sstn, oil sat., soft	10.629	62.680	0.10324	0.511840	0.000011	0.348403	0.000008	1550
	2358.5	Sand	3.425	20.039	0.10406	0.511837	0.000009	0.348399	0.000006	1565
	2363.5	Sand/silt	3.813	22.769	0.10194	0.511536	0.000010	0.348399	0.000006	1923
	2365.0	Laminated mudstn/siltstn.	22.527	132.527	0.10347	0.511316	0.000009	0.348390	0.000008	2233
	2366.3	Sand	2.461	14.491	0.10339	0.511377	0.000009	0.348406	0.000006	2152
	2371.6	Choc brown, coaly mudstone	1.434	7.782	0.11218	0.511871	0.000007	0.348415	0.000004	1635
	2376.5	Mud/sand	5.005	29.062	0.10484	0.511818	0.000009	0.348397	0.000008	1601
	2381.8	M. massive sstn.	4.430	26.382	0.10223	0.511833	0.000008	0.348406	0.000007	1546
	2388.6	Sand	5.850	35.334	0.10079	0.511421	0.000007	0.348402	0.000006	2050
	2392.0	Massive, M. sstn, oil sat.	4.097	22.723	0.10974	0.511426	0.000010	0.348404	0.000007	2208
	2396.5	Sand	5.485	30.784	0.10847	0.511499	0.000007	0.348406	0.000006	2085
	2398.6	Dark grey muddy siltstn.	6.864	39.953	0.10458	0.511477	0.000008	0.348401	0.000006	2044
	2402.2	Sand	3.281	19.977	0.09998	0.511510	0.000007	0.348400	0.000006	1924
	2410.9	Indurated, muddy siltstone, tectonised	1.504	7.633	0.11998	0.511683	0.000007	0.348426	0.000005	2045
	2419.2	Poorly sorted C. sstn, altered fspar, oil sat.	1.592	9.396	0.10317	0.511578	0.000010	0.348407	0.000008	1889
	2421.7	Sand	1.610	9.764	0.10039	0.511630	0.000009	0.348410	0.000006	1779
	2425.4	Silt	38.012	208.790	0.11082	0.511440	0.000007	0.348397	0.000007	2211
	2430.0	Sand	1.551	8.837	0.10687	0.511555	0.000009	0.348405	0.000008	1981
	2439.5	Sand	4.933	29.454	0.10194	0.511546	0.000008	0.348401	0.000006	1910
	2444.5	Dark grey, soapy muddy siltstn.	9.653	53.746	0.10933	0.511475	0.000007	0.348395	0.000006	2133
	2447.6	M–F sstn, crumbling, oil sat.	2.143	11.605	0.11240	0.511594	0.000007	0.348402	0.000005	2027
	2452.3	Sand	2.422	13.342	0.11051	0.511636	0.000007	0.348413	0.000006	1934
	2458.9	Dark grey massive mudstn/siltstn.	1.765	9.307	0.11542	0.511780	0.000007	0.348393	0.000006	1817
	2464.1	M–C sstn poor cementation, oil sat.	1.265	8.000	0.09628	0.511672	0.000007	0.348406	0.000006	1669
	2468.0	Sand	2.583	14.938	0.10524	0.511483	0.000008	0.348394	0.000006	2047
	2471.8	Laminated F. sstn with rippled shale lamenae	7.843	45.014	0.10607	0.511548	0.000008	0.348406	0.000005	1976
	2478.5	M–C poorly sorted sstn. oil sat.	2.127	12.470	0.10384	0.511655	0.000015	0.348404	0.000012	1799
	2482.6	Sand	2.088	12.074	0.10526	0.511530	0.000009	0.348405	0.000006	1986
	2486.0	Mud/silt	7.919	43.278	0.11139	0.511531	0.000009	0.348393	0.000006	2096
	2487.6	Sand	0.868	4.961	0.10644	0.511556	0.000008	0.348415	0.000006	1972
	2493.5	Mud	15.234	87.816	0.10561	0.511702	0.000008	0.348408	0.000006	1765
	2499.0	Fine sand; brown	4.115	23.476	0.10670	0.511420	0.000009	0.348405	0.000006	2157
	2504.2	Sand	1.196	7.035	0.10346	0.511467	0.000008	0.348413	0.000006	2037
	2509.7	Grey, F-sstn, w-mica, cemented	5.356	31.150	0.10467	0.511531	0.000007	0.348409	0.000006	1974
	2518.0	Silt/sand	6.201	36.145	0.10444	0.511419	0.000010	0.348409	0.000006	2117
	2527.0	Silt/sand	5.519	31.104	0.10801	0.511525	0.000008	0.348400	0.000007	2041

Well	Depth (m)	Lithology	Sm (p.p.m.)	Nd (p.p.m.)	$^{147}Sm/^{144}Nd$	$^{143}Nd/^{144}Nd$	±2 SE	$^{145}Nd/^{144}Nd$	±2 SE	Prov. Age (Ma)
	2531.0	Grey massive mudstone	8.398	47.006	0.10875	0.511484	0.000006	0.348404	0.000006	2110
	2535.2	Silt/sand	2.919	17.013	0.10444	0.511663	0.000008	0.348411	0.000006	1798
	2542.3	Sand	2.242	12.968	0.10523	0.511558	0.000009	0.348402	0.000006	1949
	2546.1	Brown mudstone with grey/green redu. spots	4.838	27.159	0.10843	0.511636	0.000008	0.348403	0.000004	1899
	2548.5	Silt/sand	3.207	18.508	0.10550	0.511502	0.000007	0.348399	0.000006	2027
	2552.8	Massive, pale brown M. sstn, mod cementation	2.343	13.386	0.10656	0.511599	0.000008	0.348406	0.000006	1917
	2555.2	Sand	4.111	23.580	0.10613	0.511564	0.000007	0.348414	0.000008	1956
	2558.2	Brown muddy siltstn, slickensides	9.574	54.684	0.10658	0.511533	0.000009	0.348400	0.000006	2005
	2559.2	Mud/silt	16.224	48.745	0.20261	0.511547	0.000009	0.348402	0.000006	
	2590.5	Silt/sand	8.044	45.170	0.10841	0.511941	0.000007	0.348412	0.000007	1485
	2604.6	Mud/silt	5.415	29.770	0.11074	0.511957	0.000007	0.348408	0.000006	1494
34/10-C-6	2825.3	Medium-grained red–white speckled sandstone	3.101	17.070	0.11058	0.511948	0.000007	0.348416	0.000006	1505
	2830.9	Red–brown siltstone with blotchy reduction spots	7.865	43.899	0.10907	0.511911	0.000007	0.348413	0.000007	1535
	2839.5	Medium-grained red–white speckled sandstone	3.553	19.683	0.10990	0.511959	0.000007	0.348413	0.000005	1480
	2850.0	Medium-grained red–white speckled sandstone	4.900	25.168	0.11852	0.511967	0.000007	0.348414	0.000007	1593
	2852.0	Red–brown, massive siltstone	6.957	38.900	0.10889	0.511942	0.000008	0.348418	0.000006	1490
	2912.4	Med-gr, laminated, brown, friable sandstone	5.644	30.705	0.11191	0.511963	0.000007	0.348416	0.000008	1502
	2924.7	Medium-grained, grey sandstone	9.449	57.578	0.11153	0.511963	0.000008	0.348398	0.000008	1497
	2933.3	Red–brown mudstone, slickensides	8.345	46.017	0.11040	0.511938	0.000008	0.348396	0.000007	1516
	3025.0	Dark red–brown indurated mudstone	10.345	55.559	0.11336	0.511944	0.000011	0.348395	0.000007	1549
	3035.5	Coarse-grained red–white speckled sandstone	2.258	12.908	0.10648	0.511902	0.000009	0.348411	0.000008	1512
	3043.3	Medium-grained red–white speckled sandstone	4.216	22.945	0.11186	0.511947	0.000010	0.348388	0.000009	1524
	3050.7	Dark red–brown indurated mudstone	10.427	59.810	0.10613	0.511921	0.000007	0.348410	0.000008	1482
	3073.2	Medium-grained red–white speckled sandstone	5.144	30.492	0.10270	0.511909	0.000008	0.348417	0.000007	1454
34/10-C-13	2542.0	> 250 μm: mudstone chips; cuttings	2.175	10.717	0.12356	0.512078	0.000006	0.348408	0.000006	1502
	2551.0	125–250 μm: felsic sand; cuttings	1.388	7.353	0.11489	0.511949	0.000011	0.348392	0.000008	1565
	2551.0	> 250 μm: grey mudstone chips; cuttings	1.874	9.122	0.12504	0.512100	0.000008	0.348415	0.000006	1490
	2575.0	125–250 μm: felsic sand; cuttings	2.366	8.736	0.16488	0.512388	0.000009	0.348403	0.000006	1771
	2585.4	Impure M-sstn, massive, carb. lam, oil sat.	2.332	13.129	0.10816	0.511818	0.000016	0.348418	0.000011	1648
	2588.5	Impure, pale-gr–br silty sstn, rubble	6.878	37.585	0.11139	0.511381	0.000007	0.348411	0.000006	2304
	2594.5	Impure M-sstn, rubble	2.527	15.116	0.10175	0.511467	0.000007	0.348409	0.000007	2008
	2598.2	Massive, M-sstn, oil sat, crumbly	1.391	8.318	0.10075	0.511850	0.000011	0.348423	0.000007	1519
	2601.5	Grey mudstn, friable	2.631	12.798	0.12517	0.511787	0.000011	0.348401	0.000008	1991
	2605.9	M. sstn, massive, friable	1.442	8.150	0.10774	0.511870	0.000007	0.348399	0.000007	1572
	2609.0	Dark choc brown mudstone	2.954	16.651	0.10801	0.511645	0.000007	0.348413	0.000005	1879
	2613.8	Impure M–F silty sstn, pale grey, cemented	14.767	89.737	0.10017	0.511496	0.000013	0.348401	0.000006	1945
	2617.8	M. sstn, massive, v. friable, pale br	1.803	10.445	0.10506	0.511455	0.000015	0.348409	0.000008	2081

Well	Depth (m)	Lithology	Sm (p.p.m.)	Nd (p.p.m.)	$^{147}Sm/^{144}Nd$	$^{143}Nd/^{144}Nd$	±2SE	$^{145}Nd/^{144}Nd$	±2SE	Prov. Age (Ma)
	2623.3	M–C sstn, pale br, v. friable	2.126	12.402	0.10436	0.511613	0.000008	0.348413	0.000006	1862
	2629.4	Grey lam siltstn, cemented	7.689	43.656	0.10721	0.511581	0.000008	0.348403	0.000006	1952
	2633.2	M–C graded sstn, v. friable, pale br	2.177	12.729	0.10413	0.511581	0.000007	0.348402	0.000006	1900
	2635.6	Pale gr–br mudstone, bit frible, slickensides	6.288	35.187	0.10879	0.511476	0.000006	0.348400	0.000005	2122
	2638.5	Impure F–sstn, pale col. massive	1.495	7.764	0.11724	0.511510	0.000006	0.348405	0.000005	2245
	2644.4	M–C sstn, altered fspar, 2nd por, massive	1.562	8.957	0.10612	0.511643	0.000010	0.348393	0.000006	1851
	2656.7	M–C graded sstn, brown, friable	1.578	8.980	0.10699	0.511661	0.000013	0.348406	0.000080	1841
	2660.9	Grey mudstone, soapy	7.514	41.606	0.10994	0.511586	0.000008	0.348398	0.000006	1993
	2663.9	M–C sstn, oil sat rubble	1.528	8.775	0.10597	0.511702	0.000006	0.348403	0.000006	1771
	2666.6	M. sstn, oil sat. rubble	4.152	28.818	0.10611	0.511787	0.000010	0.348402	0.000005	1660
	2669.3	X lam, pale sstn/dark siltstn.	35.714	195.310	0.11132	0.511664	0.000012	0.348400	0.000017	1910
	2675.5	Massive, M. sstn, oil sat.	4.950	29.969	0.10054	0.511601	0.000008	0.348404	0.000006	1818
	2679.8	Massive, dark grey mudstone	5.793	32.937	0.10706	0.511518	0.000006	0.348403	0.000005	2034
	2681.9	Impure, silty F. sstn, cemented, pale gr–br	3.454	20.200	0.10410	0.511495	0.000008	0.348415	0.000005	2012
	2689.7	Impure, silty F. sstn, massive, cemented, w-mica	10.695	62.558	0.10850	0.511570	0.000007	0.348399	0.000006	1914
	2691.8	Grey-brown siltstone, slickensides	10.084	56.577	0.10241	0.511528	0.000012	0.348392	0.000010	2046
	2696.7	Red-brown mudstone, green reduction zones	11.447	68.041	0.10898	0.511548	0.000007	0.348406	0.000006	1915
	2701.6	Black mudstone/c-sstn mixed fac.	7.594	42.421	0.11013	0.511788	0.000011	0.348396	0.000008	1701
	2706.5	Choc. brown mudstone, friable	8.602	47.544	0.10586	0.511463	0.000008	0.348414	0.000004	2165
	2708.1	F, impure, white sstn.	4.926	28.329	0.10531	0.511438	0.000008	0.348396	0.000006	2118
	2711.5	M–C massive sstn, cemented with 2nd por	1.090	6.303	0.11041	0.511453	0.000007	0.348416	0.000006	2088
	2725.0	> 500 μm: grey and brown mudstone chips; cuttings	6.160	33.962	0.10672	0.511648	0.000005	0.348406	0.000006	1916
	2725.0	250–355 μm: felsic sand(70) mudstone(30); cuttings	2.064	11.772	0.10884	0.511445	0.000005	0.348403	0.000006	2125
	2734.0	> 500 μm: mudstone chips; cuttings	5.980	33.445	0.11238	0.511598	0.000005	0.348398	0.000006	1957
	2767.0	> 500 μm: mudstone chips; cuttings	7.652	41.448	0.11644	0.511577	0.000007	0.348402	0.000006	2051
	2779.0	125–250 μm: felsic sand; cuttings	1.017	5.319	0.10469	0.511887	0.000030	0.348408	0.000020	1678
34/10-C-4X	2815.4	M. sandstone, unconsolidated	2.949	16.763	0.10709	0.511908	0.000006	0.348416	0.000006	1512
	2820.5	Unconsolidated, M–C sstn, oil sat.	1.545	8.904	0.10565	0.511752	0.000007	0.348415	0.000006	1699
	2825.5	Grey, indurated siltstone	7.574	43.234	0.10666	0.511936	0.000012	0.348410	0.000008	1469
	2830.8	M-grained sandstone, unconsolidated	0.842	4.712	0.10883	0.511906	0.000010	0.348404	0.000006	1538
	2830.8	Repeat				0.511895	0.000007	0.348412	0.000006	
	2833.6	Poorly sorted, F–C sstn, friable	2.865	16.389	0.10641	0.511470	0.000018	0.348384	0.000012	2086
	2838.0	Unconsolidated, oil sat. m. sstn.	2.163	12.981	0.10143	0.511354	0.000007	0.348410	0.000007	2146
	2843.7	Very coarse sandstone, friable	1.048	6.167	0.10348	0.511549	0.000008	0.348413	0.000004	1931
	2849.4	F-grained sandstn/siltstn. carbon laminae	4.242	23.534	0.10974	0.511761	0.000007	0.348409	0.000005	1750
	2854.5	C-grained sandstone, friable	0.867	5.017	0.10516	0.511791	0.000011	0.348402	0.000008	1641
	2860.3	Impure F. sstn, rootlets, w-mica, indurated	3.443	17.731	0.11822	0.511757	0.000008	0.348406	0.000006	1901
	2863.5	M-grained massive sandstone, friable	1.572	9.140	0.10469	0.511575	0.000007	0.348419	0.000005	1917

Well	Depth (m)	Lithology	Sm (p.p.m.)	Nd (p.p.m.)	$^{147}Sm/^{144}Nd$	$^{143}Nd/^{144}Nd$	±2 SE	$^{145}Nd/^{144}Nd$	±2 SE	Prov. Age (Ma)
	2868.7	Very coarse sandstone, friable	1.736	10.504	0.10062	0.511539	0.000010	0.348402	0.000008	1898
	2872.3	Grey-black, muddy siltstone, slickensides	18.262	114.454	0.09713	0.511443	0.000012	0.348411	0.000008	1962
	2876.5	Very coarse sandstone, friable	1.897	10.961	0.10537	0.511554	0.000008	0.348415	0.000008	1956
	2879.9	M-C grained, massive, oil sat. sstn, soft	1.445	8.544	0.10296	0.511579	0.000008	0.348407	0.000006	1884
	2882.8	Choco brown, mudstone, slickensides	3.252	18.588	0.10648	0.511507	0.000009	0.348393	0.000008	2038
	2885.5	F-grained, massive sandstone	4.742	25.895	0.11147	0.511424	0.000009	0.348411	0.000006	2246
	2892.0	M-F-grained, lamin. sandstone, cemented	4.962	27.114	0.11140	0.511528	0.000008	0.348419	0.000007	2100
	2893.6	M. sstn, massive, oil sat.	1.612	9.727	0.10086	0.511556	0.000007	0.348410	0.000005	1880
	2898.4	C-grained sandstone, rubble	1.782	10.701	0.10136	0.511636	0.000010	0.348407	0.000007	1786
	2908.0	M-F-grained, graded sandstone, friable	1.840	10.412	0.10758	0.511893	0.000012	0.348407	0.000008	1539
	2911.5	Interbedded, ripple laminated, mudstone-sstn.	7.597	43.457	0.10643	0.511770	0.000008	0.348400	0.000006	1687
	2917.5	Very coarse sandstone, friable	1.201	6.717	0.10884	0.511619	0.000008	0.348424	0.000007	1929
	2921.4	Grey siltstone, massive	3.311	18.342	0.10990	0.511704	0.000007	0.348400	0.000006	1831
	2925.0	F-Grained, massive, grey sandstone	15.356	96.023	0.09735	0.511411	0.000007	0.348419	0.000006	2005
	2930.4	M-C oil sat, sstn, friable	1.402	8.132	0.10495	0.511550	0.000009	0.348411	0.000008	1954
	2932.7	M-grained, massive, friable sandstone	2.125	11.726	0.11031	0.511419	0.000006	0.348414	0.000006	2229
	2937.0	Grey-brown, laminated, muddy siltstone	5.575	30.779	0.11026	0.511490	0.000007	0.348413	0.000008	2131
	2940.0	Red-grey-brown mudstone	4.585	26.464	0.10546	0.511530	0.000008	0.348413	0.000007	1989
	2945.5	Mottled grey-green mudstone	3.090	16.417	0.11459	0.511679	0.000008	0.348407	0.000006	1947
	2949.5	Red-brown muddy siltstone, fractures	4.292	25.062	0.10424	0.511593	0.000008	0.348415	0.000007	1886
	2952.4	Very pale, massive, M. sstn.	3.942	23.427	0.10243	0.511414	0.000009	0.348415	0.000012	2087
	2960.0	Very coarse friable sandstone	1.613	9.506	0.10328	0.511560	0.000009	0.348416	0.000007	1914
	2963.7	Red-brown friable muddy siltstone	15.618	95.413	0.09993	0.511422	0.000010	0.348420	0.000007	2034
	2974.4	Very coarse friable sandstone	1.019	5.892	0.10528	0.511697	0.000007	0.348408	0.000006	1766
	2977.4	Coarse, friable, pale sstn, 2nd porosity	3.203	19.101	0.10200	0.511467	0.000008	0.348408	0.000006	2013
	2978.7	Red-brown-grey, muddy siltstone	6.497	39.254	0.10075	0.511382	0.000009	0.348393	0.000006	2099
	2981.4	Yellowy-grey coarse sandstone	2.485	14.235	0.10625	0.511469	0.000011	0.348418	0.000008	2084
	2983.5	Grey laminated, sandy siltstn, w-mica, indurated	6.694	38.407	0.10611	0.511531	0.000009	0.348407	0.000006	1999
	2987.9	Yellowy-brown M-C sandstone	1.634	9.270	0.10731	0.511536	0.000008	0.348410	0.000007	2014
	2991.9	Brown, indurated mudstone, slickensides	7.372	42.038	0.10676	0.511541	0.000006	0.348405	0.000006	1997
	2997.8	Grey medium sandstone, well cemented	9.849	56.387	0.10577	0.511655	0.000006	0.348409	0.000008	1830
	3003.9	Red-brown very friable mudstone	5.686	30.833	0.11227	0.511706	0.000008	0.348421	0.000008	1868
	3010.0	Red-brown very friable mudstone	12.097	71.435	0.10310	0.511753	0.000009	0.348424	0.000008	1661
	3017.6	Grey-brown medium sandstone	4.875	27.394	0.10833	0.511933	0.000008	0.348412	0.000008	1495
	3026.3	Grey medium sandstone	2.112	17.937	0.10561	0.511931	0.000008	0.348410	0.000007	1462
	3030.7	Very friable mudstone	2.898	13.671	0.12905	0.511976	0.000008	0.348427	0.000006	1762
	3040.9	Red-brown mudstone	11.644	61.372	0.11551	0.511982	0.000008	0.348410	0.000007	1526
	3052.1	Red-brown friable mudstone	11.537	62.899	0.11167	0.511963	0.000012	0.348408	0.000008	1499
	3064.5	Red-brown friable mudstone	10.076	54.336	0.11290	0.511985	0.000011	0.348417	0.000008	1485

Luminescence dating of Quaternary sediments

H. M. RENDELL

Geography Laboratory, University of Sussex, Falmer, Brighton BN1 9QN, UK

Abstract: Luminescence techniques have the potential to date Quaternary sediments from a range of depositional environments. Recent developments of optical dating complement thermoluminescence dating techniques, extending their applicability. The age range of samples that can be dated shows a strong dependence on the material involved. Recent work on both dune sands and loess indicates that accurate dating to beyond 500 ka may be possible in certain circumstances. The age range limits for other material may be closer to 100 ka. Examples are given of luminescence dating of material from a range of sedimentary environments.

Luminescence dating techniques are based on dating the last exposure to light of sediment grains prior to burial. Such techniques potentially have wide application to Quaternary sediments which do not contain material that can be dated by other methods, such as ^{14}C or uranium series disequilibrium. The purpose of this paper is to review briefly the current state-of-the-art within luminescence dating, and to discuss the age range limits of the techniques and their applicability to a range of sedimentary environments.

Although the first important paper on luminescence dating of sediments was on the thermoluminescence dating of ocean sediments (Wintle & Huntley 1979), subsequent work has mostly concentrated on aeolian sediments (for recent reviews see Berger 1988; Wintle 1990, 1993; Zöller & Wagner 1991). The strength of luminescence dating, in that the material being dated is the sediment itself, is also its weakness. The accuracy of these luminescence dates can only be determined by comparison with independently dated material from the same profile or context. Independent dating is a problem precisely because appropriate materials (e.g. organic carbon or inorganic carbonate) may either not be available, or may be beyond the range of the independent technique concerned, e.g. ^{14}C.

Luminescence dating techniques

The techniques are based upon the fact that grains of quartz and feldspar are natural radiation dosimeters. When quartz and feldspar grains are exposed to a flux of ionizing radiation, electrons (e^-) and holes (h^+) become trapped at defects at the atomic level within these natural crystals. The trapped charges can be liberated by heating the crystal, and a proportion of the trapped charges recombine with the emission of light to give thermoluminescence (TL). Trapped charge can also be released by exposing the crystal to light of a particular wavelength and monitoring light output, again as a result of recombination, at a different wavelength. This process is known as optically-stimulated luminescence (OSL).

In order to obtain a date, both the radiation dose absorbed by the sample since its last exposure to light [i.e. the equivalent dose (ED)] and the annual radiation dose that the sample grains would have received since burial need to be evaluated. The way in which an age estimate may be obtained for a particular sediment is summarized schematically in Fig. 1.

Luminescence dating has concentrated almost exclusively on the dating of quartz and feldspar grains. The grain size ranges used are polarized between the large grain size fraction (125–300 μm) and fine grains (2–10 μm). A narrow range of grain sizes (e.g. 150–180 μm) is normally chosen from within the potential size range for the large grains. Heavy liquids are used to separate the different mineral fractions of the large grains. Fine-grain mixtures are generally polymineral, but etching with fluorosilicic acid may be used to isolate the fine-grain quartz component. The different ionizing radiations, i.e. alpha, beta, gamma and cosmic, that make up the natural radiation flux have different degrees of penetration. The penetration distances of alpha particles are relatively short. The alpha particles emitted from radionuclides in the uranium and thorium series have mean ranges of 23 and 28 μm, respectively (Aitken 1985). Large grains are therefore generally etched with hydrofluoric acid to remove the outer, alpha-irradiated portion of each grain.

From Dunay, R. E. & Hailwood, E. A. (eds), 1995, *Non-biostratigraphical Methods of Dating and Correlation*
Geological Society Special Publication No. 89, pp. 223–235

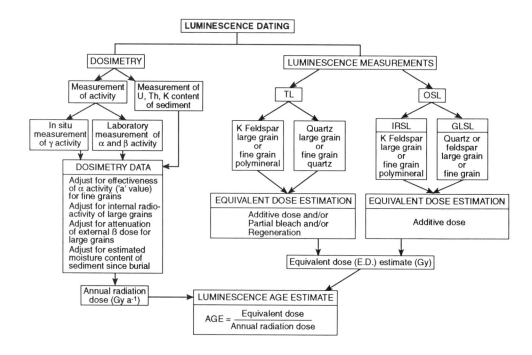

Fig. 1. Schematic flow chart for luminescence dating of sediments. TL, thermoluminescence; OSL, optically-stimulated luminescence; IRSL, infra-red stimulated luminescence; GLSL, green light stimulated luminescence. Sample activity typically measured by *in situ* gamma spectrometry, thick source alpha counting and thick source beta counting. U, Th, K measurement by neutron activation analysis, gamma spectrometry, alpha spectrometry or by I.C.P.M.S. K analysis by A.A.S. or flame photometry.

TL dating was originally developed for dating pottery and burnt stone, and only began to be applied to sediments in a systematic way in the late 1970s and early 1980s, when it became apparent that exposure of quartz and feldspar grains to light was effective in resetting the TL signal (Aitken 1985). The revolutionary step of using optical rather than thermal stimulation was made by Huntley *et al.* (1985), who used an argon-ion laser (emission 514 nm) to stimulate emission in quartz and feldspar. Later, important work by Hutt *et al.* (1988) showed that infrared stimulation, in the wavelength range 800–950 nm, could be used to excite emission from feldspars. Currently, OSL work appears to be polarized between the use of green light to produce green light stimulated luminescence (GLSL), using a 514 nm laser, a halogen or xenon lamp with filters, for either quartz or feldspar, and the use of infrared diodes to produce infrared stimulated luminescence (IRSL) for feldspar.

The zeroing mechanism

An important difference between the effectiveness of solar resetting for TL and OSL in sediments is shown schematically in Fig. 2. It has become clear that, irrespective of the length of time of light exposure or of the wavelength range of light used, a residual TL signal will always remain. This residual signal is sometimes referred to as the 'unbleachable' component of TL. By contrast, very short light exposures, of the order of minutes, are sufficient to zero completely OSL signals from both quartz and feldspar grains (Godfrey Smith *et al.* 1988). This critical difference is particularly important in two respects: first, for the dating of sedimentary environments where light exposure may be relatively short or of low intensity, and secondly, for geologically 'young' sediments where TL signal levels are likely to be low and therefore barely distinguishable from the TL 'residual' level.

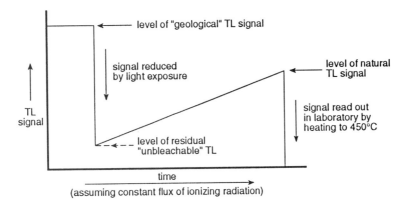

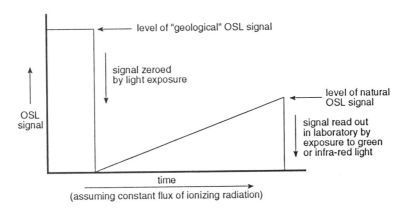

Fig. 2. Schematic representation of changes in thermoluminescence (TL) and optically stimulated luminescence (OSL) signals with time.

The problems associated with optical resetting of TL signals in sediments are discussed in some detail by Huntley (1985) and Berger (1990), with particular reference to the choice of appropriate methods of ED determination. As far as TL is concerned, there is considerable evidence that optical/solar resetting is more effective for feldspars than for quartz. But, in either case, the nature of the depositional environment may militate against complete resetting (Kronberg 1983; Gemmell 1985; Ditlefson 1992; Forman & Ennis 1992). Berger (1988, 1990) has advocated the use of the partial bleach technique of ED estimation and the use of longer wavelengths of light (550 nm) for laboratory bleaching of sediments, particularly where deposition was subaqueous. However, it is clear that partial bleach techniques for TL remain problematic, both in terms of the wavelength range of light used for laboratory bleaching and in terms of the extrapolation techniques employed for growth

curves (Berger & Eyles 1994). A recent field experiment designed to investigate underwater bleaching of TL and OSL for both quartz and feldspar, at a range of different water depths, showed that, after a 3 h exposure, IRSL and OSL signals from feldspar and quartz were effectively reset down to water depths of 10 m, whereas TL signals showed much less resetting (Rendell *et al.* 1994). Interestingly, this study also showed that quartz OSL signals were reset more rapidly than feldspar IRSL signals, but that the reverse was true for the TL signals. These results are in broad agreement with data on subaerial exposure of quartz and feldspar grains.

Estimation of absorbed radiation dose

The methods of obtaining ED estimates for both TL and OSL are shown schematically in Fig. 3. The accurate estimation of the radiation dose

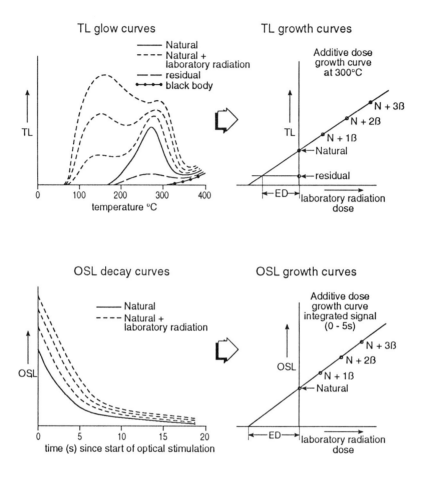

Fig. 3. Schematic representation of Equivalent Dose determination by the Additive Dose method for both TL and OSL. Note: OSL curves are measured after removal of an unstable component of the signal either by pre-heating or long-term storage.

absorbed by the sample grains, since their last exposure to light, is obviously of fundamental importance in luminescence dating. A range of techniques was developed in TL sediment dating during the 1980s, but all generally involve the measurement of the natural TL signal in samples, the measurement of the TL output per unit of laboratory radiation dose, and some strategy for estimating the level of residual signal at deposition (Aitken 1985; Berger 1988). All the ED estimation techniques concentrate on TL emissions in the higher temperature part of the TL glow curve (i.e. 270–400°C). The reason for this choice of temperature range is that this is the region of stable storage of trapped charge at ambient temperatures (Mejdahl 1988). Some of the methods of ED determination appear to be

prone to problems. In particular, the regeneration method has been shown to underestimate sample age in a consistent manner.

In contrast to the TL measurements, giving temperature-resolved information, OSL information is limited to a series of decay curves (see Fig. 3). OSL measurements have to be made after any thermally-unstable component of the signal has been removed. This removal is normally accomplished by either pre-heating sample aliquots in the laboratory or by long-term storage. The additive dose method has tended to be favoured in OSL, partly because measurements for quartz show a change in dose–response after laboratory bleaching (Smith *et al.* 1990), and similar changes in sensitivity are also reported for feldspar (Duller 1992). The recent

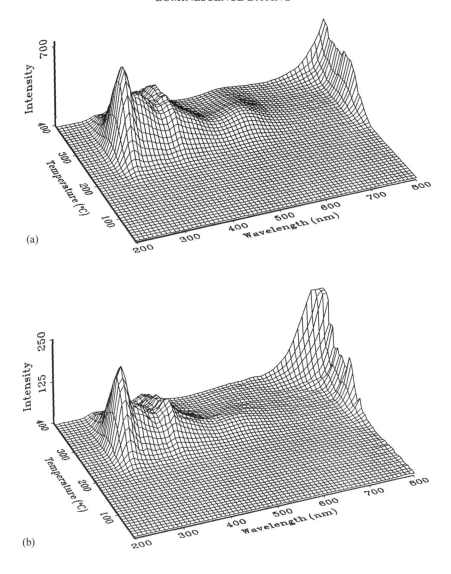

Fig. 4. Natural TL emission spectra for potassium feldspar separates from fluvial sands from the Tajo valley, near Toledo, central Spain. (a) ARG2.1; (b) TO2L.

development of automated TL/OSL systems allowed Duller (1991) to introduce a single aliquot technique for ED estimation using IRSL from potassium feldspars. This technique has considerable advantages, both in terms of the rapidity with which age determination can be made and in terms of improved precision compared to conventional dating procedures.

In addition to the choice of ED estimation technique, the choice of the wavelength range of TL emissions to be monitored continues to be the subject of debate. TL emissions are con-

ventionally monitored by a photomultiplier tube via broad-band optical filters. Until recently, the choice of filters has been based on an imperfect knowledge of the TL emission spectrum and it has become apparent that ED values obtained for large-grained potassium feldspars, for example, can vary as a function of the choice of emission wavelengths (Balescu & Lamothe 1992). High-sensitivity measurements of TL emission spectra of some of the potassium feldspar samples analysed by Balescu & Lamothe (1992), using equipment designed and built at

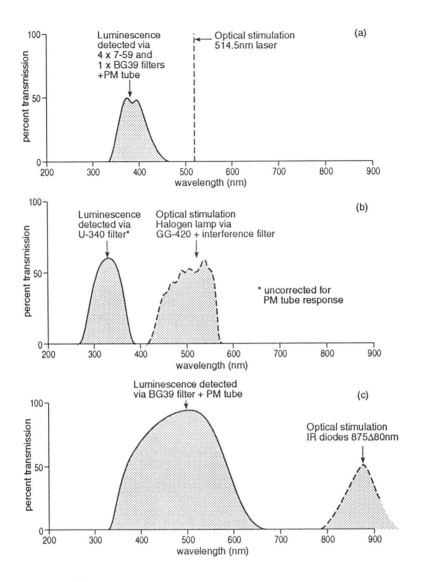

Fig. 5. Stimulation and detection wavelengths in OSL (**a**) green laser stimulation (source: Spooner 1992); (**b**) broad band green light stimulation using a halogen lamp and optical filters (source: Bøtter-Jensen & Duller 1993) and (**c**) infrared stimulation using light emitting diodes (sources: Duller 1992; Spooner 1992).

Sussex University by Luff & Townsend (1993), have revealed the presence of three distinct emission bands at 280, 340 and 400 nm in the temperature range used for ED determination (Rendell *et al.* 1993). These emission bands appear to be common to potassium feldspar separates from a range of environments and, as an example, natural TL emission spectra for two feldspar separates from samples of fluvial sands from Central Spain are shown in Fig. 4. The

right choice of filters may, therefore, be critical to the success of feldspar dating, since it is apparent that problems may ensue if particular combinations of the emission bands are 'sampled' via broad-band filters.

The choice of detection wavelengths for OSL measurements is constrained by the need to shield the PM tube from the stimulating wavelength(s). The various options are illustrated in Fig. 5. It is clear that the choice of detection wavelength range is much more flexible in the case of IRSL than it is for GLSL, where emissions are generally monitored in the blue or blue–UV. Whether such a choice is significant in affecting the ED estimation has still to be established.

Upper age limits for luminescence dating

The ultimate limits of luminescence dating are determined by the availability of defects at which the e^- and h^+ produced by the passage of ionizing radiation through the crystals, may be trapped and the stability of storage within those traps (Mejdahl 1988). The effect of these limitations is that, eventually, the dose–response curves for quartz and feldspar reach saturation. All other things being equal, the oldest ages are obtained where there is a combination of low sensitivity to dose on the part of the natural dosimeter and a low environmental dose rate. Such a combination of circumstances appears to exist in the case of quartz grains extracted from a dune sequence associated with raised beaches in southeast Australia (Huntley et al. 1985, 1993a). The environmental dose rate was low, of the order of $0.50–0.60\,\mathrm{Gy\,ka^{-1}}$, compared to more typical values of $2.0–4.5\,\mathrm{Gy\,ka^{-1}}$. TL ages going back to beyond 800 ka, in reasonably good agreement with oxygen isotope data, were obtained from relating times of dune formation to sea-level changes. In an interesting addition to this work, Huntley et al. (1993b) have also reported IRSL dates for feldspar inclusions within some of these dune sand samples, and the dates obtained, in the range 0–300 ka, are in broad agreement with the isotope chronology, although with a tendency to be systematically younger than the TL dates for the same sequence.

Although severe underestimation of TL ages of loesses from NW Europe appears to occur, particularly if the regeneration method of ED determination is used, age estimates of up to 300 ka have been obtained for loess deposits in central and eastern Europe (Zöller & Wagner 1991). Berger et al. (1992), however, have reported TL dates for loess samples from sites with reasonably good independent age control in Alaska and New Zealand, back to 800 ka, using the partial bleach and additive dose (total bleach) methods of ED determination. Unlike the study by Huntley et al. (1993a), the dose rates for these loess samples were in the range $1.18–4.30\,\mathrm{Gy\,ka^{-1}}$, and therefore were not unusually low. However, it should be noted that the errors reported for this data set are high, particularly for the oldest sample, which has a

TL age of 730 ± 250 ka (i.e. an error of 34%). Indeed, almost all of the other TL dates in excess of 300 ka have errors in the range 15–25%. Wintle et al. (1993) criticized this work on the basis, amongst other things, of the problems associated with the extrapolation of non-linear growth curves in order to establish a value for the ED. However, Berger (1993) has defended both the extrapolations used and the claims to be able to date loess of up to 800 ka.

One particular obstacle to the dating of older material is the loss of signal from supposedly thermally-stable traps in feldspars. Mejdahl (1988) discusses long-term fading problems in potassium feldspar and suggests a potential method for correcting dates older than 10 ka. Fading of signal at ambient temperatures may also prove to be a serious problem for OSL/IRSL dating of feldspars (Spooner 1992), with the potential to affect all age determinations.

Lower age limits in luminescence dating

The tremendous advantage of OSL age determinations over TL ones is that the problem of the residual signal is removed, and therefore geologically young material can be dated with increased accuracy and precision (Smith et al. 1990; Edwards 1993). Somewhat strangely, much of the initial work on OSL dating of sediments has focused on aeolian sediments, whereas the particular utility of OSL is of special importance where the bleaching history of the sediment grains must be presumed to have been short prior to deposition and burial. The partial bleach method for TL ED estimation has tended to be employed precisely because it has the potential to overcome problems of short or 'incomplete' bleaching prior to deposition, but OSL methods now present an alternative solution to this problem.

Problems in the evaluation of annual radiation dose

Determination of the annual dose rate experienced by sediment grains since burial may, at first sight, seem to be far more straightforward than the luminescence techniques for ED determination. One of the fundamental assumptions underlying dose-rate calculations is that secular equilibrium exists in the long decay chains of $^{238}\mathrm{U}$, $^{235}\mathrm{U}$ and $^{232}\mathrm{Th}$. Unfortunately, this assumption may not always hold. Problems of disequilibrium may arise from the escape of thoron or radon, or from post-depositional changes affecting the uranium or thorium contents of the sediment, such as solution and leaching of material (Aitken 1985), weathering processes, and the deposition of grain coatings (Questiaux 1991). Post-depositional changes may also have radical effects on any feldspars present, and on the potassium content of the sediments. Again, this type of change introduces uncertainty into the dose rate. In addition, Parish (1992) has noted that the weathering of feldspar grains *in situ* may also have the effect of changing their luminescence stability.

In addition to differences which may arise between the results of the range of methods for evaluating the radioactivity within the sediments, there is another complicating factor involving time-dependent changes in sediment water content.

The presence of water in the pore spaces of a sediment attenuates the external radiation dose to sediment grains, therefore some estimate of the mean water content since burial is required before the annual radiation dose can be calculated. This introduces a major source of uncertainty into dose-rate estimation. Although it is clear that the end-points, as far as dose rates are concerned, involve assuming that the sample was either saturated with water or completely dry, the problem for many subaerial sediments lies in deducing the likely range of values of the natural water contents.

Selected examples of luminescence dating with independent age control in different sedimentary environments

Recent developments in luminescence dating of sediments have naturally been reflected in the changing contents of proceedings of the last four international seminars on TL dating, held in 1984, 1987, 1990 and 1993. A glance at these proceedings reveals that studies have been undertaken predominantly on sediments from aeolian environments, particularly sand dunes and loess, with sediments from fluvial, lacustrine, marine and colluvial environments meriting comparatively little attention. Despite the development of OSL techniques, TL still dominated OSL as the preferred research approach in the proceedings of the 1990 meeting. However, this dominance changed at the international seminar in Krems, Austria, in July 1993, where the focus of effort appeared to have shifted to optical dating.

The following examples give some idea of the range of current applications of luminescence dating.

Dating of aeolian sands and peats or ash layers

Intercalation of sands and peats allows inter-comparison between luminescence dates and ^{14}C dates. Recent examples from the Netherlands using OSL (Stokes 1991) and New Zealand using TL (Shepherd & Price 1990) show agreement within the errors of the luminescence techniques used.

Dating of intertidal sediments

Holocene intertidal clastic sediments intercalated with peat were investigated by Plater & Poolton (1992) using both OSL and TL techniques. The OSL dates showed good stratigraphic agreement with ^{14}C dates for the peat layers. The rapidity of bleaching of the OSL signal was reported as a major advantage for dating these particular sediments.

Dating of glaciolacustrine and other glaciogenic sediments

Data from Berger et al. (1987) show good agreement with independent age estimates, but only when the settling velocity of the sediments was sufficiently slow to allow some bleaching to occur as the grains settled to the lake bed. The partial bleach method was used for ED estimation. Work on waterlain sediments from Spitsbergen, reported by Forman & Ennis (1992), also highlighted problems of incomplete resetting for these materials. These types of sedimentary environment are an obvious testing ground for OSL techniques.

Examples of luminescence dating with limited independent age control

The following examples of work undertaken in the Sussex Laboratory show what can be achieved using these dating techniques, even though independent age control may be limited. Given that these results are from just one laboratory, the range of sedimentary environments is inevitably limited, but the results are instructive.

Dating of fluvio-glacial and aeolian sands, Chelford, UK

TL age determinations on the quartz and feldspar components of the sand deposits at Chelford have already been reported (Rendell *et*

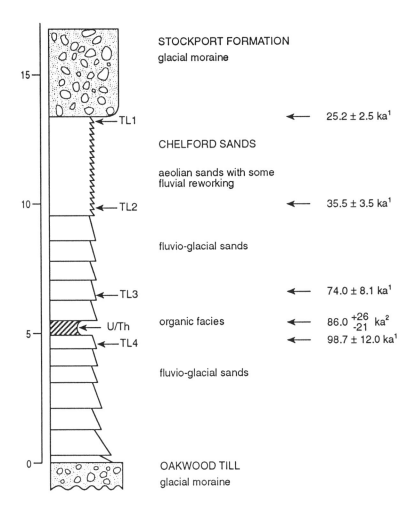

Fig. 6. Section through fluvio-glacial and organic deposits at Chelford. Sources: [1] Rendell 1992; [2] Heijnis & van der Plicht 1992.

al. 1991; Rendell 1992). The results of subsequent and independent uranium series disequilibrium dating of material from the Chelford Interstadial organic horizon have now been published (Heijnis & van der Plicht 1992) and show good agreement with the TL age determinations (Fig. 6). Earlier attempts to date the organic horizon by ^{14}C simply resulted in age determinations which indicated that material was beyond the range of this particular technique, i.e. > 45 ka (see Rendell et al. 1991).

Correlation of loessic deposits in Pakistan and Kashmir

TL dating has proved to be a useful tool for

the correlation of loess sequences within both northern Pakistan and Kashmir, even though problems have arisen with the dating of older material. In Pakistan, for example, loess sequences studied in the Potwar area contained no material that could be independently dated (Rendell 1989). Samples were dated using the additive dose and partial bleach methods, and examples of the results are shown in Fig. 7. Data for the longest profile, Riwat R89, show age estimates increasingly monotonically with depth down the profile. By contrast, the regeneration method of ED determination was employed for the TL dating of the Kashmir loess, owing to problems with the additive dose method of ED determination, and the ages appear to go into

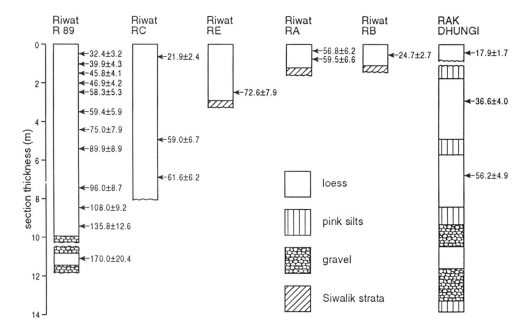

Fig. 7. TL dating results for loess sections in northern Pakistan. Source: Rendell 1989.

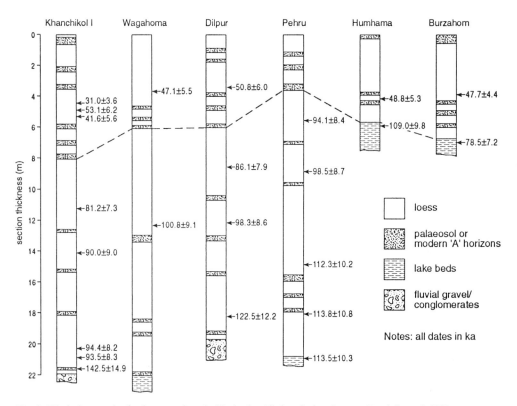

Fig. 8. TL dating results for loess sections in Kashmir. All data in ka. Source: Rendell *et al.* 1989.

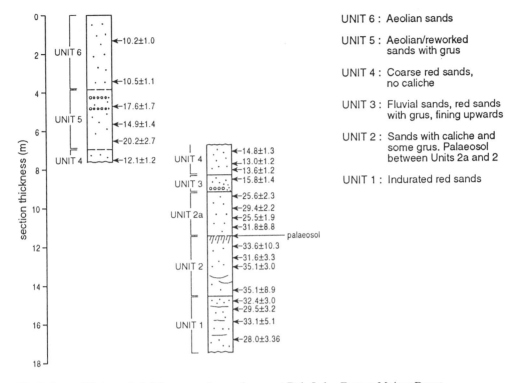

Fig. 9. Quartz TL dates (in ka) for a complex sand ramp at Dale Lake, Eastern Mojave Desert.

saturation at *c.* 100–120 ka (Singhvi *et al.* 1987; Rendell *et al.* 1989). In the case of the Kashmir sequence, although the TL dates provide a guide to relative rather than absolute age of the older loess deposits (see Fig. 8), the dates do provide an independent basis for correlation between the different sections.

Correlation of loessic deposits in the UK

Samples from 35 sites in southeast England were analysed using standard TL techniques (Parks & Rendell 1992). The results allowed sample sites to be grouped by age and revealed distinct differences between a group of Late Pleistocene loessic deposits (10–22 ka) and older deposits dating to the last glacial cycle and possibly to the penultimate glacial period.

Dating complex climbing and falling dunes in the Mojave Desert (colluvial/aeolian mixtures)

Luminescence dating techniques are currently being applied to a series of complex sand ramps in the Eastern Mojave Desert, California. The sand ramps hold the key to dating periods of sand movement in the Mojave during the late Quaternary and are therefore a potentially valuable source of palaeoenvironmental information. Unfortunately, luminescence dating can only be as good as the sediment being dated, and it has become clear during the present study that mixtures of grains with different light-exposure histories appear to be present in some parts of these complex ramps. Data for quartz TL ages for the Dale Lake sand ramp are given in Fig. 9. The data show a distinct stratigraphic break in sequence between units 2a and 3, but otherwise the degree of scatter is relatively high. Work on the optical dating of samples from this sequence is currently in progress.

Conclusions

Luminescence dating techniques have the potential to date the last exposure to light of sediment grains from a range of depositional environments. Far more effort has been expended on dating sediments from aeolian environments than all other types of depositional environment. The recent development of optical dating, both GLSL and IRSL, should help to redress this balance since these techniques are particularly applicable to environments in which light exposures have been fairly short.

The financial support of NERC and SERC for the luminescence work at Sussex University is gratefully acknowledged.

References

AITKEN, M. J. 1985. *Thermoluminescence Dating*. Academic Press, London.

BALESCU, S. & LAMOTHE, M. 1992. The blue emission of K-feldspar coarse grains and its potential for overcoming TL age underestimation. *Quaternary Science Reviews*, **11**, 45–51.

——, PACKMAN, S. C. & WINTLE, A. G. 1991. Chronological separation of interglacial raised beaches for northwestern Europe using thermoluminescence. *Quaternary Research*, **35**, 91–102.

BERGER, G. W. 1988. Dating Quaternary events by luminescence. *In:* EASTERBROOK, D. J. (ed.) *Dating Quaternary Sediments*. Geological Society of America Special Paper, **227**, 13–50.

—— 1990. Effectiveness of natural zeroing of the thermoluminescence in sediments. *Journal of Geophysical Research*, **95** B8, 12 375–12 397.

—— 1993. Dating loess up to 800 ka by thermoluminescence: reply. *Geology*, **21**, 569.

——, CLAGUE, J. J. & HUNTLEY, D. J. 1987. Thermoluminescence dating applied to glaciolacustrine sediments from central British Columbia. *Canadian Journal of Earth Sciences,* **24**, 425–434.

—— & EYLES, N. 1994. Thermoluminescence chronology of Toronto-area Quaternary sediments and implications for the extent of the midcontinent ice sheet(s). *Geology*, **22**, 31–34.

——, PILLANS, B. J. & PALMER, A. S. 1992. Dating loess up to 800 ka by thermoluminescence. *Geology*, **20**, 403–406.

BØTTER-JENSEN, L. & DULLER, G. A. T. 1992. A new system for measuring OSL of quartz samples. *Nuclear Tracks and Radiation Effects*, **20**, 549–553.

DITLEFSON, C. 1992. Bleaching of K-feldspars in turbid water suspensions: a comparison of photo- and thermoluminescence signals. *Quaternary Science Reviews*, **11**, 33–38.

DULLER, G. A. T. 1991. Equivalent dose determinations using single aliquots. *Nuclear Tracks and Radiation Effects*, **18**, 371–378.

—— 1992. *Luminescence chronology of raised marine terraces, south-west North Island, New Zealand*. PhD thesis, University of Wales, UK.

EDWARDS, S. R. 1993. Luminescence dating of sand from the Kelso dunes, California. *In:* PYE, K. (ed.) *The Dynamics and Environmental Context of Aeolian Sedimentary Systems*. Geological Society, London, Special Publication, **72**, 59–68.

FORMAN, S. L. & ENNIS, G. 1992. Limitations of thermoluminescence to date waterlain sediments from glaciated environments of western Spitsbergen, Svalbard. *Quaternary Science Reviews*, **11**,

61–70.

GEMMELL, A. M. D. 1985. Zeroing of the TL signal of sediment undergoing fluvial transport: a laboratory experiment. *Nuclear Tracks and Radiation Measurements*, **10**, 695–702.

GODFREY-SMITH, D. I., HUNTLEY, D. J. & CHEN, W. H. 1988. Optical dating studies of quartz and feldspar sediment extracts. *Quaternary Science Reviews*, **7**, 373–380.

HEIJNIS, H. & VAN DER PLICHT, J. 1992. Uranium/thorium dating of Late Pleistocene peat deposits in NW Europe, uranium/thorium isotope systematics and open system behaviour of peat layers. *Chemical Geology*, **94**, 161–171.

HUNTLEY, D. J. 1985. On the zeroing of the thermoluminescence of sediments. *Physics and Chemistry of Minerals*, **12**, 122–127.

——, GODFREY-SMITH, D. I. & THEWALT, M. L. W. 1985. Optical dating of sediments. *Nature*, **313**, 105–107.

——, HUTTON, J. T. & PRESCOTT, J. R. 1985. South Australian sand dunes: a TL sediment test sequence: preliminary results. *Nuclear Tracks and Radiation Measurements*, **10**, 757–758.

——, —— & —— 1993a. The stranded beachdune sequence of south-east South Australia: a test of thermoluminescence dating, 0–800 ka. *Quaternary Science Reviews*, **12**, 1–20.

——, —— & —— 1993b. Optical dating using inclusions within quartz grains. *Geology*, **21**, 1087–1090.

HUTT, G., JAEK, I. & TCHONKA, J. 1988. Optical dating: K-feldspars optical response stimulation spectra. *Quaternary Science Reviews*, **7**, 381–385.

KRONBERG, C. 1983. Preliminary results of age determination by TL of interglacial and interstadial sediments. *P.A.C.T.*, **9**, 595–605.

LUFF, B. J. & TOWNSEND, P. D. 1993. High sensitivity thermoluminescence spectrometer. *Measurement Science Technology*, **4**, 65–71.

MEJDAHL, V. 1988. Long-term stability of the TL signal in alkali feldspars. *Quaternary Science Reviews*, **7**, 357–360.

PARISH, R. 1992. IRSL dating and feldspar weathering. Geography Discussion Paper 2, University of Sussex.

PARKS, D. A. & RENDELL, H. M. 1992. Thermoluminescence dating and geochemistry of loessic deposits in SE England. *Journal of Quaternary Science*, **7**, 99–107.

PLATER, A. J. & POOLTON, N. R. J. 1992. Interpretation of Holocene sea level tendency and intertidal sedimentation in the Tees estuary using sediment luminescence techniques: a viability study. *Sedimentology*, **39**, 1–15.

QUESTIAUX, D. 1991. Optical dating of loess – comparisons between different grain-size fractions for infra-red and green excitation wavelengths. *Nuclear Tracks and Radiation Measurements*, **18**, 133–139.

RENDELL, H. M. 1989. Loess deposition during the Late Pleistocene in Northern Pakistan. *Zeitschrift für Geomorphologie N.F., Supplementband*, **76**, 247–255.

—— 1992. A comparison of TL age estimates from different mineral fractions of sands. *Quaternary Science Reviews*, **11**, 79–83.

——, GARDNER, R. A. M., AGRAWAL, D. P. & JUYAL, N. 1989. Chronology and stratigraphy of Kashmir loess. *Zeitschrift für Geomorphologie N.F., Supplementband*, **76**, 213–223.

——, TOWNSEND, P. D., LUFF, B. J., WINTLE, A. G. & BALESCU, S. 1993. Spectral analysis of TL in the dating of potassium feldspars. *Physica Status Solidi*, **138**, 335–341.

——, WEBSTER, S. E. & SHEFFER, N. A. 1994. Underwater bleaching of signals from sediment grains: new experimental data. *Quaternary Geochronology*, **13**, 433–435.

——, WORSLEY, P., GREEN, F. & PARKS, D. 1991. Thermoluminescence dating of the Chelford Interstadial. *Earth and Planetary Science Letters*, **103**, 182–189.

SHEPHERD, M. J. & PRICE, D. M. 1990. Thermoluminescence dating of late Quaternary dune sand, Manawatu/Horowhenua area, New Zealand: a comparison with ^{14}C age determinations. *New Zealand Journal of Geology and Geophysics*, **33**, 535–539.

SINGHVI, A. K., BRONGER, A., PANT, R. K. & SAUER, W. 1987. Thermoluminescence dating and its implications for the chronostratigraphy of loess–paleosol sequences in the Kashmir Valley (India). *Chemical Geology*, **65**, 45–56.

SMITH, B. W., RHODES, E. J., STOKES, S., SPOONER, N. A. & AITKEN, M. J. 1990. Optical dating of sediments: initial quartz results from Oxford. *Archaeometry*, **32**, 19–31.

SPOONER, N. A. 1992. Optical dating: preliminary results on the anomalous fading of luminescence of feldspars. *Quaternary Science Reviews*, **11**, 139–145.

STOKES, S. 1991. Quartz-based optical dating of Weichselian coversands from the eastern Netherlands. *Geologie en Mijnbouw*, **70**, 327–337.

WINTLE, A. G. 1990. A review of current research on TL dating of loess. *Quaternary Science Reviews*, **9**, 385–397.

—— 1993. Luminescence dating of aeolian sands: an overview. *In:* PYE, K. (ed.) *The Dynamics and Environmental Context of Aeolian Sedimentary Systems.* Geological Society, London, Special Publication, **72**, 49–58.

—— & HUNTLEY, D. J. 1979. Thermoluminescence dating of a deep-sea sediment core. *Nature*, **279**, 710–712.

——, QUESTIAUX, D. G., ROBERTS, R. G. & SPOONER, N. A. 1993. Dating loess up to 800 ka by thermoluminescence: comment. *Geology*, **21**, 568.

ZÖLLER L. & WAGNER, G. A. 1991. Thermoluminescence dating of loess – recent developments. *Quaternary International*, **1**, 61–64.

Wireline log-cyclicity analysis as a tool for dating and correlating barren strata: an example from the Upper Rotliegend of The Netherlands

CHANG-SHU YANG & WIM F. P. KOUWE

International Geo Consultants (IGC) BV, Kapelstraat 5B, 2161 HD Lisse, The Netherlands

Abstract: The stratigraphic subdivision and correlation of the Upper Rotliegend Group in The Netherlands offshore is difficult due to lack of biostratigraphic data. A regional study of 150 released wells and core data has revealed the preservation of five supersequences, 12 sequences and short-term cyclicities in the Upper Rotliegend deposits. They are interpreted to reflect long- and short-term climate changes, which controlled the deposition of Rotliegend sequences.

The supersequences (second order) serve best for regional correlation and prediction of sand distribution. The sequences (third order) are best suited for detailed block- or field-scale correlation and distribution of (un)favourable lithofacies. The short-term (higher order) cycles represent orbital-forcing (Milankovitch) cycles and are used as quantitative parameters in wireline log cyclicity analysis. This includes the calibration of sequence boundaries and maximum flooding surfaces, the determination of the absolute age, the estimation of the duration time and the net accumulation rate of the preserved rock succession, and the pattern recognition of the trend of net accumulation rate variations within the third-order sequence. The method has revealed a new possibility of dating and correlating strata such as the Upper Rotliegend.

The Upper Rotliegend Group is a succession of barren siliciclastic and evaporitic red-bed sediments (Fig. 1a; Glennie 1986, 1990; Nederlandse Aardolie Maatschappij B.V. & Rijks Geologische Dienst 1980; Van Wijhe *et al.* 1980). The Upper Rotliegend Group has been assigned an Early Permian age (e.g. Harland *et al.* 1990). A recent study, however, suggests that the deposition of the Upper Rotliegend Group may have continued into Late Permian (Menning 1994). The Upper Rotliegend deposits extend in a belt from Poland to the UK, called the Southern Permian Basin (Fig. 1b). The basin is delimited to the north by the Mid-North Sea High and Ringkobing–Fyn High, and to the south by the Variscan orogen. The London–Brabant Massif and Rhenish Massif are the two southern bounding elements in the Netherlands. The Upper Rotliegend deposits have been the main target for petroleum exploration in the southern North Sea since the discovery of the giant Groningen gas field in 1959. This is mainly due to the favourable stratigraphic combination of the thick, gas sourcing Carboniferous coal measures below, the sealing salts of the Upper Permian Zechstein above, and the often excellent reservoirs of the Rotliegend sandstones in between.

Although the deposits have been studied extensively, it has remained impossible to find a method to correlate them reliably. The main problem in dealing with the Upper Rotliegend Group has been its barren nature. This has thus far prevented detailed stratigraphic correlations and estimations of the absolute age and time span of the Upper Rotliegend Group.

In 1991, InterGeos completed a non-exclusive regional study of the Upper Rotliegend Group in the Netherlands' offshore (Intergeos 1991). This study involved 150 wells, of which 50 contained cored intervals with a total length of 1200 m. The sequence stratigraphic principles (Vail *et al.* 1977; Haq *et al.* 1987; Posamentier *et al.* 1988; Posamentier & Vail 1988; Van Wagoner *et al.* 1990) were adapted to the Rotliegend Basin and new concepts and tools were developed in this study (Yang & Nio 1994; Yang & Baumfalk 1994). This high-resolution sequence stratigraphy method is an integration of sequence stratigraphic analysis and cyclicity analysis of wireline logs. It is based on the intrinsic composite cyclic nature of the Upper Rotliegend deposits, which have been proven to be made up of depositional cycles of various durations (Intergeos 1991; Yang & Nio 1994; Yang & Baumfalk 1994). The recognition and determination of high-frequency cycles within the third-order sequences is the basis of these new concepts. These high-frequency cycles have been recognized as Milankovitch cycles, and they

From Dunay, R. E. & Hailwood, E. A. (eds), 1995, *Non-biostratigraphical Methods of Dating and Correlation*
Geological Society Special Publication No. 89, pp. 237–259

show an intimate relationship with the development of the third-order sequences. This paper is intended to illustrate the potential of wireline log-cyclicity analysis in dating and correlating barren strata such as the Upper Rotliegend.

Milankovitch cycles

Milankovitch cycles reflect variations in the orbital parameters of the Earth, which causes significant variations in the distribution of solar radiation over the Earth's surface over periods of 10 ka to several thousands of years. Milankovitch cyclicities are compound (e.g. Hays *et al.* 1976; Imbrie & Imbrie 1979). The principal Milankovitch cycles for the Recent are (Berger *et al.* 1989): (1) **Eccentricity**: the variation of the eccentricity of the Earth's orbit. One of the principal periods of eccentricity is 100 ka. It affects total insolation; (2) **Obliquity**: the variation of the obliquity of the Earth's rotation axis with principal periods of 54 and 41 ka. It affects seasonality and insolation, with greatest effect in high latitudes; (3) **Precession**: the precession of the equinoxes with principal periods of 23 and 19 ka. It affects radiation intensity for each season, mainly at low latitudes.

Table 1. *Estimated periods of the Milankovitch cycles in the Permian (after Berger* et al. *1989)*

A	100 ka	Eccentricity
A'	67 ka	
B	44.3 ka	Obliquity
C	35.1 ka	Obliquity
D	30 ka	
E	21.0 ka	Precession
F	17.6 ka	Precession

It is known from orbital dynamics that the obliquity and precession cycles had shorter periods in the geological past (Berger & Loutre 1994). This reflects the slowing down of the Earth's rotation rate caused by tidal friction and the increase of the Earth–Moon distance. The periods of the cycles in geological time have been calculated by Berger *et al.* (1989). The periods of the Milankovitch cycles in the Permian were estimated as 100 (eccentricity), 44.3 and 35.1 (obliquity), and 21 and 17.6 ka (precession) (Fig. 2). These periods have been observed in the Upper Rotliegendes (Yang & Baumfalk 1994). In addition, some interference cycles can also be recognized in the Upper Rotliegendes, such as the 67 and 30 ka cycles (Yang & Baumfalk 1994). For convenience, the Permian cycles 100, 67, 44.3, 35.1, 30, 21 and 17.6 ka are named

A, A', B, C, D, E and F cycles, respectively (Table 1).

Climate and sedimentary facies cycles of the Upper Rotliegend

Milankovitch cycles have been recognized in various sedimentary sequences (InterGeos database; Melnyk & Smith 1989; Smith 1989; Worthington 1990; Fischer & Bottjer 1991; Nio *et al.* 1993; de Boer & Smith 1994). The orbital-forced Milankovitch cycles influence the insolation (solar radiation) incident upon the surface of the Earth, mainly in relative terms over the seasons. These changes cause periodic widespread climatic changes and sea-level/base-level fluctuations, which in turn are important controls on sequence development and sedimentary facies variations as recorded in wireline log data. How these climatic changes are reflected in the sedimentary record will depend on the latitude, the general depositional setting, the importance of seasonality in the sediment supply, and the depositional processes in the receiving basin.

During the deposition of the Upper Rotliegend the southern Permian Basin was located at a palaeolatitude similar to that of the present-day Sahara (Glennie 1986, 1990), under arid to semi-arid climatic conditions. The Southern Permian Basin can be subdivided into three belts, each of which displays its own sedimentary succession. The basin centre is characterized by an alternation of halites with lacustrine claystones. In the basin fringe area a succession of aeolian (dune and sand flat) sandstones alternated with fluvial sandstones was deposited. Lacustrine deposits can be intercalated in marginal sub-basins (e.g. Off Holland Low). In the transition zone an intertonguing of the aeolian and fluvial sandstones from the basin fringe and lacustrine mudstones from the basin centre is observed (Fig. 1a).

In the past these deposits have been studied extensively, and a good sedimentological understanding of virtually all encountered lithofacies has been reached. Overviews of Upper Rotliegend lithofacies associations are given by Glennie (1990), Reijers *et al.* (1993, fig. 4), and Yang & Nio 1994.

The observed alternation of halites and claystones in the basin centre was interpreted to have been caused by climatic variations (Trusheim 1971; Hedemann *et al.* 1984). Only recently these climatic variations are also believed to be preserved in the lithological columns of the basin fringe and transitional zones. The linking of the basinal and the transitional

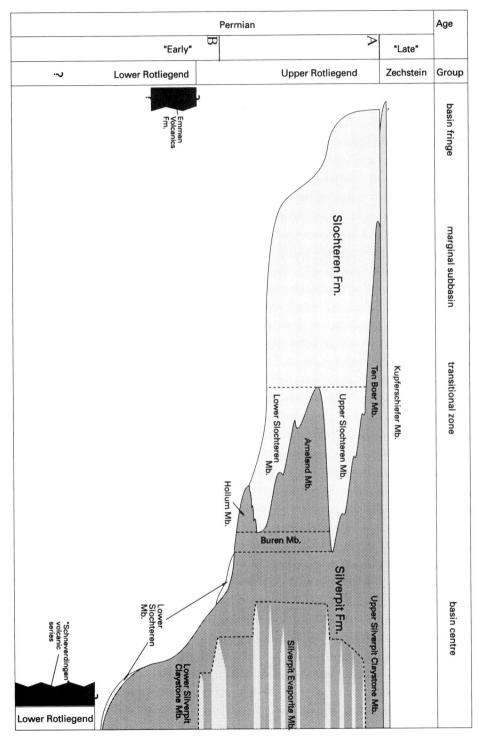

Fig. 1. (a) Lithostratigraphic subdivision of the Upper Rotliegend Group. A, Early Permian–Late Permian boundary of Harland *et al.* (1990); B, the same boundary of Menning (1994).

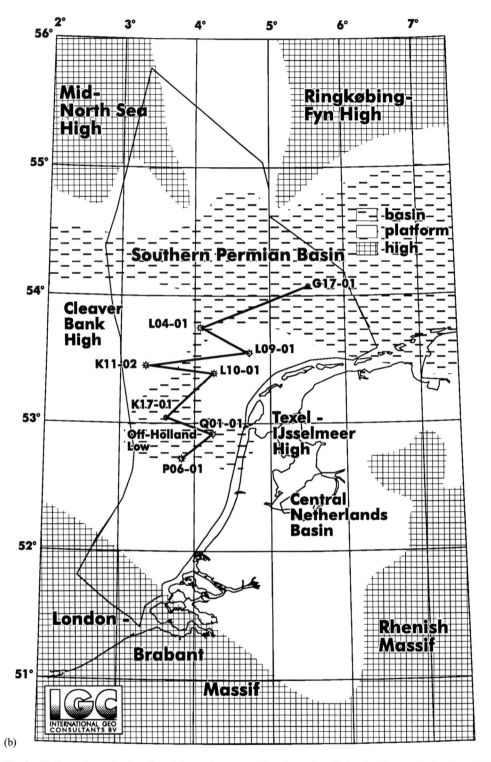

(b)

Fig. 1. **(b)** General structural setting of the study area and locations of studied wells. The correlation line of Fig. 10 is also indicated.

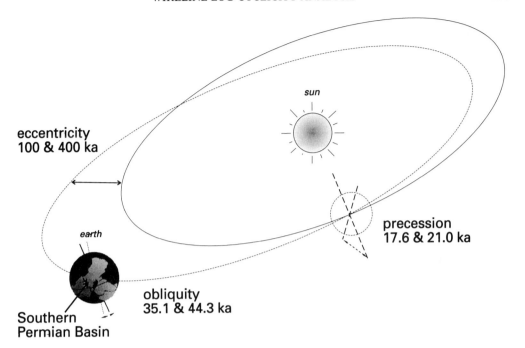

Fig. 2. Milankovitch cycles for the Upper Rotliegend (Early Permian). The Milankovitch cycle values are after Berger *et al.* (1989).

provinces remained a matter of debate (Gralla 1988 v. Gast 1988, 1991).

Extensive core observations in InterGeos' (1991) Rotliegend study show distinct cyclic sedimentary facies changes in the Upper Rotliegend succession (Intergeos 1991; Yang & Nio 1994). Such cyclic sedimentary facies changes are also observed in the Upper Rotliegend in Germany (Gast 1988, 1991). These sedimentary facies changes occur as second-, third- or higher-order cycles.

Typical facies changes of the Upper Rotliegend in the basin centre

Figure 3 shows the typical facies changes of the Upper Rotliegend in Well G17-01 in the basin centre. Most third-order cycles show an upward facies change from halites to lacustrine clay-stones. Third-order cycles can be grouped into second-order cycles, which are dominated by halites in the lower part and lacustrine clay-stones in the upper part. The time duration estimation was based on the recognition of the Milankovitch cycles in this well using interval spectral analysis, which will be discussed later. As shown in Fig. 3, most third-order cycles have a duration of 0.4–1.7 Ma, and most second-order cycles a duration of 1.3–2.3 Ma.

Typical facies changes of the Upper Rotliegend in the basin margin

Figure 4 illustrates the typical facies changes of part of the Upper Rotliegend Group in Well K11-02 in the basin margin. The cored interval is characterized by an alternation of aeolian dune-slip-face, dry sandflat, wadi and mudflat deposits.

The aeolian dune-slip-face facies consists of well-sorted, fine to medium sands. The sands are porous and display distinct planar cross-bedded foresets (Fig. 4) with steeply-dipping and in-verse-graded sandflow laminae. These sandflow laminae show clear grain size segregation and alternation of finer-grained (dark) and coarse-grained (light) laminae with thicknesses of a few millimetres to 1 cm. These sandflow laminae indicate avalanching of non-cohesive sands on the dune-slip-face.

The dry sandflat facies consists of light red, very fine to medium sand with some silt. The sands are moderate to well sorted with fair to good porosity. Dry sandflat often display homogeneous sand, horizontal planar bedding, or wind-ripple laminations (Fig. 4). The dry sandflats were formed in an aeolian setting as sand sheets. They were also formed in a fluvial terminal fan.

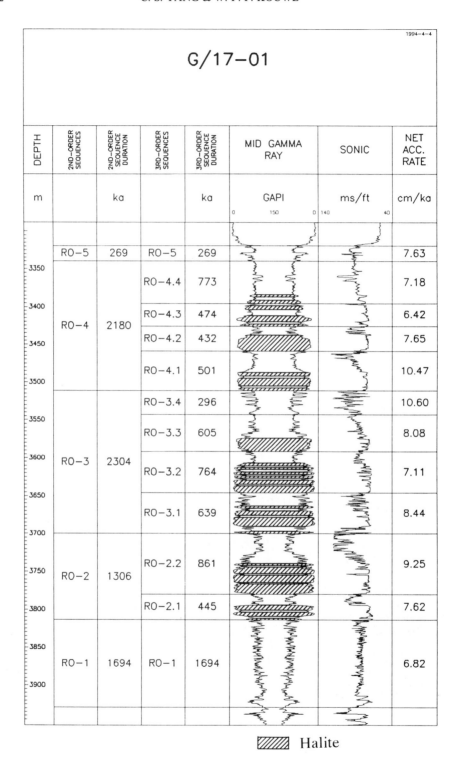

Fig. 3. Facies changes and time durations of the second- and third-order cycles in the Upper Rotliegend in Well G17-01 of The Netherlands offshore.

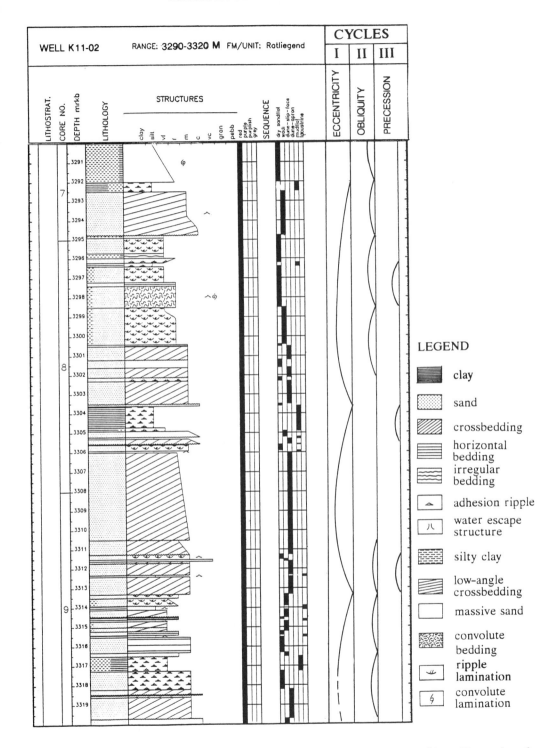

Fig. 4. Core description of the Upper Rotliegend in Well K11-02 of The Netherlands offshore. Three orders of sedimentary cycles are indicated. They can be compared to the Milankovitch eccentricity, obliquity and precession cycles.

Milankovitch climate cycle (4th order)	climate	sedimentary facies		
		fringe	transition	basin
	humid	sabkha	lake	lake
	subhumid	damp sandflat / wadi	sabkha / fluvial fan	lake
	semi-arid	wadi / dry sandflat	damp - dry sandflat	lake / saline lake
	arid	dry sandflat / dunes	dunes / dry sandflat	evaporite lake
	semi-arid	dry sandflat / wadi	dry sandflat / fluvial fan	saline lake
	subhumid	braided river / sandflat	fluvial fan / sabkha	saline lake
	humid	sabkha/ pond	sabkha lake	lake

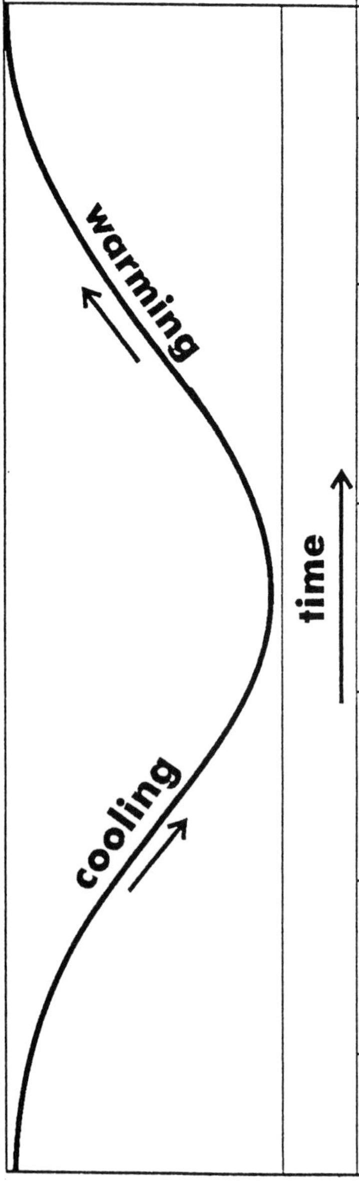

Fig. 5. (a) The climate-forcing sedimentary facies model. The Milankovitch cycles cause periodic climate and base-level fluctuations, controlling both sequence development and sedimentary facies patterns.

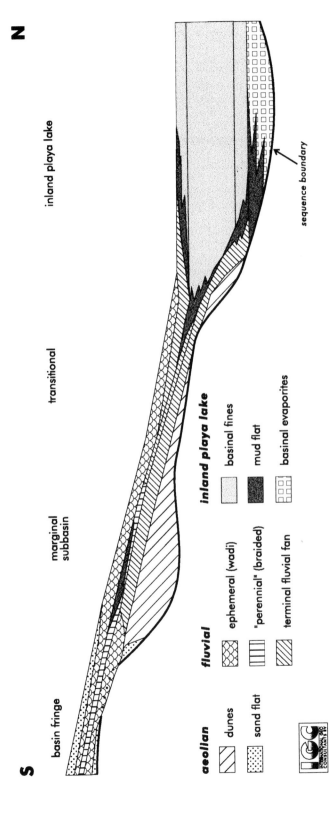

Fig. 5. (b) Sedimentary facies changes of the Upper Rotliegend through a single sequence. The sedimentary facies were based on extensive core observations and wireline log analysis.

The wadi facies consists of poorly sorted sand and clayey sand. Wadi facies often display 0.5–2 m thick fining-upwards sequences. Each fining-upwards sequence has an erosional base, covered by clay pebbles. This is overlain by homogeneous, parallel-bedded, low-angle cross-bedded, or ripple laminated, sands (Fig. 4). The wadi successions were deposited by highly ephemeral fluvial systems characterized by sporadic floods of highly suspended-load flows.

The mudflat facies consists of dark red clay, often with intercalations of thin silt and very fine sand layers. They show abundant adhesion ripples and ripple laminations (Fig. 4). The mudflat facies was deposited by sheetflood process in inland sabkha areas.

The vertical facies changes show clear cycles. Three types of sedimentary cycles can be observed from this cored interval, as indicated in Fig. 4. The type I cycle has a wavelength of 10–11.5 m, and is characterized by an upward facies change from dune-slip-face to wadi or mudflat. Type II cycles have a wavelength of 3.3–3.5 m and shows an upward facies change from dune-slip-face or dry sandflat to wadi or mudflat. A type II cycle may also show a facies change from wadi to mudflat, or a break in dune deposits. Type III cycles have a wavelength of 2–2.2 m and displays an upward facies change from dune-slip-face or dry sandflat to wadi or mudflat. All these cyclic facies changes reflect climate cycles from arid to relative humid conditions. Since the cored Upper Rotliegend interval in this well has a net accumulation rate of 9.4–10.9 cm per ka, the three types of sedimentary cycles represent the time periods of 100, 33–35 and 20–22 ka, respectively, which match very well with the Milankovitch eccentricity, obliquity and precession cycles in the Permian. This example clearly supports the hypothesis that the sedimentary facies changes in the Upper Rotliegend indeed reflect the climate changes related to the Milankovitch cycles.

Climate-forcing sedimentary facies model

The typical cyclic sedimentary facies changes observed in the core are summarized in Fig. 5a. The basin centre was characterized by evaporite (halite) and saline lake facies during arid climate periods, and lacustrine deposits during relatively humid periods. The basin margin was dominated by dune and aeolian dry sandflat facies during arid periods, which changed to damp sandflat, fluvial (wadi) and sabkha facies as the climate became relatively humid. The evaporite/

claystone cycles found in the basin centre correspond very well with cyclic lithofacies stacking patterns in the transitional and basin fringe provinces. Such cyclic sedimentary facies changes are often observed at a scale of fourth- or higher-order cycles (time duration 10 s to 400 ka, e.g. Fig. 4), but also very distinct at a scale of second- and third-order cycles (Fig. 3; see also Yang & Nio 1994). The regional study has revealed the presence of 12 third-order sequences in the Netherlands, which can be grouped into five supersequences. They could correspond to Gondwana glaciation cycles.

The climate-forcing sedimentary facies model is different from the traditional sedimentary facies interpretation based on the uniformitarian concept of Lyell (1830). The climate-forcing sedimentary facies model is based on observations that the Upper Rotliegend sedimentary facies changed as a response to periodic widespread climatic variations, and not as a result of progradation of stable depositional systems (Yang & Nio 1994). Consequently, the vertical succession of sedimentary facies records mainly the history of climatic changes, not the succession of lateral adjacent facies. Therefore, the observed vertical facies cycles were not related to an autocyclic mechanism caused by the natural redistribution of energy within a depositional system. They have an allocyclic nature controlled by some external cause. The climate changes, associated with fluctuations of the lake level and groundwater table, are the driving force in creating the strata patterns and sedimentary facies variations during the deposition of the Upper Rotliegend (Fig. 5b). The climate-forcing sedimentary facies model is an important modification of the existing concepts of sequence stratigraphy when applied to the Upper Rotliegend continental basin (Yang & Nio 1994). The major modifications involve the following aspects:

(1) Exxon sequence stratigraphy approach was developed in marine basins. The Upper Rotliegend Group of The Netherlands was deposited in a continental basin. The base level in the Rotliegend continental basin was the lake level/groundwater table, which was not directly controlled by global eustatic sea level, and at times this base level was c. 250 m below the eustatic sea level (Smith 1970, 1979; Ziegler 1982; Glennie 1990).

(2) Climate change is the driving force causing changes in a number of parameters in a continental basin, such as sediment supply, fluvial discharge and types (e.g.

braided river or wadi), wind speed and direction, and the position of the lake level and the groundwater table. These factors controlled the sequence stratigraphic development and sedimentary facies distribution in the Rotliegend continental basin.

(3) The vertical and lateral facies relationship of the Upper Rotliegendes in The Netherlands displays abrupt facies changes, which reflect important climate changes. These climate changes occurred throughout the southern Permian Basin and can also be recognized in the Upper Rotliegend in northwestern Germany (Gast 1988, 1991). As a result of climate changes, the major facies belts not only shifted in their position, but also changed in their character (Fig. 5b). The rate of climate-induced sedimentary facies changes exceeded the progradation rate of sediment facies belts. This is in contrast to the uniformitarian approach of the Exxon sequence stratigraphy.

(4) Since the climate-induced sedimentary facies changes are basinwide and synchronous in geological terms, they provide a sound base for sequence stratigraphic correlation throughout the Rotliegend Basin (Fig. 5b). The key terms used in the sequence stratigraphic study of the Upper Rotliegend Basin are similar to those used in marine basins, although the exact meaning has been modified (Yang & Nio 1994). The sequence boundary is interpreted to represent the maximum rate of increasing aridity causing rapid fall of lake level and groundwater table, followed by a period of lowstand of the lake. The lowstand systems tract is bounded by the sequence boundary at the base and the regional flooding surface at the top. It is the lowstand of the lake in a continental basin, and not the lowstand in its usual meaning (i.e. sea level). This lowstand systems tract is characterized by evaporite successions in the basin centre, and extensive aeolian (e.g. dune-slip-face, dune apron and dry sandflat) deposits in the basin flank and margin. Both facies reflect an arid climate. The preservation of the aeolian deposits depends on accommodation space related to basin subsidence. Substantial aeolian dune deposits could be preserved in areas of sufficient accommodation space, whereas relatively thin dry sandflats, or even aeolian deflation surfaces, mark areas of lower subsidence. The

overlying transgressive systems tract consists of lacustrine facies in the basin centre, extensive sabkha facies (damp sandflat and mudflat) in the basin flank, and ephemeral fluvial (wadi) and dry-damp sandflat along the basin margin which indicates a change from arid to relatively humid (semi-arid) climate and an expansion of the lake. The maximum flooding surfaces, which separate the transgressive systems tract below from the highstand systems tract above, represent the maximum rate of increasing humidity causing rapid rise of the lake level and groundwater table, followed by a period of highstand of the lake. The maximum flooding surface is marked by expanded lake facies in the basin centre and flank, and extensive inland sabkha and fluvial facies in the marginal area. Lacustrine conditions could also be established in some marginal sub-basins with relatively high subsidence rates. The overlying highstand systems tract is dominated by lake facies in the basin centre, and wadi, fluvial terminal fan (dry-damp sandflat) and sabkha (mudflat) facies in the basin flank and margin which reflect a period of relatively humid climate, followed by a period of decreasing humidity.

Despite these modifications, apparent similarities exist between the depositional cycles of the Upper Rotliegend and the Exxon sequences: (1) The effect of lake-level variations on strata stacking patterns in the Upper Rotliegendes sequences is comparable to the effect of sea-level fluctuations in sequence stratigraphy; (2) The third-order sequences in the Upper Rotliegend Group have a time duration of 0.7–1.4 Ma (Yang & Baumfalk 1994), comparable to the time durations of the Exxon third-order sequences; (3) The same climate change was the driving force causing both the eustatic sea-level fluctuations in the marine basins and the base-level fluctuations in the continental basins. Both could correspond to Gondwana glaciation cycles during the deposition of the Upper Rotliegend. The sequence stratigraphic framework established in the continental basin should be comparable to the sequence stratigraphy established in the marine basins.

Cyclicity analysis methods

Good summaries of general spectral analysis principles have been given by Weedon (1991),

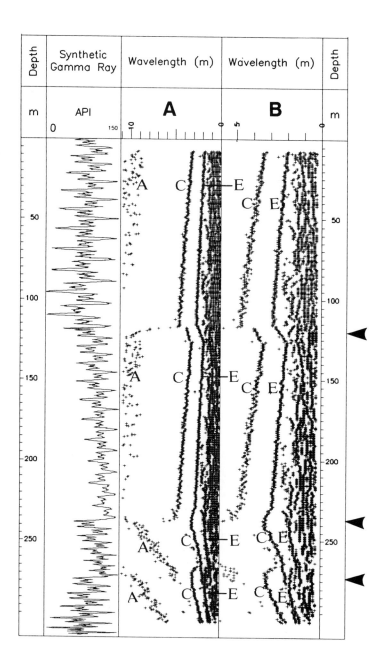

(a)

Fig. 6. (a) Sliding window maximum entropy spectral analysis applied to a synthetic GR log. The spectrum bands or spectrum peaks reveal a blurred or interrupted pattern related to discontinuity in the log (arrowed), and a distinct shift of the spectral bands related to the changing net accumulation rate (see text for explanation). In (a) and (b) A, C and E indicate the 100, 35.1 and 21 ka cycles, respectively.

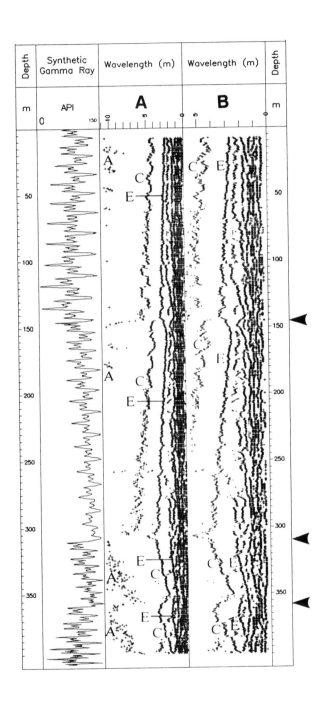

(b)

Fig. 6. (b) Sliding window maximum entropy spectral analysis applied to a decompacted synthetic GR log. The spectrum peaks display a shift towards the lower frequency range after the decompaction, and this shift becomes more distinct in the clay-dominated intervals (see text for explanation).

Schwarzacher (1991), and ten Kate & Sprenger (1992). In order to reveal the climate and sedimentary cycles in the Upper Rotliegend Group, wireline log-cyclicity analysis was made using Fourier and maximum entropy spectral analysis.

Fourier spectral analysis

Fourier spectral analysis assumes that any continuous, single valued function can be considered as the sum of a set of sinusoidal functions (harmonics). Usually, the analysed signal is a function of time, and the composing harmonics can be characterized by their wavelengths (or frequencies) and amplitudes. The amplitude and phase of each harmonic are estimated from the data.

The spectral analysis produces so-called spectrograms, showing specific harmonic values plotted v. the relative contribution of the variance of each harmonic to the total variance of the analysed signal, which is a measure for the relative importance of each harmonic. If the original signal contains cyclic patterns, the power spectrum derived from the data will show pronounced peaks corresponding with the frequencies of these cyclicities.

Maximum entropy spectral analysis

Maximum entropy spectral analysis is based on the autocorrelations calculated from the data. Maximum entropy has the advantage of high resolution in the low-frequency range. It produces a continuous spectrum. In addition, this method can be used to analyse short sequences. Presence of only a few cycles in the 'original' data is sufficient to get a reliable estimation of the cycle length and power. Since most third-order sequences in the Upper Rotliegend are rather short, maximum entropy spectral analysis is extensively used in the cyclicity analysis of the Upper Rotliegend sequences.

Wireline log-cyclicity analysis

In a wireline log spectral analysis, the analysed signal is a single valued function of depth. The wavelength in the wireline log-cyclicity analysis corresponds to the thickness of electrofacies (i.e. wireline log facies) cycle, while the frequency equals the number of electrofacies cycles found in the analysed interval.

The application of spectral analysis to wireline logs involves two techniques. Each has its own objective. First, a sliding window spectral analysis is done for the recognition and calibration of discontinuities, and determination of intervals of 'continuous' deposition. Secondly, interval spectral analysis is made to determine the absolute age and estimate the duration and the net accumulation rate of the designated intervals.

Sliding window spectral analysis

The depth in a stratigraphic succession can be regarded as a function of time. Different geological processes could cause various distortions of the initial time–depth relationships in a stratigraphic section (Weedon 1991; Schwarzacher 1991).

The basic assumption is that the data points of the wireline log signal, which are equally spaced in terms of depth, can also be considered as equally spaced in terms of time if: (1) the net accumulation rate is constant; and (2) the sedimentary succession has a relatively uniform lithology and therefore has a similar thickness variation after the compaction and diagenesis.

The basic assumptions of a constant net accumulation rate and a relatively uniform compaction are not always true. As a result of changing net accumulation rate and/or differential compaction, the same Milankovitch cycles will be recorded as electrofacies cycles with changing thickness. This will inevitably influence the cyclicity patterns. In addition, hiatuses will produce discontinuities, which may break the continuous cyclicity pattern. Moreover, autocyclic variations of depositional process may cause the thickness deviation of the electrofacies cycles from regular wavelength, thus add noise into the cyclicity pattern.

Changing net accumulation rate and hiatus

In fact, the net accumulation rate of sediment is not always constant and the sedimentation may be discontinuous. In order to examine the effect of changing net accumulation rate or discontinuity on wireline log-cyclicity patterns, a synthetic gamma ray (GR) log is constructed using the three main Milankovitch cycles (the 100, 35.1 and 21 ka cycles) as major parameters (Fig. 6a). This synthetic GR log is constructed under conditions of changing net accumulation rate and discontinuous sedimentation. It consists of a stacking of four synthetic GR logs. The lowermost one is constructed with a fining-up and thickening-up trend, followed by a coarsening-up and thickening-up log. This is overlain by a coarsening-up and thinning-up log. The uppermost part is a fining up and thinning-up log.

The cyclicity patterns of the synthetic logs have been investigated with a sliding window spectral analysis technique (Fig. 6a). In this analysis a window of 15 m was moved down the synthetic wireline log data. Since the window contains only a very short segment of the synthetic log, the 'gradual change' of the net accumulation rate within the window can be neglected. The 'abrupt change' of the net accumulation rate within the window (e.g. the hiatus), however, will produce distinct disturbances. For each step in the analysis, the window is shifted down the log by 0.6 m. In each step a power spectrum is calculated for the window using maximum entropy spectral analysis. The high-frequency part of the spectrogram is plotted next to the synthetic GR log and positioned at the middle depth-point of the window (Fig. 6a). The power spectra are plotted as peak crests. In column A of Fig. 6a the high-frequency part (75%) of the spectrogram is plotted to give an overview of the spectrum peaks (including eccentricity, obliquity and precession peaks), whereas in column B the high-frequency part (40%) of the spectrogram is enlarged to show details of the spectrum peaks in the high-frequency range (the obliquity and precession peaks).

The results of the sliding window spectral analysis reveal interesting patterns related to the discontinuity or the changing net accumulation rate: (1) In intervals of 'continuous sedimentation', spectrum crests or spectrum peaks of successive windows may partly overlap. This will produce a pattern of continuous spectrum bands; (2) In intervals where cyclicities are missing, or blurred, the sliding window spectral analysis shows a blank pattern. If the window includes a discontinuity, the spectrum pattern will be interrupted, such as the discontinuities indicated by arrows in Fig. 6a. It has turned out that the blurred intervals are generally related to discontinuities in the sedimentary record, e.g. sequence boundaries; (3) The changing net accumulation rate will cause a distinct shift of the spectral band. The spectral bands will shift to the left (the lower-frequency range) if the net accumulation rate increases (thickening-up succession), and to the right (the higher-frequency range) if the net accumulation rate decreases (thinning-up succession).

Compaction

The synthetic GR log shown in Fig. 6a is decompacted to examine the effect of differential compactions on wireline log-cyclicity patterns. The decompaction calculation used a decom-paction factor, which varies according to the clay : sand ratio of each sampling depth. Consequently the clay dominated intervals were decompacted more than the sand-dominated intervals. As a result, the data points, which were originally equally spaced in terms of depth, became unequally spaced after the decompaction, and a process of resampling is necessary to obtain a new set of data points, which are equally-spaced in terms of depth.

The cyclicity patterns of the decompacted synthetic logs are studied with the same sliding window spectral analysis technique (Fig. 6b). The spectrum bands display a shift towards the lower-frequency range after the decompaction, and this shift (i.e. increase of cycle thickness) becomes more distinct in the clay-dominated intervals, which have been decompacted more than the sand-dominated intervals. In addition, the spectrum bands become less regular after the decompaction. This may reflect the noise introduced into the data by the resampling process after the decompaction.

Sliding window spectral analysis of the Upper Rotliegend in Well L09-01

Figure 7 shows an example of the sliding window maximum entropy spectral analysis as applied to the GR log of the Upper Rotliegend in Well L09-01. In this analysis a window of 22.5 m is moved down the GR data. At each step, the window is shifted down the GR log by 0.9 m. For each step a power spectrum is calculated for the window using maximum entropy spectral analysis. The high-frequency part (70%) of the spectrogram is plotted next to the GR log and positioned at the middle depth point of the window. In this example, the power spectra are plotted as stacked full spectra (labelled as GR-log spectrum band in Fig. 7), or as peak crests only (labelled as GR-log spectrum peaks in Fig. 7). The 'spectrum band' display shows details of the continuous power spectrum, while the 'spectrum peak' display can show a more accurate depth position of the spectrum.

The spectrum bands in Fig. 7 illustrate several interesting features in the Upper Rotliegend of Well L09-01: (1) the sequence boundaries are characterized by breaks in the spectrum bands; (2) in most sequences (sequence RO2, RO3.1, RO3.2, RO3.3, RO4.1, RO4.3, RO4.4 and RO5) the spectrum bands show a thinning-upward trend (spectrum bands shift to the right), indicating an upward decrease of the net accumulation rate within these sequences; (3) in sequences RO3.4 and RO4.2, the spectrum

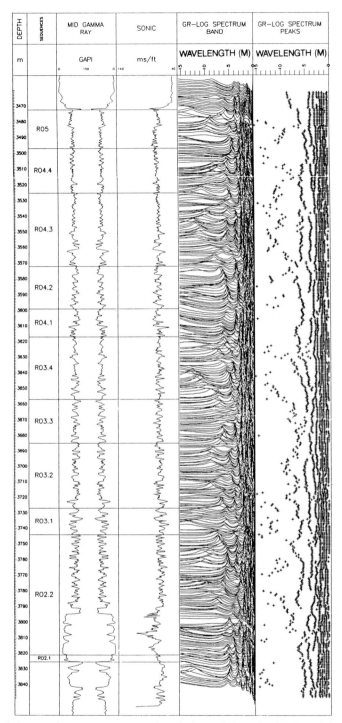

Fig. 7. Sliding window maximum entropy spectral analysis applied to the GR log of the Upper Rotliegend sequences in Well L09-01. The power spectra can be plotted as stacked full spectra (labelled as GR-log spectrum band on the left-hand side), or as peak crests only (labelled as GR-log spectrum peaks on the right-hand side). The GR log is plotted in the mirror image display (MID) format to enhance the pattern of the sequences. See text for explanation.

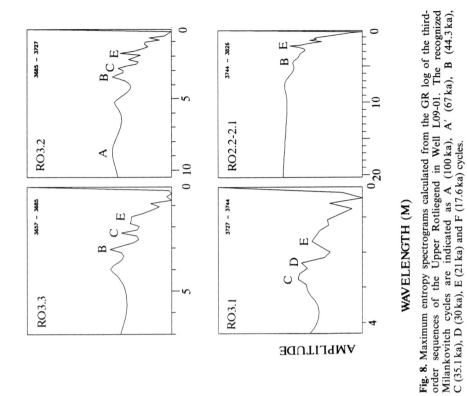

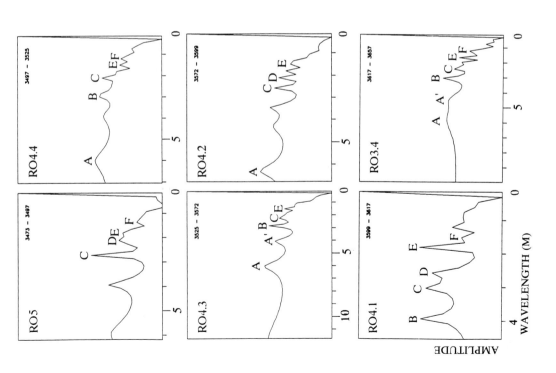

WAVELENGTH (M)

Fig. 8. Maximum entropy spectrograms calculated from the GR log of the third-order sequences of the Upper Rotliegend in Well L09-01. The recognized Milankovitch cycles are indicated as A (100 ka), A' (67 ka), B (44.3 ka), C (35.1 ka), D (30 ka), E (21 ka) and F (17.6 ka) cycles.

bands show a thickening-upward trend (spectrum bands shift to the left), indicating an upward increase of the net accumulation rate within these sequences. These features can be used to calibrate sequence boundaries and to help well-to-well correlations. The net accumulation rates of preserved sediments were controlled by basin subsidence, base-level fluctuations, sediment supplies, and compactions. For wells of similar subsidence (e.g. at the same fault block), same depositional systems and similar lithology, the pattern recognition in the spectral analysis may reveal similar trend of net accumulation rate variations. On the other hand, the differences in the trend of the net accumulation rate variations between neighbouring wells may reflect the differences in subsidence, sediment supply and/or lithology.

Interval spectral analysis

In interval spectral analysis we are mainly interested in the ratios between the dominant frequencies in the data. The ratios between the most significant frequency peaks can be determined from the power spectrum, and compared with the known ratios between the Milankovitch cycles. Our basic assumption in a spectral analysis of wireline log data is that the cyclicities present in the log data represent Milankovitch cycles if their ratios are close to the ratios of the Milankovitch cyclicities. This allows estimation of the absolute age, the duration time and the net accumulation rate of sequences.

Figure 8 shows the spectrograms calculated from the GR log of the third-order Upper Rotliegend sequences in Well L09-01. As most sequences are rather short, the spectrograms were calculated using maximum entropy spectral analysis, which produces a continuous spectrum. The spectrograms show the wavelengths of the frequencies plotted v. the relative contribution of the variance of each frequency to the total variance of the analysed signal, which is a measure of the relative importance of each frequency. The spectrograms reveal a limited number of pronounced peaks with wavelengths in the range of 1–10 m, indicating that the original signal contains cyclic patterns. Each peak can be characterized by its wavelength (or frequency) and amplitude, which are estimated from the data. The ratios between these peaks (i.e. the peaks labelled with letters in Fig. 8) are comparable with the ratios between the Milankovitch cycles for the Permian. Based on the similarity of the ratios, some of the major peaks revealed in the spectrograms of the Upper

Rotliegend sequences can be recognized as the principal Milankovitch cycles (Fig. 8). In some Upper Rotliegend sequences (e.g. sequences RO4.4, RO4.3, RO3.4 and RO3.2), a more or less complete set of Milankovitch cycles (cycles A–F) can be identified. In some other sequences (e.g. RO2.2-2.1 and RO3.1), however, only a few Milankovitch cycles can be detected.

Age determination

The periods of the Milankovitch cycles have not been constant, but have changed through geological history. Berger *et al.* (1989) calculated the theoretical values of the Milankovitch cycles in the geological past. Each geological stage was characterized by a set of typical ratios between the Milankovitch cycles (Table 2). For age determination, the ratios of the cycles detected from the wireline log spectral analysis are compared to the typical ratios of the Milankovitch cycles at various geological times. For example, Fig. 9 shows the spectrum diagram calculated from the GR log of the Upper Rotliegend sequence RO4.4 in Well L04-01. There are six peaks in the spectrum diagram. Their wavelengths are 7.01, 4.63, 3.11, 2.45, 1.49 and 1.22 m, respectively. The ratios of these cycles with reference to the eccentricity are therefore 1.0, 0.66, 0.444, 0.350, 0.212 and 0.174. These ratios match very well with the typical ratios of the Milankovitch cycles as estimated for the Permian [i.e. 1.0 (A), 0.67 (A'), 0.433 (B), 0.351 (C), 0.21 (E) and 0.176 (F); see Table 2]. By comparison of the ratios, it can be judged that the analysed sequence was deposited during the Permian (*c.* 270 Ma). It should be noted that the Milankovitch cycles as calculated by Berger *et al.* (1989) are theoretical values, and other views exist (e.g. Walker & Zahnle 1986; ten Kate & Sprenger 1993). The accuracy of peak recognition and age determination depends on: (1) the understanding and accurate evaluation of the theoretical values of the Milankovitch cycles in the geological past; and (2) good quality wireline log data with minor distortions from different geological processes. Given these conditions an accuracy of *c.* 15% can be expected for age determination.

Estimation of sequence duration and net accumulation rate

Based on the comparison of the ratios, individual peaks in the wireline log spectrum can be

Table 2. *Specific ratios of the Milankovitch cycles at various geological times* (after Berger *et al.* 1989)

Period	Age	Ratios (with reference to eccentricity)				
	(Ma)	A	B	C	E	F
Holocene	0	1.0	0.540	0.410	0.230	0.190
Lower Cretaceous	72	1.0	0.512	0.394	0.225	0.186
Permian	270	1.0	0.443	0.351	0.210	0.176
Lower Carboniferous	298	1.0	0.429	0.343	0.207	0.174
Middle Devonian	380	1.0	0.395	0.321	0.199	01.68
Early Silurian	440	1.0	0.372	0.305	0.193	0.164

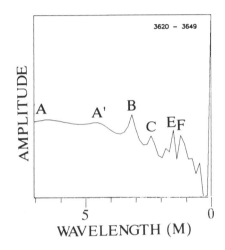

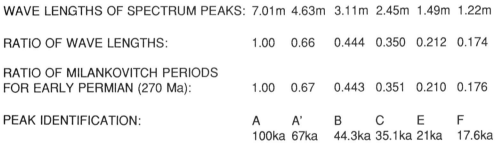

WAVE LENGTHS OF SPECTRUM PEAKS: 7.01m 4.63m 3.11m 2.45m 1.49m 1.22m

RATIO OF WAVE LENGTHS: 1.00   0.66   0.444 0.350 0.212 0.174

RATIO OF MILANKOVITCH PERIODS
FOR EARLY PERMIAN (270 Ma):   1.00   0.67   0.443 0.351 0.210 0.176

PEAK IDENTIFICATION: A   A'   B   C   E   F
                                              100ka 67ka   44.3ka 35.1ka 21ka   17.6ka

Fig. 9. Age determination using cyclicity analysis. The ratios of the dominant cycles in the GR log spectrogram of the Upper Rotliegend sequence RO4.4 in Well L04-01 match very well with the specific ratios of the Milankovitch cycles which are characteristic of the Early Permian. By comparison of the ratios, it can be judged that the analysed sequence was deposited during the Early Permian (*c.* 270 Ma).

assigned to specific Milankovitch cycles. As the periods of these Milankovitch cycles have been reconstructed for the Permian, the number of cycles contained in the sequence gives us the time span necessary to lay down the corresponding sedimentary interval. By dividing the thickness of the studied sequence by its duration, a net accumulation rate of the sequence is obtained. Given: (1) a good understanding and

accurate evaluation of the theoretical values of the Milankovitch cycles in the geological past; and (2) good quality wireline log data with minor distortions from different geological processes, an accuracy of *c.* 10% can be expected for the estimation of sequence durations and net accumulation rates.

The duration times and net accumulation rates of the Upper Rotliegend sequences in studied wells are shown in Fig. 10. Most sequences have a duration time of 280–861 ka; the net accumulation rates are in the range of 6–14 cm ka^{-1}.

Although most peaks have a ratio very similar to the ratio of Milankovitch cycles, some peaks show discrepancies with Milankovitch cycles. Several factors are considered to have contributed to the discrepancies: (1) There was always a certain amount of 'noise' generated by autocyclic processes or by wireline logging and data sampling process; (2) Some peaks may represent additional components of the Milankovitch cycles or the interference pattern of the Milankovitch cycles; (3) Milankovitch cycles may display varying amplitudes throughout geological times and in different parts (latitudes) of the Earth's surface. Therefore, not all Milankovitch cycles were recorded everywhere; (4) Milankovitch cycles (in terms of time) were recorded as cyclic patterns in stratigraphic sections (in terms of depth) only when the net accumulation rate was constant. Changes in net accumulation rates or presence of hiatuses could contribute to the discrepancies between the Milankovitch cycles and the stratigraphic cycles; (5) Compaction had different effect on sandy and muddy sections. This causes distortions in the wavelengths of various cycles.

Application to correlation of the Upper Rotliegend

An example of a well-to-well correlation of the Upper Rotliegend sequences supported by cyclicity analysis is shown in Fig. 10. The GR sliding window spectral analysis and interval spectral analysis, as made in the Upper Rotliegend sequences of Well L09-01 were also made for the Upper Rotliegend in other wells. The sliding window spectral bands, as well as the time durations and net accumulation rates, estimated for the third-order sequences of selected wells are displayed in Fig. 10.

Based on the sliding window spectral analysis, it is possible to recognize patterns of the trend of net accumulation rate variations within the third-order sequence in order to calibrate sequence boundaries and to help well-to-well correlations. Sequence RO1 is the lowermost sequence in the study area. It is penetrated only in Well G17-01. Sliding window spectral bands indicate several internal breaks within this sequence. Sequence RO2.1 is a short sequence; it shows a thinning-up trend. The overlying sequence RO2.2 is characterized by an overall thinning-up trend in Well G17-01. A similar

pattern can also be observed in the same sequence in Wells L09-01 and L04-01.

In the overlying sequences, sliding window spectral bands revealed several important breaks at the sequence boundaries of RO3.1, RO3.3, RO4.1, RO4.3 and RO5, which are distinct in most wells (Fig. 10) These sequence boundaries are marked by the occurrence of halite (Well G17-01) or aeolian deposits (most other wells) (InterGeos 1991). Above these sequence boundaries, the spectrum bands often show a thinning-upward trend (spectrum bands shift to the right), indicating an upward decrease of the net accumulation rate.

In addition to the patterns of the sliding window spectrum bands, Fig. 10 shows also the lateral variations of the time duration and net accumulation rate of each third-order sequence. Most sequences have a time duration of 200–861 ka and a net accumulation of 6–14 cm ka^{-1}.

The method can also be used to obtain estimates of maximum sequence duration. By adding the maximum duration estimations of all sequences together, a minimum duration for the time span in which the Upper Rotliegend was deposited could be calculated. It turns out that the Upper Rotliegend in the Netherlands offshore had a total time duration of 10.7 Ma (Yang & Baumfalk 1994; Yang & Nio 1994). An error of c. 10% is assumed for this estimation due to the presence of hiatuses and differential compaction in the Upper Rotliegend sequences.

Figure 11 displays a chronostratigraphic overview of the Upper Rotliegend sequences in studied wells. The estimated time durations of each sequence in the studied wells were plotted in a chronstratigraphic chart. The deposits of each sequence were subsequently characterized using wireline log-cluster analysis, which defines electrofacies based on wireline logs. These electrofacies were subsequently calibrated with extensive core observations and the corresponding sedimentary facies were interpreted. Thus, the sedimentary record of each sequence was subdivided into slices, each with a specific main sedimentary facies which was deposited under a specific climate condition and could be assigned to a specific period of time within the third-order sequence. The distribution of the sedimentary facies shows distinct cycles of widespread facies changes (Fig. 11), which reflect climate cycles from arid to relatively humid conditions. During arid periods an evaporite lake occupied the basin centre, while aeolian dunes and aeolian sandflats developed in the basin margins. In the following relatively humid period the basin centre was dominated by lake, whereas fluvial (wadi) and sabkha facies became important in the basin

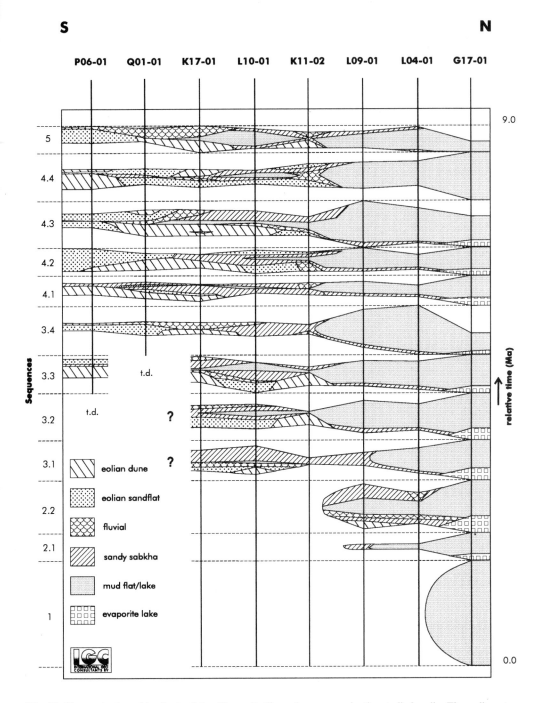

Fig. 11. Chronostratigraphic chart of the Upper Rotliegend sequences in the studied wells. The sedimentary facies are based on wireline log-cluster analysis and calibrated with extensive core observations. The chart shows lateral facies variations from basin centre (Well G17-01) to basin margin (Well P06-01). The vertical facies changes do not represent lateral facies relations, but reflect mainly cyclic climate changes (see text for explanation).

margin. Therefore, the vertical facies changes do not represent lateral facies relations, but reflect mainly cyclic climate changes. Figure 11 shows also that considerable time durations of each sequence were not preserved in the sedimentary record in most wells due to erosion and/or hiatuses.

Discussion

Milankovitch cyclicities have been recognized in virtually all types of sediments (InterGeos database; Melnyk & Smith 1989; Smith 1989; Worthington 1990; Fischer & Bottjer 1991; Nio *et al.* 1993; de Boer & Smith 1994; and many others). The widespread preservation of the Milankovitch cyclicities in sedimentary records gives us a new approach to the study of sedimentation and preservation of the sedimentary record. If depositional cycles were formed by autocyclic processes, the resulting sedimentary record would consist of a pattern of randomly stacked cycles. These would not display significant periodicity peaks in spectral analysis. The fact that specific periods are found implies that the preserved sedimentary record is not produced by a random autocyclic process, but by an allocyclic process controlled by one or more external parameters. The most obvious periodic external parameters able to regulate both climate and sedimentation are the Milankovitch orbital cyclicities. In combination with the fact that the ratios of the observed short-term cycles match those of the orbital-forcing cycles closely, this supports the interpretation that the observed cycles are in fact Milankovitch cycles.

Conclusions

Applications of cyclicity analysis may reveal: (1) the break of cyclicity bands at the discontinuity (e.g. a sequence boundary); (2) the trend of net accumulation rate variations within a third-order sequence; (3) the absolute numerical age of a third-order sequence; (4) the time duration of a third-order sequence; (5) the net accumulation rate of a third-order sequence. These are important in the sequence stratigraphic analysis and inter-well correlation of barren strata.

It is crucial to realize that the mathematical treatment and computer analysis of wireline logs is not a 'black box', but an additional tool to improve geological knowledge in a more objective way. Integrated with basin dynamics information, and relevant well information, this tool can add valuable information in the exploration, development and production stages.

The authors thank the Geological Survey of The Netherlands for providing the wireline logs of released wells for this study, and thank InterGeos B.V. for permission to publish this paper. Appreciation and thanks are due to S. D. Nio, Y. A. Baumfalk, J. J. van den Hurk, and other colleagues for helpful discussions in this study; and also to the reviewers and the editors for valuable comments and suggestions.

References

BERGER, A. & LOUTRE, M. F. 1989. Astronomical forcing through geological time. *In:* DE BOER, P. L. & SMITH, D. G. (eds) *Orbital Forcing and Cyclic Sequences.* International Association of Sedimentologists, Special Publication, **19**, 15–24.

——, —— & DEHANT, V. 1989. Astronomical frequencies for pre-Quaternary palaeoclimate studies. *Terra Nova,* **1**, 474–479.

DE BOER, P. L. & SMITH, D. G. (eds) 1994. *Orbital Forcing and Cyclic Sequences.* International Association of Sedimentologists, Special Publication, **19**.

FISCHER, A. G. & BOTTJER, D. J. (eds) 1991. *Orbital Forcing and Cyclic Sequences.* Journal of Sedimentary Petrology, Special Issue, **61**(7).

GAST, R. E. 1988. Rifting im Rotliegenden Niedersachsens. *Die Geowissenschaften,* **4**, 115–122.

—— 1991. The perennial Rotliegend saline lake in NW Germany. *Geol. Jb., Hannover,* **A119**, 25–59.

GLENNIE, K. W. 1986. Development of NW Europe's southern Permian gas basin. *In:* BROOKS, J., GOFF, J. & VAN HOORN, B. (eds) *Habitat of Palaeozoic Gas in N.W. Europe.* Geological Society, London, Special Publication, **23** 3–22.

—— 1990. Lower Permian–Rotliegend. *In:* GLENNIE, K. W. (ed.) *Introduction to the Petroleum Geology of the North Sea.* Blackwell Scientific Publications, Oxford, 120–152.

GRALLA, P. 1988. Das Oberrotliegende in NW Deutschland – Lithostratigraphie und Faziesanalyse. *Geologisches Jahrbuch, Reihe A,* **106**, 3–59.

HAQ, B. U., HARDENBOL, J. & VAIL, P. R. 1987. Chronology of fluctuating sea levels since the Triassic (250 million years ago to present). *Science,* **235**, 1156–1167.

HARLAND, W. B., ARMSTRONG, R. L., COX, A. V., CRAIG, L. E., SMITH, A. G. & SMITH, D. G. 1990. *A Geologic Time Scale 1989.* Cambridge University Press, Cambridge.

HAYS, J. D., IMBRIE, J. & SHACKLETON, N. J. 1976. Variations in earth's orbit: pacemaker the ice ages. *Science,* **194**, 1121–1132.

HEDEMANN, H. A., MASCHECK, W., PAULUS, B. & PLEIN, E. 1984. Mitteilung zur lithostratigraphischen Gliederung des Oberrotliegenden im nordwestdeutschen Becken. *Nachschr. Deutsches Geol. Gesampt,* **30**, 100–107.

IMBRIE, J. & IMBRIE, J. Z. 1979. Modelling the climate response to orbital variations. *Science,* **207**, 943–953.

INTERGEOS 1991. Rotliegend of The Netherlands offshore. Intergeos Report no. EP91046.

LYELL, C. 1830. *Principles of Geology.* Vol 1. John

Murray, London.

MELNYK, D. H. & SMITH, D. G. 1989. Outcrop to subsurface cycle correlation in the Milankovitch frequency band: Middle Cretaceous, central Italy. *Terra Nova*, **1**, 432–436.

MENNING, M. 1994. A numerical time scale for the Permian and Triassic periods: an integrated time analysis. *In:* SCHOLL, P. A., PERYT, T. H. & ULHNER-SCHOLLE, D. S. (eds) *The Permian of Northern Pangea, Vol. 1: palaeogeography, palaeoclimates and stratigraphy.* Springer-Verlag, Berlin.

NEDERLANDSE AARDOLIE MAATSCHAPPIJ B.V. & RIJKS GEOLOGISCHE DIENST 1980. *Stratigraphic Nomenclature of The Netherlands.* Het Koninklijk Nederlands Geologisch Mijnbouwkundig Genootschap, 32.

NIO, S. D., YANG, C. S. & BAUMFALK, Y. A., *ET AL.* 1993. Computer analysis of depositional sequences using wireline logs – A new method for determining rates of geologic processes. *In:* ARMENTROUT, J. M., BLOCH, R. OLSON, H. C. & PERKINS, B. F. (eds) *Rates of Geologic Processes.* Society of Economic Paleontologists and Mineralogists Foundation, Gulf Coast Section, 14th Annual Research Conference, 141–154.

POSAMENTIER, H. W., JERVEY, M. P. & VAIL, P. R. 1988. Eustatic controls on clastic deposition I – conceptual framework. *In:* WILGUS, C. K., HASTINGS, B. S., KENDALL, C. G. St C., POSAMENTIER, H. W., ROSS, C. A. & VAN WAGONER, J. C. (eds) *Sea-level Changes: An Integrated Approach. Society of Economic Paleontologists and Mineralogists, Special Publication,* **42**, 109–124.

——— & VAIL, P. R. 1988. Eustatic controls on clastic deposition II – sequence and systems tract models. *In:* WILGUS, C. K., HASTINGS, B. S., KENDALL, C. G. St C., POSAMENTIER, H. W., ROSS, C. A. & VAN WAGONER, J. C. (eds) *Sea-level Changes: An Integrated Approach. Society of Economic Paleontologists and Mineralogists, Special Publication,* **42**, 125–154.

REIJERS, T. J. A., MIJNLIEFF, H. F., PESTMAN, P. J. & KOUWE, W. F. P. 1993. Lithofacies and their interpretation: a guide to standardized description of sedimentary deposits. Mededelingen Rijks Geologische Dienst, Geological Survey of The Netherlands, **49**.

SCHWARZACHER, W. 1991. Milankovitch cycles and the measurement of time. *In:* EINSELE, G., RICKEN, W. & SEILACHER, A. (eds) *Cycles and Events in Stratigraphy.* Springer-Verlag, Berlin, 855–863.

SMITH, D. B. 1970. The palaeogeography of the British Zechstein. *In:* RAU, J. L. & DELLWIG, L. F. (eds) *Third Symposium on Salt.* Northern Ohio Geological Society, **1**, 20–23.

——— 1979. Rapid marine transgressions and regressions of the Upper Permian Zechstein Sea. *Journal of the Geological Society, London,* **136**, 155–156.

SMITH, D. G. 1989. Stratigraphic correlation of presumed Milankovitch cycles in the Blue Lias (Hettangian to earliest Sinemurian), England. *Terra Nova*, **1**, 457–460.

TEN KATE, W. G. H. Z. & SPRENGER, A. 1992. *Rhythmicity in Deep-water Sediments, Documentation and Interpretation by Pattern and Spectral Analysis.* Academisch proefschrift, Vrije Universiteit, Amsterdam.

TRUSHEIM, F. 1971. Zur Bildung der Salzlager im Rotliegenden und Mesozoikum Mitteleuropas. *Beih. Geol. Jb., Hannover,* **112**.

VAIL, P. R., MITCHUM, R. M. Jr & THOMPSON, S. III 1977. Seismic stratigraphy and global changes of sea level, part four: global cycles of relative changes of sea level. *In:* PAYTON, C. E. (ed.) *Seismic Stratigraphy – Applications to Hydrocarbon Exploration.* American Association of Petroleum Geologists, Memoir, **26**, 83–98.

VAN WAGONER, J. C., MITCHUM, R. M., CAMPION, K. M. & RAHMANIAN, V. D. 1990. *Siliciclastic Sequence Stratigraphy in Well Logs, Cores and Outcrops: Concepts for High-resolution Correlation of Time and Facies.* American Association of Petroleum Geologists, Methods in Exploration Series, **7**.

VAN WIJHE, D. H., LUTZ, M. & KAASSCHIETER, J. P. H. 1980. The Rotliegend in The Netherlands and its gas accumulations. *Geologie en Mijnbouw,* **59**, 3–24.

WALKER, J. C. G. & ZAHNLE, K. J. 1986. Lunar nodal tide and distance to the Moon during the Precambrian. *Nature*, **320**, 600–602.

WEEDON, G. P. 1991. The spectral analysis of stratigraphic time series. *In:* EINSELE, G., RICKEN, W. & SEILACHER, A. (eds) *Cycles and Events in Stratigraphy.* Springer-Verlag, Berlin, 840–854.

WORTHINGTON, P. F. 1990. Sediment cyclicity from well logs. *In:* HURST, A., LOVELL, M. A. & MORTON, A. C. (eds) *Geological Applications of Wire-line Logs.* Geological Society, London, Special Publication, **48**, 123–132.

YANG, C. S. & BAUMFALK, Y. A. 1994. Milankovitch cyclicity in the Upper Rotliegend Group of The Netherlands offshore. *In:* DE BOER, P. L. & SMITH, D. G. (eds) *Orbital Forcing and Cyclic Sequences.* International Association of Sedimentologists, Special Publication, **19**, 47–61.

——— & NIO, S. D. 1994. Applications of high-resolution sequence stratigraphy to the Upper Rotliegend in The Netherlands offshore. *In:* WEIMER, P. & POSAMENTIER, H. W. (eds) *Siliciclastic Sequence Stratigraphy: Recent Developments and Applications.* American Association of Petroleum Geologists, Memoir, **58**, 285–316.

ZIEGLER, P. A. 1982. *Geological Atlas of Western and Central Europe.* Elsevier, Amsterdam.

Index